2012 中国环境统计年报

ANNUAL STATISTIC REPORT ON ENVIRONMENT IN CHINA

中华人民共和国环境保护部　编
MINISTRY OF ENVIRONMENTAL PROTECTION
OF THE PEOPLE'S REPUBLIC OF CHINA

U0333673

中国环境出版社·北京

图书在版编目（CIP）数据

中国环境统计年报. 2012/中华人民共和国环境保护部编.
—北京：中国环境出版社，2013.12
ISBN 978-7-5111-1686-4

Ⅰ．①中…　Ⅱ．①中…　Ⅲ．①环境统计—统计资料—
中国—2012—年报　Ⅳ．①X508.2-54

中国版本图书馆 CIP 数据核字（2013）第 302370 号

出 版 人　王新程
责任编辑　殷玉婷
责任校对　唐丽虹
封面设计　彭　杉

出版发行　中国环境出版社
　　　　　（100062　北京市东城区广渠门内大街 16 号）
　　　　　网　　　址：http://www.cesp.com.cn
　　　　　电子邮箱：bjgl@cesp.com.cn
　　　　　联系电话：010-67112765（编辑管理部）
　　　　　　　　　　010-67187041（学术著作图书出版中心）
　　　　　发行热线：010-67125803，010-67113405（传真）
印　　刷　北京中科印刷有限公司
经　　销　各地新华书店
版　　次　2013 年 12 月第 1 版
印　　次　2013 年 12 月第 1 次印刷
开　　本　889×1194　1/16
印　　张　28.25
字　　数　600 千字
定　　价　196.00 元

《中国环境统计年报·2012》编委会

组织编写　环境保护部污染物排放总量控制司

中国环境监测总站

资料提供　各省、自治区、直辖市环境保护厅（局）

批　　准　中华人民共和国环境保护部

目录

9　重点城市环境统计　

10　各工业行业环境统计

IV

11　流域及入海陆源废水排放统计　307～372

12　大气污染防治重点区域废气排放统计　373～386

13 环境管理统计

14 附表

15 主要统计指标解释

综述

2012 年，中国共产党第十八次全国代表大会把生态文明建设纳入中国特色社会主义事业五位一体的总体布局，提出推进生态文明，建设美丽中国。在党中央、国务院的正确领导下，环境保护各项工作取得积极进展，主要污染物总量减排工作扎实推进，民生环境问题综合整治成效显现，重点流域区域污染防治取得新进展，农村环境保护和生态保护得到切实强化。与 2011 年相比，化学需氧量（COD）排放量下降 3.05%，氨氮排放量下降 2.62%，二氧化硫（SO_2）排放量下降 4.52%，氮氧化物（NO_x）排放量下降 2.77%。但是环境形势依然严峻，环境风险不断凸显，污染治理任务依然艰巨。

2012 年全国废水排放总量 684.8 亿吨。其中，工业废水排放量 221.6 亿吨，城镇生活污水排放量 462.7 亿吨。废水中化学需氧量排放量 2 423.7 万吨，其中工业源化学需氧量排放量为 338.5 万吨，农业源化学需氧量排放量为 1 153.8 万吨，城镇生活化学需氧量排放量为 912.8 万吨。废水中氨氮排放量 253.6 万吨，其中工业源氨氮排放量为 26.4 万吨，农业源氨氮排放量为 80.6 万吨，城镇生活氨氮排放量为 144.6 万吨。

全国废气中二氧化硫排放量 2 117.6 万吨。其中，工业二氧化硫排放量为 1 911.7 万吨，城镇生活二氧化硫排放量为 205.7 万吨。全国废气中氮氧化物排放量 2 337.8 万吨。其中，工业氮氧化物排放量为 1 658.1 万吨，城镇生活氮氧化物排放量为 39.3 万吨，机动车氮氧化物排放量为 640.0 万吨。全国废气中烟（粉）尘排放量 1 234.3 万吨。其中，工业烟（粉）尘排放量为 1 029.3 万吨，城镇生活烟（粉）尘排放量为 142.7 万吨，机动车烟（粉）尘排放量为 62.1 万吨。

全国一般工业固体废物产生量 32.9 亿吨，综合利用量 20.2 亿吨，贮存量 6.0 亿吨，处置量 7.1 亿吨，倾倒丢弃量 144.2 万吨，全国一般工业固体废物综合利用率为 61.0%。全国工业危险废物产生量 3 465.2 万吨，综合利用量 2 004.6 万吨，贮存量 846.9 万吨，处置量 698.2 万吨，倾倒丢弃量 16.1 吨，全国工业危险废物综合利用处置率为 76.1%。

全国共调查统计了城镇污水处理厂 4 628 座，设计处理能力达到 1.5 亿吨/日，全年共处理污水 416.2 亿吨；生活垃圾处理厂（场）2 125 座，全年共处理生活垃圾 1.97 亿吨，其中采用填埋方式处置的生活垃圾共 1.75 亿吨，采用堆肥方式处置的共 0.03 亿吨，采用焚烧方式处置的共 0.18 亿吨；危险废物集中处理（置）厂（场）722 座，医疗废物集中处理（置）厂（场）236 座，全年共综合利用危险废物 375.2 万吨，处置危险废物 340.7 万吨。

1

统计调查企业基本情况

ANNUAL STATISTIC REPORT ON ENVIRONMENT IN CHINA
2012

1.1 工业企业调查基本情况

2012 年，全国重点调查了 147 996 家工业企业，其中，有废水及废水污染物排放的企业有 93 283 家，有废气及废气污染物排放的企业有 111 827 家，有工业固体废物产生的企业有 105 244 家，有危险废物产生的企业有 23 249 家。对其他工业企业的污染排放量按比率进行估算。

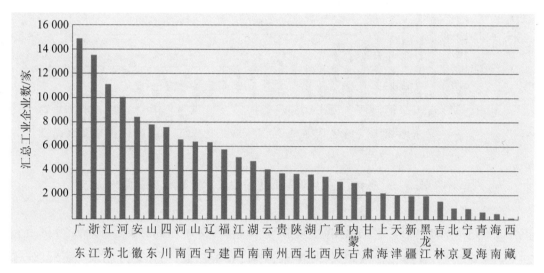

图 1-1　各地区汇总重点调查工业企业数量分布情况

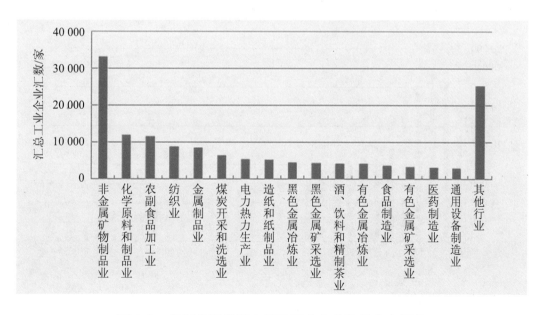

图 1-2　各行业汇总重点调查工业企业数量分布情况

重点调查工业企业中，共有 85 673 套废水治理设施，形成了 26 620 万吨/日的废水处理能力，投入运行费 667.7 亿元。共处理 527.5 亿吨工业废水，去除化学需氧量 2 092.1 万吨，氨氮 119.7 万吨，石油类 27.9 万吨，挥发酚 6.4 万吨，氰化物 0.5 万吨。

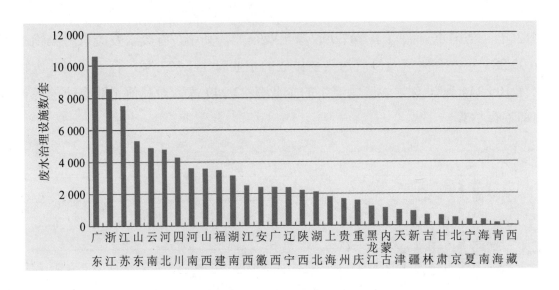

图 1-3　各地区重点调查工业企业废水治理设施数量

重点调查工业企业中，在使用的工业锅炉和窑炉数分别为 10.3 万台和 9.9 万台，共安装 225 913 套废气治理设施（其中，脱硫设施 20 968 套，脱硝设施 832 套，除尘设施 177 985 套），形成了 164.9 亿立方米/时的废气处理能力，投入运行费用 1 452.3 亿元。共去除二氧化硫 4 021.2 万吨，氮氧化物 135.2 万吨，烟（粉）尘 76 770.1 万吨。

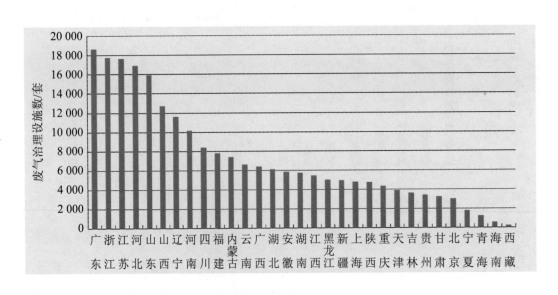

图 1-4　各地区重点调查工业企业废气治理设施数量

1.2 农业源调查基本情况

2012 年，重点调查了 135 614 家规模化畜禽养殖场，9 426 家规模化畜禽养殖小区，对种植业、水产养殖业和其他养殖专业户按产排污强度等进行了核算。

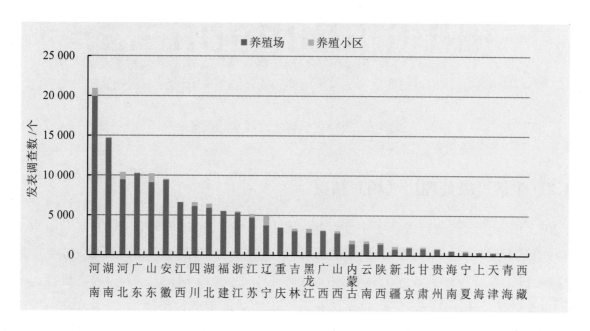

图 1-5 各地区农业源重点调查规模化养殖场和规模化养殖小区数量分布情况

1.3 集中式污染治理设施调查基本情况

1.3.1 城镇生活污水集中处理厂情况

2012 年，全国共调查统计 4 628 座城市污水处理厂，比 2011 年增加 654 座；设计处理能力为 15 313.8 万吨/日，比 2011 年新增 1 322.9 万吨/日；运行费用为 348.2 亿元，比 2011 年增加 41.0 亿元。全年共处理废水 416.2 亿吨，比 2011 年增加 13.3 亿吨，其中，处理生活污水 361.8 亿吨，占总处理水量的 86.9%。再生水生产量 18.0 亿吨，再生水利用量 12.1 亿吨。共去除化学需氧量 1 013.1 万吨，氨氮 88.6 万吨，油类 4.6 万吨，总氮 73.1 万吨，总磷 12.5 万吨，挥发酚 0.1 万吨，氰化物 0.1 万吨。污水处理厂的污泥产生量为 2 418.6 万吨，污泥处置量为 2 418.5 万吨。

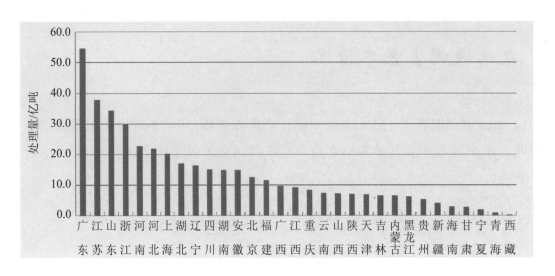

图 1-6　各地区城镇生活污水处理量

1.3.2　生活垃圾处理厂（场）情况

2012 年，全国共调查统计了生活垃圾处理厂（场）2 125 座，比 2011 年增加 86 座；填埋设计容量达 325 399 万立方米；堆肥设计处理能力达到 26 984 吨/日，比 2011 年增加 6 292 吨/日；焚烧设计处理能力达到 66 416 吨/日，比 2011 年增加 44 791 吨/日；运行费用为 98.5 亿元，比 2011 年增加 39.4 亿元。全年共处理生活垃圾 1.97 亿吨，其中采用填埋方式处置的生活垃圾共 1.75 亿吨，采用堆肥方式处置的共 0.03 亿吨，采用焚烧方式处置的共 0.18 亿吨。

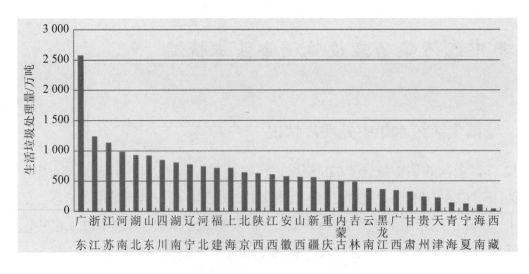

图 1-7　各地区生活垃圾处理量

1.3.3 危险废物（医疗废物）集中处理（置）厂（场）情况

2012 年，全国共调查统计危险废物集中处理（置）厂（场）722 座，比 2011 年增加 78 座；医疗废物集中处理（置）厂（场）236 座，比 2011 年减少 24 座；危险废物设计处置能力达到 59 805 吨/日；运行费用为 53.9 亿元，比 2011 年增加 5.8 亿元。全年共综合利用危险废物 375.2 万吨。全年共处置危险废物 340.7 万吨，其中工业危险废物 223.2 万吨，医疗废物 50.3 万吨。采用填埋方式处置的危险废物共 110.5 万吨，采用焚烧方式处置的 129.1 万吨。

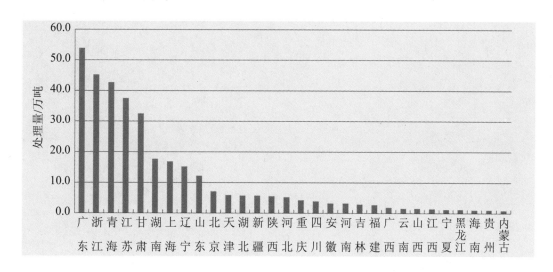

图 1-8 各地区危险废物集中处理（置）量

2

废 水

ANNUAL STATISTIC REPORT ON ENVIRONMENT IN CHINA 2012

2.1 废水及主要污染物排放情况

2.1.1 废水排放情况

2012 年，全国废水排放量 684.8 亿吨，比 2011 年增加 3.9%。

工业废水排放量 221.6 亿吨，比 2011 年减少 4.0%；占废水排放总量的 32.3%，比 2011 年减少 2.6 个百分点。

城镇生活污水排放量 462.7 亿吨，比 2011 年增加 8.1%；占废水排放总量的 67.6%，比 2011 年增加 2.7 个百分点。

集中式污染治理设施废水（不含城镇污水处理厂，下同）排放量 0.5 亿吨，占废水排放总量的 0.1%。

表 2-1　全国废水及其主要污染物排放情况

年份	排放源 / 排放量	合计	工业源	农业源	城镇生活源	集中式
2011	废水/亿吨	659.2	230.9	—	427.9	0.4
	化学需氧量/万吨	2 499.9	354.8	1 186.1	938.8	20.1
	氨氮/万吨	260.4	28.1	82.7	147.7	2.0
2012	废水/亿吨	684.8	221.6	—	462.7	0.5
	化学需氧量/万吨	2 423.7	338.5	1 153.8	912.8	18.7
	氨氮/万吨	253.6	26.4	80.6	144.6	1.9
变化率/%	废水	3.9	−4.0	—	8.1	—
	化学需氧量	−3.0	−4.6	−2.7	−2.8	—
	氨氮	−2.6	−6.0	−2.5	−2.1	—

注：1. 自 2011 年起环境统计中增加农业源的污染排放统计。农业源包括种植业、水产养殖业和畜禽养殖业排放的污染物。

2. 集中式污染治理设施排放量指生活垃圾处理厂（场）和危险废物（医疗废物）集中处理（置）厂（场）垃圾渗滤液/废水及其污染物的排放量。

2.1.2 化学需氧量排放情况

2012 年，全国废水中化学需氧量排放量 2 423.7 万吨，比 2011 年减少 3.05%。

工业废水中化学需氧量排放量 338.5 万吨，比 2011 年减少 4.6%；占化学需氧量排放总量的 14.0%，与 2011 年持平。

农业源排放化学需氧量 1 153.8 万吨，比 2011 年减少 2.7%，其中畜禽养殖业 1 099.0 万吨，比 2011 年减少 1.4%，水产养殖业 54.8 万吨，比 2011 年减少 2.8%；占化学需氧量排放总量的 47.6%，与 2011 年持平。

城镇生活污水中化学需氧量排放量 912.8 万吨，比 2011 年减少 2.8%；占化学需氧量排放总量的 37.6%，与 2011 年持平。

集中式污染治理设施废水中化学需氧量排放量 18.7 万吨，其中生活垃圾处理厂（场）18.6 万吨，危险（医疗）废物集中处理（置）厂（场）1 472.1 吨；占化学需氧量排放总量的 0.8%。

2.1.3 氨氮排放情况

2012 年，全国废水中氨氮排放量 253.6 万吨，比 2011 年减少 2.62%。

工业废水氨氮排放量 26.4 万吨，比 2011 年减少 6.0%；占氨氮排放总量的 10.4%，与 2011 年持平。

农业源氨氮排放量 80.6 万吨，比 2011 年减少 2.5%，其中种植业 15.2 万吨，与 2011 年持平，畜禽养殖业 63.1 万吨，比 2011 年减少 3.2%，水产养殖业 2.3 万吨，比 2011 年增加 1.1%；占氨氮排放总量的 31.8%，与 2011 年持平。

城镇生活污水中氨氮排放量 144.6 万吨，比 2011 年减少 2.1%；占氨氮排放总量的 57.0%，与 2011 年持平。

集中式污染治理设施废水中氨氮排放量 1.9 万吨，其中生活垃圾处理厂（场）1.9 万吨，危险（医疗）废物集中处理（置）厂（场）36.1 吨，占氨氮排放总量的 0.8%。

2.1.4 废水中其他主要污染物排放情况

2012 年，全国工业废水中石油类排放量 1.7 万吨，挥发酚排放量 1 481.4 吨，氰化物排放量 171.8 吨，分别比 2011 年减少 15.8%、38.5%、20.2%。

工业废水中重金属汞、镉、六价铬、总铬、铅及砷排放量分别为 1.1 吨、26.7 吨、70.4 吨、188.6 吨、97.1 吨和 127.7 吨，分别比 2011 年减少 8.3%、23.9%、33.7%、35.0%、35.6%、12.1%。

表 2-2 全国工业废水中重金属及其他污染物排放量　　　　　单位：吨

污染物 年份	石油类	挥发酚	氰化物	汞	镉	六价铬	总铬	铅	砷
2011	20 589.1	2 410.5	215.4	1.2	35.1	106.2	290.3	150.8	145.2
2012	17 327.2	1 481.4	171.8	1.1	26.7	70.4	188.6	97.1	127.7
变化率/%	−15.8	−38.5	−20.2	−8.3	−23.9	−33.7	−35.0	−35.6	−12.1

2.2 各地区废水及主要污染物排放情况

2.2.1 各地区废水排放情况

2012 年，废水排放量大于 30 亿吨的省份共 7 个，依次为广东、江苏、山东、浙江、河南、河北、湖南，7 个省份废水排放总量为 335.1 亿吨，占全国废水排放量的 48.9%。工业废水排放量前 3 位的是江苏、广东和山东，分别占全国工业废水排放量的 10.7%、8.4%、8.3%；城镇生活污水排放量前 3 位依次是广东、江苏、山东，分别占全国城镇生活污水排放量的 14.1%、7.8% 和 6.4%。

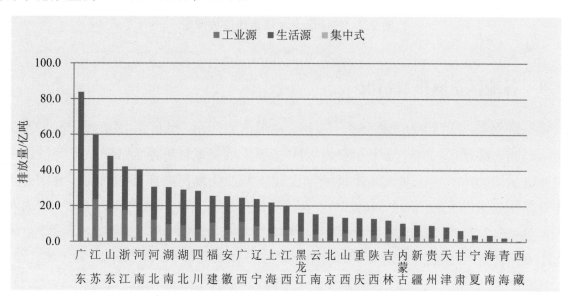

图 2-1　各地区废水排放情况

2.2.2 各地区化学需氧量排放情况

化学需氧量排放量大于 100 万吨的省份有 10 个，依次为山东、广东、黑龙江、河南、河北、辽宁、四川、湖南、江苏和湖北，10 个省份的化学需氧量排放量为 1 408.7 万吨，占全国化学需氧量排放量的 58.1%。工业化学需氧量排放量前 3 位的依次是广东、江苏和广西，分别占全国工业化学需氧量排放量的 7.0%、6.8% 和 5.9%；农业化学需氧量排放量前 3 位的依次是山东、黑龙江和河北，分别占全国农业化学需氧量排放量的 11.6%、9.1% 和 8.0%；城镇生活化学需氧量排放量前 3 位的依次是广东、四川和江苏，分别占全国城镇生活化学需氧量排放量的 10.4%、6.6%、6.3%。

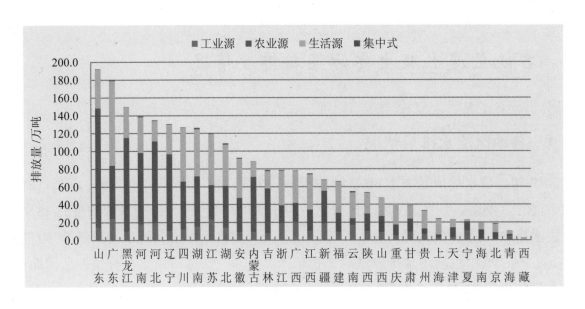

图 2-2　各地区化学需氧量排放情况

2.2.3　各地区氨氮排放情况

氨氮排放量大于 10 万吨的省份有 11 个，依次为广东、山东、湖南、江苏、河南、四川、湖北、浙江、河北、辽宁和安徽，11 个省份的氨氮排放量为 156.3 万吨，占全国氨氮排放量的 61.6%。工业氨氮排放量前 3 位的依次为湖南、江苏和河北，分别占全国工业氨氮排放量的 9.8%、6.2%和 6.0%；农业氨氮排放量前 3 位的依次为山东、河南和湖南，分别占全国农业氨氮排放量的 9.2%、7.9%和 7.8%；城镇生活氨氮排放量前 3 位的依次为广东、江苏和山东，分别占全国城镇生活氨氮排放量的 10.3%、6.7%和 5.7%。

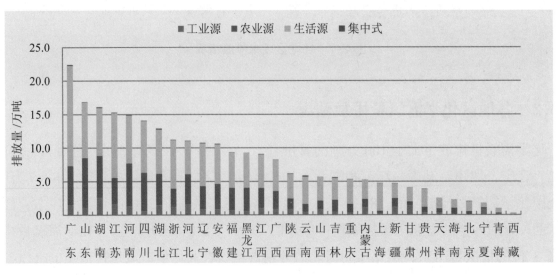

图 2-3　各地区氨氮排放情况

2.2.4 各地区工业石油类、挥发酚和氰化物排放情况

工业废水中石油类排放量大于 700 吨的省份有 10 个，依次为江苏、山西、河南、山东、河北、湖北、内蒙古、湖南、陕西和安徽，10 个省份的石油类排放量为 9 594.0 吨，占全国工业废水石油类排放量的 55.4%。

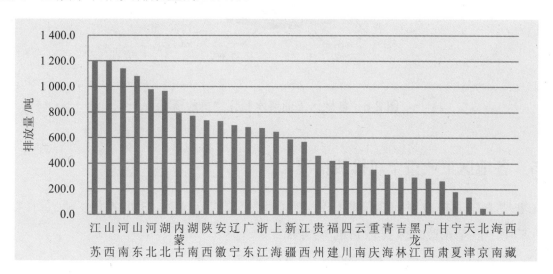

图 2-4　各地区工业废水石油类排放情况

工业废水中挥发酚排放量大于 100 吨的省份有 4 个，依次山西、内蒙古、河南、河北，4 个省份的挥发酚排放量为 1 197.4 吨，占全国工业废水挥发酚排放量的 80.8%。

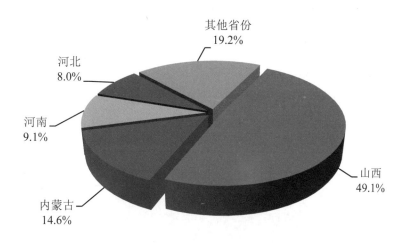

图 2-5　各地区工业废水挥发酚排放情况

工业废水中氰化物排放量大于 10 吨的省份有 6 个，依次山西、河南、江苏、河北、湖南和广东，6 个省份的氰化物排放量为 96.8 吨，占全国工业废水氰化物排放量的 56.0%。

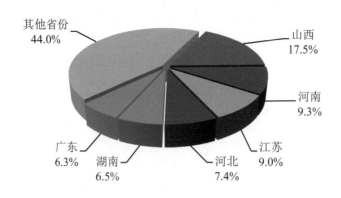

图 2-6　各地区工业废水氰化物排放情况

2.2.5　各地区工业重金属排放情况

工业废水中汞排放量前 3 位省份依次为湖南、湖北、江苏，3 个省份工业废水汞排放量为 0.5 吨，占全国废水汞排放量的 49.4%。

工业废水中镉排放量前 3 位的省份依次为湖南、江西、云南，3 个省份工业废水镉排放量为 17.3 吨，占全国工业废水镉排放量的 64.8%。

工业废水中六价铬排放量前 3 位的省份依次为江西、湖北和浙江，3 个省份工业废水六价铬排放量为 38.1 吨，占全国废水六价铬排放量的 54.1%。

工业废水中总铬排放量前 3 位的省份依次为河南、广东和浙江，3 个省份工业废水总铬排放量为 80.4 吨，占全国废水总铬排放量的 42.6%。

工业废水中铅排放量前 3 位的省份依次为湖南、云南和甘肃，3 个省份工业废水铅排放量为 54.1 吨，占全国废水铅排放量的 55.7%。

工业废水中砷排放量前 3 位的省份依次为湖南、云南和湖北，3 个省份工业废水砷排放量为 73.5 吨，占全国废水砷排放量的 57.6%。

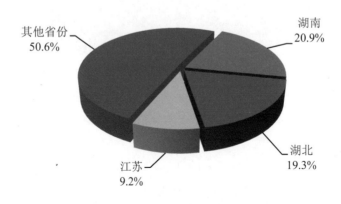

图 2-7　各地区工业废水汞排放情况

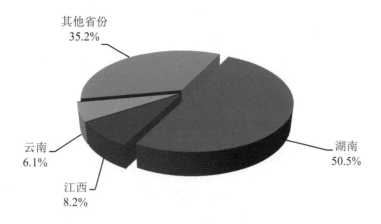

图 2-8　各地区工业废水镉排放情况

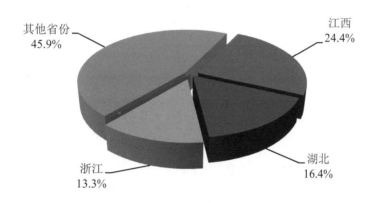

图 2-9　各地区工业废水六价铬排放情况

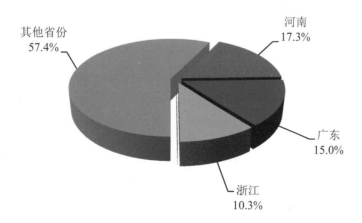

图 2-10　各地区工业废水总铬排放情况

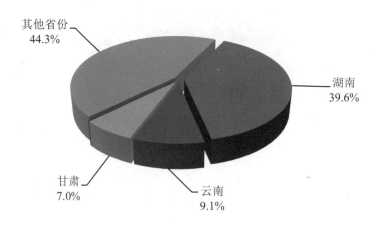

图 2-11 各地区工业废水铅排放情况

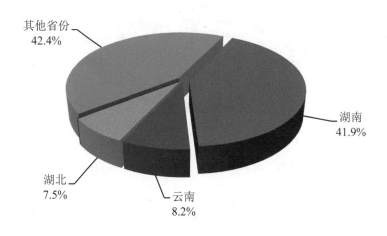

图 2-12 各地区工业废水砷排放情况

2.3 工业行业废水及主要污染物排放情况

2.3.1 行业废水排放情况

2012 年，在调查统计的 41 个工业行业中，废水排放量位于前 4 位的行业依次为造纸和纸制品业，化学原料及化学制品制造业，纺织业，农副食品加工业；4 个行业的废水排放量 101.1 亿吨，占重点调查工业企业废水排放总量的 49.7%。

表 2-3　重点行业废水排放情况　　　　　　　　　　　　　　　　　单位：亿吨

年份 \ 行业	合计	造纸和纸制品业	化学原料及化学制品制造业	纺织业	农副食品加工业
2011	104.9	38.2	28.8	24.1	13.8
2012	101.1	34.3	27.4	23.7	15.7
变化率/%	−3.6	−10.2	−4.9	−1.7	13.8

注：自 2011 年起，环境统计按《国民经济行业分类》（GB/T 4754—2011）标准执行分类统计，下同。

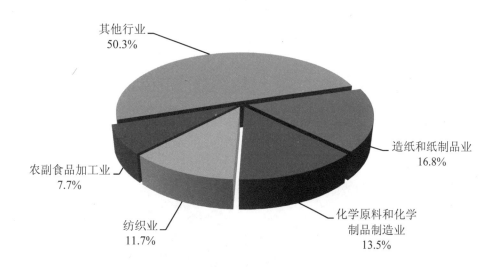

图 2-13　重点行业废水排放情况

2012 年，造纸和纸制品业废水排放量前 5 位的省份依次是浙江、广东、山东、河北和河南，5 个省份造纸和纸制品业废水排放量为 15.8 亿吨，占该行业重点调查工业企业废水排放量的 46.0%。

化学原料和化学制品制造业废水排放量前 5 位的省份依次是江苏、山东、湖北、河南和浙江，5 个省份化学原料和化学制品制造业废水排放量为 12.7 亿吨，占该行业重点调查工业企业废水排放量的 46.4%。

纺织业废水排放量前 5 位的省份依次是江苏、浙江、广东、山东和福建，5 个省份纺织业废水排放量为 19.3 亿吨，占该行业重点调查工业企业废水排放量的 81.4%。

农副食品加工业废水排放量前 5 位的省份依次是广西、山东、云南、河南和河北，5 个省份农副食品加工业废水排放量为 8.9 亿吨，占该行业重点调查工业企业废水排放量的 56.9%。

2.3.2　行业化学需氧量排放情况

2012 年，在调查统计的 41 个工业行业中，化学需氧量排放量位于前 4 位的行业依次为造纸和纸制品业，农副食品加工业，化学原料及化学制品制造业，纺织业；4 个行业的化学需氧量排放量 173.6 万吨，占重点调查工业企业排放总量的 57.1%，较 2011 年下降 2.4 个百分点。

表2-4　重点行业化学需氧量排放情况　　　　　　　　　　　　　　　　单位：万吨

年份 \ 行业	合计	造纸和纸制品业	农副食品加工业	化学原料及化学制品制造业	纺织业
2011	191.5	74.2	55.3	32.8	29.2
2012	173.6	62.3	51.0	32.5	27.7
变化率/%	−9.3	−16.0	−7.8	−0.9	−5.1

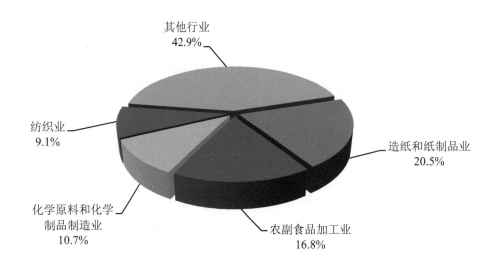

图 2-14　重点行业化学需氧量排放情况

　　2012 年，造纸和纸制品业化学需氧量前 5 位的省份依次是广东、广西、河北、湖南和宁夏，5 个省份造纸和纸制品业化学需氧量排放量为 24.3 万吨，占该行业重点调查工业企业化学需氧量排放量的 39.0%。

　　农副食品加工业化学需氧量排放量前 5 位的省份依次是云南、广西、甘肃、河北、黑龙江，5 个省份农副食品加工业化学需氧量排放量为 23.5 万吨，占该行业重点调查工业企业化学需氧量排放量的 46.1%。

　　化学原料和化学制品制造业化学需氧量排放量前 5 位的省份依次是江苏、湖北、河南、湖南和山东，5 个省份化学原料和化学制品制造业化学需氧量排放量为 12.4 万吨，占该行业重点调查工业企业化学需氧量排放量的 38.3%。

　　纺织业化学需氧量排放量较大的省份依次是浙江、广东和江苏，3 个省份纺织业化学需氧量排放量为 16.8 万吨，占该行业重点调查工业企业化学需氧量排放量的 60.4%。

2.3.3 行业氨氮排放情况

2012 年，在调查统计的 41 个工业行业中，氨氮排放量位于前 4 位的行业依次为化学原料及化学制品制造业，造纸和纸制品业，农副食品加工业，纺织业；4 个行业的氨氮排放量 14.4 万吨，占重点调查工业企业排放总量的 59.2%，较 2011 年下降 1.3 个百分点。

表 2-5　重点行业氨氮排放情况　　　　　　　　　　　　　　　　单位：万吨

行业 年份	合计	化学原料及化学制品制造业	造纸和纸制品业	农副食品加工业	纺织业
2011	15.9	9.3	2.5	2.1	2.0
2012	14.4	8.4	2.1	1.9	1.9
变化率/%	−10.1	−9.7	−16.0	−9.5	−5.0

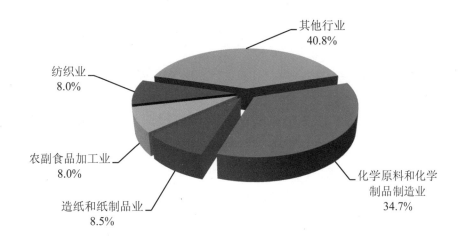

图 2-15　工业行业氨氮排放情况

2012 年，化学原料和化学制品制造业氨氮排放量前 5 位的省份依次是湖南、新疆、湖北、甘肃和安徽，5 个省份化学原料和化学制品制造业氨氮排放量为 4.1 万吨，占该行业重点调查工业企业氨氮排放量的 48.5%。

造纸和纸制品业氨氮前 5 位的省份依次是河北、山东、湖南、广东和浙江，5 个省份造纸和纸制品业氨氮排放量为 0.9 万吨，占该行业重点调查工业企业氨氮排放量的 43.4%。

农副食品加工业氨氮排放量前 5 位的省份依次是河北、广西、河南、黑龙江和山东，5 个省份农副食品加工业氨氮排放量为 0.7 万吨，占该行业重点调查工业企业氨氮排放量的 35.0%。

纺织业氨氮排放量较大的省份依次是浙江、广东和江苏，3 个省份纺织业氨氮排放量为 1.2 万吨，占该行业重点调查工业企业氨氮排放量的 63.4%。

2.3.4 行业石油类排放情况

2012 年，石油类排放量位于前 4 位的行业依次为黑色金属冶炼和压延加工业，煤炭开采和洗选业，化学原料和化学制品制造业，石油加工、炼焦和核燃料加工业；4 个行业石油类排放量为 9 401.0 吨，占重点调查工业企业石油类排放量的 54.3%。

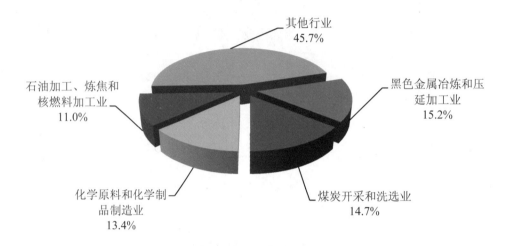

其他行业
45.7%

石油加工、炼焦和
核燃料加工业
11.0%

黑色金属冶炼和压
延加工业
15.2%

化学原料和化学制
品制造业
13.4%

煤炭开采和洗选业
14.7%

图 2-16 工业行业石油类污染物排放情况

2012 年，黑色金属冶炼和压延加工业石油类排放量前 5 位的省份依次是内蒙古、山西、河北、云南和湖北，5 个省份黑色金属冶炼和压延加工业石油类排放量为 1 299.0 吨，占该行业重点调查工业企业石油类排放量的 49.2%。

煤炭开采和洗选业石油类排放量前 5 位的省份依次是山西、河南、贵州、安徽和山东，5 个省份煤炭开采和洗选业石油类排放量为 1 172.8 吨，占该行业重点调查工业企业石油类排放量的 46.2%。

化学原料和化学制品制造业石油类排放前 5 位的省份依次是江苏、河南、湖南、福建和陕西，5 个省份化学原料和化学制品制造业行业石油类排放量为 900.3 吨，占该行业重点调查工业企业石油类排放量的 38.8%。

石油加工、炼焦和核燃料加工业石油类排放量前 5 位的省份依次是山西、河南、辽宁、河北和内蒙古，5 个省份石油加工、炼焦和核燃料加工业石油类排放量为 1 036.6 吨，占该行业重点调查工业企业石油类排放量的 54.6%。

2.3.5 行业挥发酚排放情况

2012 年，挥发酚排放量最大的行业为石油加工、炼焦和核燃料加工业，挥发酚排放量为 1 234.9 吨，占重点调查工业企业挥发酚排放量的 83.4%；其次为化学原料和化学制品制造业，挥发酚排放量为 91.1 吨，占重点调查工业企业挥发酚排放量的 6.1%。

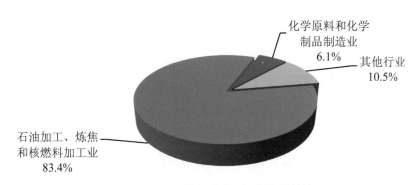

图 2-17　工业行业挥发酚排放情况

2012 年，石油加工、炼焦和核燃料加工业挥发酚排放量较大的省份为山西和内蒙古，其中山西石油加工、炼焦和核燃料加工业挥发酚排放量为 711.6 吨，内蒙古为 209.5 吨，分别占行业重点调查工业企业挥发酚排放量的 57.6%和 17.0%。

化学原料和化学制品制造业挥发酚排放量前较大的省份依次为江苏、浙江，其中江苏化学原料和化学制品制造业挥发酚排放量为 24.2 吨，浙江 21.8 吨，分别占该行业重点调查工业企业挥发酚排放量的 26.6%和 24.0%。

2.3.6　行业氰化物排放情况

2012 年，氰化物排放量位于前 4 位的行业依次为化学原料和化学制品制造业，石油加工、炼焦和核燃料加工业，金属制品业，黑色金属冶炼和压延加工业；4 个行业石油类排放量为 155.8 吨，占该行业重点调查工业企业氰化物排放量的 90.7%。

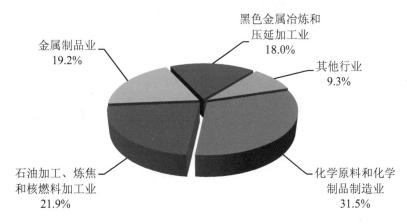

图 2-18　工业行业氰化物排放情况

2012 年，化学原料和化学制品制造业氰化物排放量前 5 位的省份依次为山西、湖南、河南、山东和江苏，这 5 个省份化学原料和化学制品制造业氰化物排放量为 27.7 吨，占该行业重点调查工业企业氰化物排放量的 51.2%。

石油加工、炼焦和核燃料加工业氰化物排放量较大的省份依次为山西和河北，其中山西该行业氰化物排放量为 15.5 吨，河北为 8.5 吨，分别占该行业重点调查工业企业氰化物排放量的 41.2%和 22.6%。

金属制品业氰化物排放量较大省份依次是广东、浙江和江苏，其中广东该行业氰化物排放量为 7.3 吨，浙江为 6.7 吨，江苏为 5.8 吨，分别占该行业重点调查工业企业氰化物排放量的 22.0%、20.2%和 17.6%。

黑色金属冶炼和压延加工业氰化物排放量前 5 位的省份依次是河南、山西、河北、湖南和湖北，5 个省份黑色金属冶炼和压延加工业石氰化物放量为 15.9 吨，占该行业重点调查工业企业氰化物排放量的 51.3%。

2.3.7　行业重金属污染物排放情况

2012 年，重金属（汞、镉、六价铬、总铬、铅、砷）排放量位于前 4 位的行业依次为金属制品业、有色金属冶炼和压延加工业，皮革、毛皮、羽毛及其制品和制鞋业，化学原料和化学制品制造业；4 个行业重金属排放量为 377.1 吨，占重点调查工业企业重金属排放量的 73.7%。

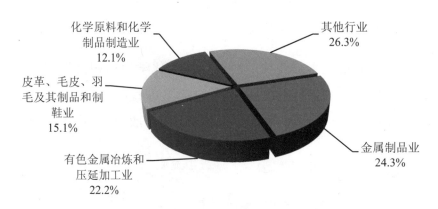

图 2-19　工业行业重金属排放情况

汞排放量前 3 位的行业依次是化学原料和化学制品制造业，有色金属矿采选业和有色金属冶炼和压延加工业；这 3 个行业汞排放量为 0.9 吨，占重点调查工业企业汞排放量的 86.5%。化学原料和化学制品制造业汞排放量前 3 位的省份依次是湖北、湖南和江苏，3 个省份化学原料和化学制品制造业汞排放量为 361.3 千克，占该行业重点调查工业企业汞排放量的 78.6%；有色金属矿采选业汞排放量前 3 位的省份依次是内蒙古、江西和四川，3 个省份有色金属矿采选业汞排放量为 179.9 千克，占该行业重点调查工业企业汞排放量的 60.4%；有色金属冶炼和压延加工业汞排放量前 3 位的身份依次是湖南、甘肃和陕西，3 个省份有色金属冶炼和压延加工业汞排放量为 135.9 千克，占该行业重点调查工业企业汞排放量的 76.5%。

镉排放量前 3 位的行业依次是有色金属冶炼和压延加工业，有色金属矿采选业，化学原料和化学制品制造业；这 3 个行业镉排放量为 24.7 吨，占重点调查工业企业镉排放量的 92.5%。有色金属冶炼和压延加工业镉排放量前 3 位的省份依次是湖南、云南和江西，这 3 个省份有色金属冶炼和压延加工业的镉排放量为 14.6 吨，占该行业重点调查工业企业镉排放量的 71.2%；有色金属矿采选业镉排放量前 3 位的省份依次是江西、湖南和广西，这 3 个省份有色金属矿采选业镉排放量为 1.6 吨，占该行业重点调查工业企业镉排放量的 61.1%；化学原料和化学制品制造业镉排放量前 3 位的省份依次是湖南、甘肃和湖北，这 3 个省份化学原料和化学制品制造业镉排放量为 1.4 吨，占该行业重点调查工业企业镉排放量的 86.6%。

铅排放量前 3 位的行业依次是有色金属冶炼和压延加工业，有色金属矿采选业，化学原料和化学制品制造业；这 3 个行业铅排放量为 86.0 吨，占重点调查工业企业铅排放量的 88.5%。有色金属冶炼和压延加工业铅排放量前 3 位的省份依次是湖南、江西和云南，这 3 个省份有色金属冶炼和压延加工业铅排放量为 35.3 吨，占该行业重点调查工业企业铅排放量的 68.7%；有色金属矿采选业铅排放量前 3 位的省份依次是湖南、广西和内蒙古，这 3 个省份有色金属矿采选业铅排放量为 13.6 吨，占该行业重点调查工业企业铅排放量的 52.4%；化学原料和化学制品制造业铅排放量为湖南、甘肃和福建，这 3 个省份化学原料和化学制品制造业行业铅排放量为 6.7 吨，占该行业重点调查工业企业铅排放量的 78.7%。

砷排放量前 3 位的行业依次是化学原料和化学制品制造业，有色金属冶炼和压延加工业，有色金属矿采选业，这 3 个行业砷排放量为 110.6 吨，占重点调查工业企业砷排放的 86.6%。化学原料和化学制品制造业砷排放量前 3 位的省份依次是湖南、湖北和安徽，3 个省份化学原料和化学制品制造业砷排放量 40.1 吨，占该行业重点调查工业企业砷排放量的 91.0%；有色金属冶炼和压延加工业砷排放量前 3 位的省份依次是湖南、江西和云南，3 个省份有色金属冶炼和压延加工业砷排放量为 23.8 吨，占该行业重点调查工业企业砷排放量的 66.1%；有色金属矿采选业砷排放量前 3 位的省份依次是云南、广西和湖南，3 个省份有色金属矿采选业砷排放量为 17.1 吨，占该行业重点调查工业企业砷排放量的 55.7%。

总铬、六价铬排放量较大的行业为金属制造业，皮革、毛皮、羽毛及其制品和制鞋业和汽车制造业。金属制造业总铬、六价铬排放量为 74.0 吨和 44.7 吨，分别占重点调查工业企业总铬、六价铬排放量的 41.0% 和 63.6%，主要分布在广东、江西和浙江 3 个省份，3 个省份金属制造业总铬、六价铬排放量为 48.2 吨和 31.0 吨，分别占重点调查工业企业总铬、六价铬排放量的 62.3% 和 69.3%。皮革、毛皮、羽毛及其制品和制鞋业主要污染物为总铬，该行业总铬排放量最高省份为河南，其次为湖南，分别占 41.7% 和 20.6%。

表2-6 工业行业废水重金属污染物排放情况

污染物	排放量总计/吨	主要行业及所占比例	排放量大的地区及所占比例
汞	1.1	化学原料和化学制品制造业 41.8%，有色金属矿采选业 27.1%，有色金属冶炼和压延加工业 16.1%	湖南 20.4%，湖北 18.9%，江苏 9.0%
镉	26.7	有色金属冶炼和压延加工业 76.4%，有色金属矿采选业 9.7%，化学原料和化学制品制造业 6.0%	湖南 50.5%，江西 8.2%，云南 6.1%
铅	97.1	有色金属冶炼和压延加工业 52.9%，有色金属矿采选业 26.8%，化学原料和化学制品制造业 8.8%	湖南 39.6%，云南 9.1%，甘肃 7.0%
砷	127.7	化学原料和化学制品制造业 34.5%，有色金属冶炼和压延加工业 28.2%，有色金属矿采选业 24.0%	湖南 41.9%，云南 8.2%，湖北 7.5%
六价铬	70.4	金属制品业 63.5%，汽车制造业 14.5%，黑色金属冶炼和压延加工业 5.5%	江西 24.4%，湖北 16.4%，浙江 13.3%
总铬	188.6	金属制品业 41.0%，皮革、毛皮、羽毛及其制品和制鞋业 39.2%，汽车制造业 5.7%	河南 17.3%，广东 15.0%，浙江 10.3%

2.4 重点流域废水及主要污染物排放与治理情况

根据《重点流域水污染防治"十二五"规划》中的流域分区，松花江、辽河、海河、黄河中上游、淮河、长江中下游、太湖、巢湖、滇池、三峡库区及其上游、丹江口库区及其上游等重点流域总体排放情况如下：

表2-7 2012 年重点流域废水及废水中污染物总体排放情况

污染物 流域	废水/亿吨	化学需氧量/万吨	氨氮/万吨	工业石油类/吨	工业挥发酚/吨	工业氰化物/吨	工业重金属/吨
松花江	23.9	201.4	12.8	464.9	14.2	1.7	1.0
辽河	18.1	126.0	9.8	424.1	17.7	3.8	4.6
海河	79.1	284.9	24.9	1 996.6	387.9	36.8	26.1
黄河中上游	39.9	173.4	17.5	2 549.7	605.3	25.7	41.1
淮河	66.0	262.9	28.7	1 752.8	153.6	19.0	20.2
长江中下游	124.2	381.1	48.8	4 101.4	62.9	37.7	221.7
太湖	34.5	34.9	5.4	229.6	9.4	6.7	10.4
巢湖	4.3	12.3	1.4	44.9	0.0	0.0	0.1
滇池	3.9	0.8	0.3	46.8	0.1	0.1	2.6
三峡库区	55.5	202.6	24.1	1 270.9	11.6	3.4	23.0
丹江口库区	4.5	21.9	2.9	152.7	1.7	0.7	17.0

注：本年报中重点流域是根据《重点流域水污染防治"十二五"规划》中流域分区汇总得出，下同。

2012 年，重点流域的废水排放总量为 453.9 亿吨，占全国废水排放总量的 66.3%。其中，松花江、辽河、海河、黄河中上游、淮河、长江中下游、太湖、巢湖、滇池、三峡库区及其上游、丹江口库区及其上游流域的废水排放量分别为 23.9 亿吨、18.1 亿吨、79.1 亿吨、39.9 亿吨、66.0 亿吨、124.2 亿吨、34.5 亿吨、4.3 亿吨、3.9 亿吨、55.5 亿吨和 4.5 亿吨，分别占重点流域排放总量的 5.3%、4.0%、17.4%、8.8%、14.5%、27.4%、7.6%、0.9%、0.9%、12.2%和 1.0%。

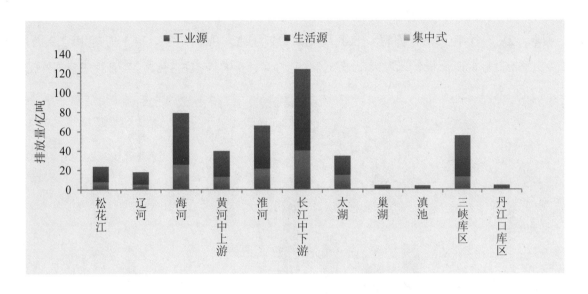

图 2-20 2012 年重点流域废水排放情况

重点流域的化学需氧量排放总量为 1 702.2 万吨，占全国化学需氧量排放总量的 70.2%。其中，松花江、辽河、海河、黄河中上游、淮河、长江中下游、太湖、巢湖、滇池、三峡库区及其上游、丹江口库区及其上游流域的化学需氧量排放量分别为 201.4 万吨、126.0 万吨、284.9 万吨、173.4 万吨、262.9 万吨、381.1 万吨、34.9 万吨、12.3 万吨、0.8 万吨、202.6 万吨和 21.9 万吨，分别占重点流域排放总量的 11.8%、7.4%、16.7%、10.2%、15.4%、22.4%、2.1%、0.7%、0.05%、11.9%、1.3%。

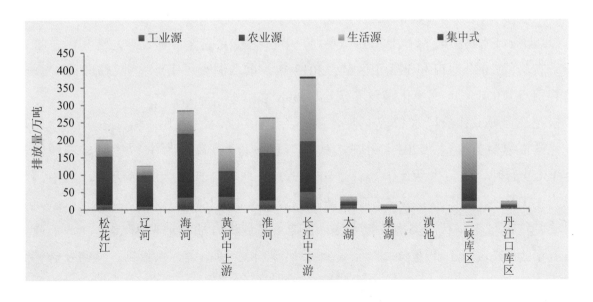

图 2-21 2012 年重点流域化学需氧量排放情况

重点流域的氨氮排放总量为 176.7 万吨，占全国氨氮排放总量的 69.7%。其中，松花江、辽河、海河、黄河中上游、淮河、长江中下游、太湖、巢湖、滇池、三峡库区及其上游、丹江口库区及其上游流域的氨氮排放量分别为 12.8 万吨、9.8 万吨、24.9 万吨、17.5 万吨、28.7 万吨、48.8 万吨、5.4 万吨、1.4 万吨、0.3 万吨、24.1 万吨和 2.9 万吨，分别占重点流域排放总量的 67.2%、5.5%、14.1%、9.9%、16.3%、27.6%、3.1%、0.8%、0.2%、13.6% 和 1.7%。

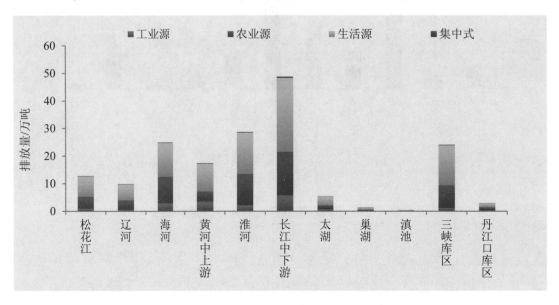

图 2-22　2012 年重点流域氨氮排放情况

2.4.1　松花江流域

（1）废水及污染物排放情况

《重点流域水污染防治"十二五"规划》中松花江流域含吉林、黑龙江和内蒙古自治区 3 个省份的 26 个地市和 173 个区县。2012 年，重点调查了工业企业 2 762 家，规模化畜禽养殖场 4 978 家，规模化畜禽养殖小区 598 家。

松花江流域共排放废水 23.9 亿吨，其中，工业废水 8.0 亿吨，城镇生活污水 15.8 亿吨。化学需氧量排放量为 201.4 万吨，其中，工业化学需氧量为 13.9 万吨，农业化学需氧量 138.9 万吨，城镇生活化学需氧量 47.4 万吨。氨氮排放量为 12.8 万吨，其中，工业氨氮为 1.0 万吨，农业氨氮 4.3 万吨，城镇生活氨氮 7.4 万吨。

松花江流域工业石油类排放量为 464.9 吨，工业挥发酚排放量为 14.1 吨，工业氰化物排放量为 1.7 吨，工业废水中 6 种重金属（包括铅、镉、汞、六价铬、总铬及砷）排放总量为 1.0 吨。

表 2-8　松花江流域废水及主要污染物排放情况

污染物 年份	废水/亿吨			化学需氧量/万吨				氨氮/万吨			
	工业源	生活源	集中式	工业源	农业源	生活源	集中式	工业源	农业源	生活源	集中式
2011	6.56	15.88	0.01	14.31	143.93	49.41	1.17	0.98	4.48	7.68	0.13
2012	8.03	15.84	0.01	13.95	138.94	47.40	1.13	0.96	4.28	7.40	0.12
变化率/%	22.5	−0.3	—	−2.6	−3.5	−4.1	—	−1.9	−4.3	−3.6	—

（2）废水及主要污染物在各地区的分布

2012 年，松花江流域废水、化学需氧量和氨氮排放量最大的均是黑龙江省，分别占该流域各类污染物排放量的 60.4%、62.2% 和 60.6%。其中，工业废水、化学需氧量和氨氮排放量最大的均是黑龙江省，分别占该流域工业排放总量的 64.1%、57.0% 和 54.9%。农业化学需氧量和氨氮排放量最大的均是黑龙江省，分别占该流域农业排放总量的 62.7% 和 61.2%。城镇生活废水、化学需氧量和氨氮排放量最大的均是黑龙江省，分别占该流域生活排放总量的 58.5%、63.5% 和 61.7%。

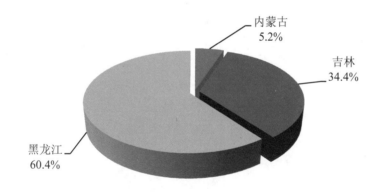

图 2-23　松花江流域废水排放区域构成

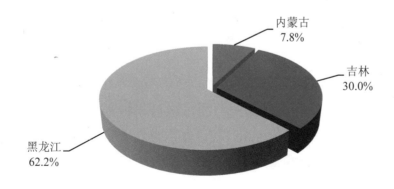

图 2-24　松花江流域化学需氧量排放区域构成

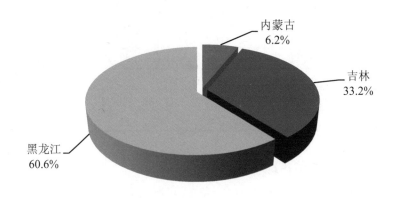

内蒙古
6.2%

吉林
33.2%

黑龙江
60.6%

图 2-25　松花江流域氨氮排放区域构成

（3）废水及主要污染物在行业的分布

2012 年，在调查统计的 41 个工业行业中，松花江流域废水排放量位于前 4 位的行业依次为煤炭开采和洗选业，农副食品加工业，造纸和纸制品业，化学原料及化学制品制造业；4 个行业的废水排放量为 3.8 亿吨，占重点调查工业企业废水排放总量的 49.8%。

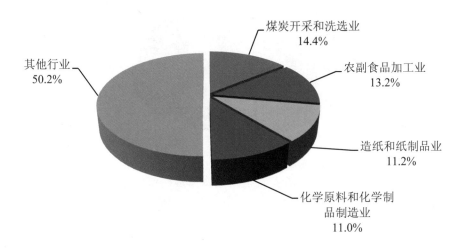

煤炭开采和洗选业
14.4%

其他行业
50.2%

农副食品加工业
13.2%

造纸和纸制品业
11.2%

化学原料和化学制品制造业
11.0%

图 2-26　松花江流域工业废水排放量行业构成

2012 年，在调查统计的 41 个工业行业中，松花江流域化学需氧量位于前 4 位的行业依次为农副食品加工业，造纸和纸制品业，酒、饮料和精制茶制造业，煤炭开采和洗选业；4 个行业的化学需氧量排放量为 9.0 万吨，占重点调查工业企业化学需氧量排放总量的 72.2%。

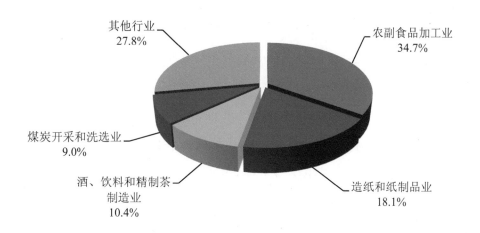

图 2-27　松花江流域工业化学需氧量排放量行业构成

2012 年，在调查统计的 41 个工业行业中，松花江流域氨氮排放量位于前 4 位的行业依次为农副食品加工业，石油加工、炼焦和核燃料加工业，化学原料和化学制品制造业，造纸和纸制品业；4 个行业的氨氮排放量为 0.56 万吨，占重点调查工业企业氨氮排放总量的 63.4%。

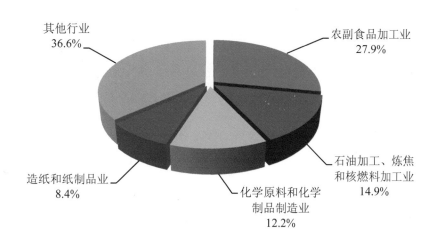

图 2-28　松花江流域工业氨氮排放量行业构成

（4）废水及污染物治理情况

2012 年，松花江流域纳入统计的污水处理厂 106 座，形成了 524 万吨/日的处理能力，年运行费用达 10.2 亿元，共处理污水 11.4 亿吨，其中生活污水 10.3 亿吨。去除化学需氧量 29.6 万吨、氨氮 2.4 万吨、油类 0.1 万吨、总氮 1.7 万吨、总磷 0.3 万吨。

松花江流域重点调查工业企业共有废水治理设施 1 492 套，形成了 1 319 万吨/日的废水处理能力，年运行费用达 29.6 亿元，共处理了 11.8 亿吨工业废水。去除工业化学需氧量 156.7 万吨、氨氮 5.3 万吨、石油类 4.1 万吨、挥发酚 5 472.6 吨、氰化物 102.1 吨。

2.4.2 辽河流域

（1）废水及污染物排放情况

《重点流域水污染防治"十二五"规划》中辽河流域含辽宁、吉林和内蒙古自治区 3 个省份的 16 个地市和 106 个区县。2012 年，重点调查了工业企业 5 219 家，规模化畜禽养殖场 3 926 家，规模化畜禽养殖小区 1 153 家。

辽河流域共排放废水 18.1 亿吨，其中，工业废水 5.3 亿吨，城镇生活污水 12.8 亿吨。化学需氧量排放量为 126.0 万吨，其中，工业化学需氧量为 8.8 万吨，农业化学需氧量 90.3 万吨，城镇生活化学需氧量 26.2 万吨。氨氮排放量为 9.8 万吨，其中，工业氨氮为 0.6 万吨，农业氨氮 3.4 万吨，城镇生活氨氮 5.8 万吨。

辽河流域工业石油类排放量为 424.1 吨，工业挥发酚排放量为 17.7 吨，工业氰化物排放量为 3.8 吨，工业废水中六种重金属（包括铅、镉、汞、六价铬、总铬及砷）排放总量为 4.6 吨。

表 2-9　辽河流域废水及主要污染物排放情况

污染物 年份	废水/亿吨			化学需氧量/万吨				氨氮/万吨			
	工业源	生活源	集中式	工业源	农业源	生活源	集中式	工业源	农业源	生活源	集中式
2011	5.57	11.72	0.01	9.53	94.85	27.51	0.80	0.66	3.49	5.88	0.11
2012	5.35	12.76	0.01	8.79	90.27	26.23	0.66	0.60	3.36	5.76	0.09
变化率/%	−4.1	8.9	—	−7.8	−4.8	−4.7	—	−9.3	−3.9	−2.0	—

（2）废水及主要污染物在各地区的分布

2012 年，辽河流域废水、化学需氧量和氨氮排放量最大的均是辽宁省，分别占该流域各类污染物排放量的 82.4%、77.7% 和 79.2%。其中，工业废水、化学需氧量和氨氮排放量最大的均是辽宁省，分别占该流域工业排放总量的 83.4%、64.5% 和 50.7%。农业化学需氧量和氨氮排放量最大的均是辽宁省，分别占该流域农业排放总量的 76.9% 和 79.6%。城镇生活废水、化学需氧量和氨氮排放量最大的均是辽宁省，分别占该流域生活排放总量的 82.0%、84.9% 和 81.8%。

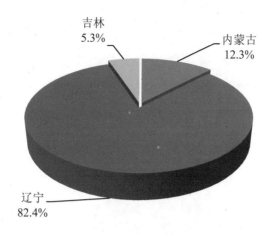

图 2-29   辽河流域废水排放区域构成

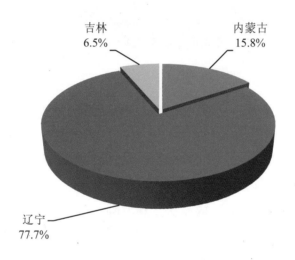

图 2-30   辽河流域化学需氧量排放区域构成

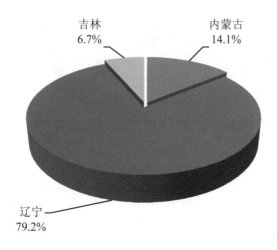

图 2-31   辽河流域氨氮排放区域构成

（3）废水及主要污染物在行业的分布

2012 年，在调查统计的 41 个工业行业中，辽河流域废水排放量位于前 5 位的行业依次为化学原料和化学制品制造业，农副食品加工业，煤炭开采和洗选业，造纸和纸制品业，黑色金属矿采选业，5 个行业的废水排放量为 2.2 亿吨，占重点调查工业企业废水排放总量的 44.5%。

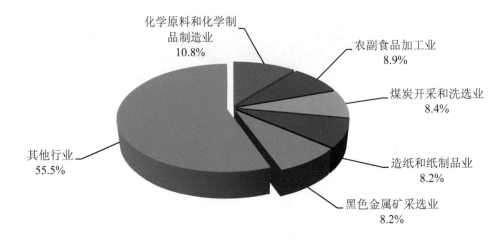

图 2-32　辽河流域工业废水排放量行业构成

2012 年，在调查统计的 41 个工业行业中，辽河流域化学需氧量位于前 4 位的行业依次为农副食品加工业，造纸和纸制品业，化学原料和化学制品制造业，酒、饮料和精制茶制造业；4 个行业的化学需氧量排放量为 5.0 万吨，占重点调查工业企业化学需氧量排放总量的 62.9%。

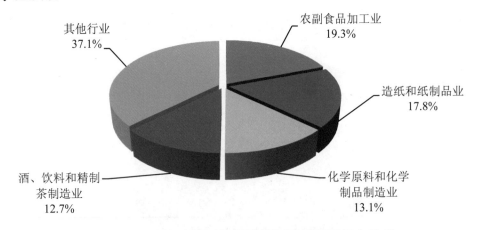

图 2-33　辽河流域工业化学需氧量排放量行业构成

2012 年，在调查统计的 41 个工业行业中，辽河流域氨氮排放量位于前 4 位的行业依次为农副食品加工业，化学原料和化学制品制造业，酒、饮料和精制茶制造业，食品制造业；4 个行业的氨氮排放量为 0.35 万吨，占重点调查工业企业氨氮排放总量的 62.9%。

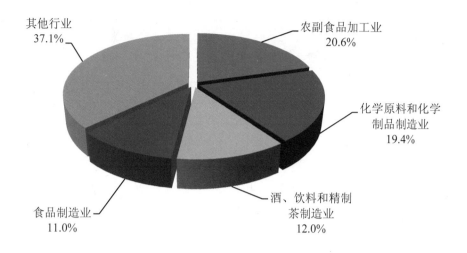

其他行业
37.1%

农副食品加工业
20.6%

化学原料和化学
制品制造业
19.4%

酒、饮料和精制
茶制造业
12.0%

食品制造业
11.0%

图 2-34　辽河流域工业氨氮排放量行业构成

（4）废水及污染物治理情况

2012 年，辽河流域纳入统计的污水处理厂 107 座，形成了 506 万吨/日的处理能力，年运行费用达 9.8 亿元，共处理污水 13.2 亿吨，其中生活污水 11.5 亿吨。去除化学需氧量 29.9 万吨、氨氮 1.9 万吨、油类 0.2 万吨、总氮 1.9 万吨、总磷 0.1 万吨。

辽河流域重点调查工业企业共有废水治理设施 1 758 套，形成了 1 084 万吨/日的废水处理能力，年运行费用达 23.0 亿元，处理了 20.7 亿吨工业废水。去除工业化学需氧量 36.7 万吨、氨氮 1.6 万吨、石油类 1.3 万吨、挥发酚 1 213.0 吨、氰化物 146.8 吨。

2.4.3　海河流域

（1）废水及污染物排放情况

《重点流域水污染防治"十二五"规划》中海河流域含北京、天津、河北、山西、内蒙古、山东、河南 7 个省份的 64 个地市和 335 个区县。2012 年，重点调查了工业企业 19 335 家，规模化畜禽养殖场 16 553 家，规模化畜禽养殖小区 1 862 家。

海河流域共排放废水 79.1 亿吨，其中，工业废水 25.8 亿吨，城镇生活污水 53.3 亿吨。化学需氧量排放量为 284.9 万吨，其中，工业化学需氧量为 34.8 万吨，农业化学需氧量 183.0 万吨，城镇生活化学需氧量 65.3 万吨。氨氮排放量为 24.9 万吨，其中，工业氨氮为 3.0 万吨，农业氨氮 9.5 万吨，城镇生活氨氮 12.3 万吨。

海河流域工业石油类排放量为 1 996.6 吨，工业挥发酚排放量为 387.9 吨，工业氰化物排放量为 36.8 吨，工业废水中 6 种重金属（包括铅、镉、汞、六价铬、总铬及砷）排放总量为 26.1 吨。

表 2-10   海河流域废水及主要污染物排放情况

年份 \ 污染物	废水/亿吨			化学需氧量/万吨				氨氮/万吨			
	工业源	生活源	集中式	工业源	农业源	生活源	集中式	工业源	农业源	生活源	集中式
2011	26.02	48.20	0.04	35.75	188.83	69.89	1.79	3.25	9.82	12.80	0.16
2012	25.76	53.34	0.04	34.78	183.01	65.28	1.85	2.98	9.48	12.32	0.16
变化率/%	-1.0	10.7	—	-2.7	-3.1	-6.6	—	-8.3	-3.4	-3.8	—

（2）废水及主要污染物在各地区的分布

2012 年，海河流域废水、化学需氧量和氨氮排放量最大的均是河北省，分别占该流域各类污染物排放量的 36.3%、45.0%和 42.7%。其中，工业废水、化学需氧量和氨氮排放量最大的均是河北省，分别占该流域工业排放总量的 43.4%、51.4%和 49.6%。农业化学需氧量和氨氮排放量最大的均是河北省，分别占该流域农业排放总量的 47.8%和 45.8%。城镇生活废水、化学需氧量和氨氮排放量最大的均是河北省，分别占该流域生活排放总量的 32.9%、33.9%和 38.7%。

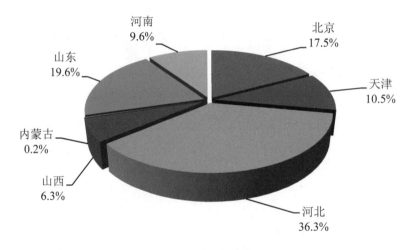

图 2-35   海河流域废水排放区域构成

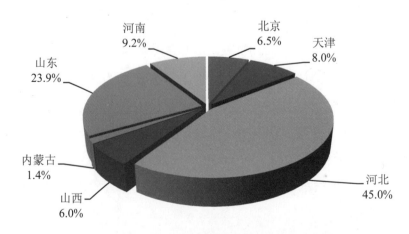

图 2-36   海河流域化学需氧量排放区域构成

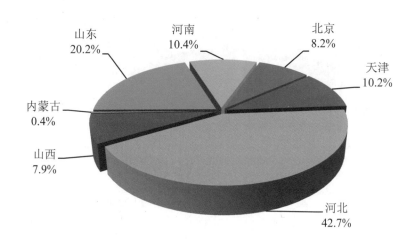

图 2-37　海河流域氨氮排放区域构成

（3）废水及主要污染物在行业的分布

2012 年，在调查统计的 41 个工业行业中，海河流域废水排放量位于前 4 位的行业依次为造纸和纸制品业，化学原料和化学制品制造业，纺织业，农副食品加工业；4 个行业的废水排放量为 12.9 亿吨，占重点调查工业企业废水排放总量的 54.8%。

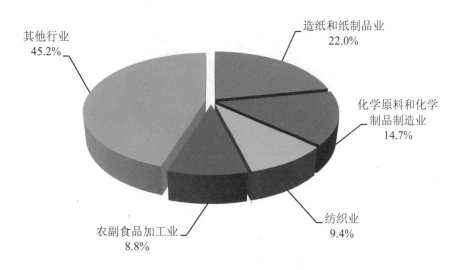

图 2-38　海河流域工业废水排放量行业构成

2012 年，在调查统计的 41 个工业行业中，海河流域化学需氧量位于前 4 位的行业依次为造纸和纸制品业，农副食品加工业，化学原料和化学制品制造业，纺织业；4 个行业的化学需氧量排放量为 19.0 万吨，占重点调查工业企业化学需氧量排放总量的 60.7%。

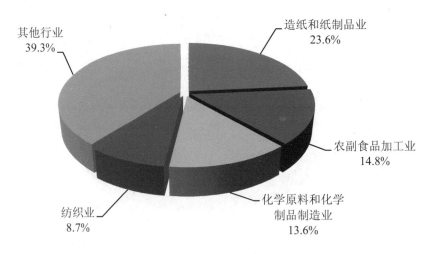

图 2-39　海河流域工业化学需氧量排放量行业构成

2012 年，在调查统计的 41 个工业行业中，海河流域氨氮排放量位于前 4 位的行业依次为化学原料和化学制品制造业，造纸和纸制品业，农副食品加工业，纺织业；4 个行业的氨氮排放量为 1.6 万吨，占重点调查工业企业氨氮排放总量的 61.3%。

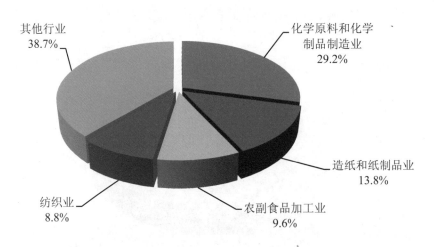

图 2-40　海河流域工业氨氮排放量行业构成

（4）废水及污染物治理情况

2012 年，海河流域纳入统计的污水处理厂 609 座，形成了 2 098 万吨/日的处理能力，年运行费用达 48.0 亿元，共处理污水 57.6 亿吨，其中生活污水 47.9 亿吨。去除化学需氧量 171.4 万吨、氨氮 17.8 万吨、油类 0.9 万吨、总氮 14.2 万吨、总磷 3.0 万吨。

海河流域重点调查工业企业共有废水治理设施 9 991 套，形成了 5 203 万吨/日的废水处理能力，年运行费用达 101.9 亿元，处理了 98.3 亿吨工业废水。去除工业化学需氧量 344.1 万吨、氨氮 16.8 万吨、石油类 4.1 万吨、挥发酚 1.6 万吨、氰化物 694.7 吨。

2.4.4 黄河中上游流域

（1）废水及污染物排放情况

《重点流域水污染防治"十二五"规划》中黄河中上游流域含山西、内蒙古、河南、陕西、甘肃、青海、宁夏 7 个省份的 52 个地市和 339 个区县。2012 年，重点调查了工业企业 12 548 家，规模化畜禽养殖场 7 697 家，规模化畜禽养殖小区 1 216 家。

黄河中上游流域共排放废水 39.9 亿吨，其中，工业废水 13.2 亿吨，城镇生活污水 26.6 亿吨。化学需氧量排放量为 173.4 万吨，其中，工业化学需氧量为 36.7 万吨，农业化学需氧量 74.0 万吨，城镇生活化学需氧量 61.7 万吨。氨氮排放量为 17.5 万吨，其中，工业氨氮为 3.5 万吨，农业氨氮 3.6 万吨，城镇生活氨氮 10.2 万吨。

黄河中上游流域工业石油类排放量为 2 549.7 吨，工业挥发酚排放量为 605.3 吨，工业氰化物排放量为 25.7 吨，工业废水中 6 种重金属（包括铅、镉、汞、六价铬、总铬及砷）排放总量为 41.1 吨。

表 2-11 黄河中上游流域废水及主要污染物排放情况

污染物 年份	废水/亿吨			化学需氧量/万吨				氨氮/万吨			
	工业源	生活源	集中式	工业源	农业源	生活源	集中式	工业源	农业源	生活源	集中式
2011	13.35	24.12	0.01	38.65	82.92	63.15	1.00	3.34	3.70	10.67	0.12
2012	13.24	26.61	0.01	36.67	74.00	61.74	1.01	3.54	3.58	10.21	0.12
变化率/%	−0.8	10.3	—	−5.1	−10.8	−2.2	—	6.1	−3.4	−4.2	—

（2）废水及主要污染物在各地区的分布

2012 年，黄河中上游流域废水、化学需氧量和氨氮排放量在各地区的分布相对比较平均。废水、化学需氧量和氨氮排放量最大的均是陕西省，分别占该流域各类污染物排放量的 24.0%、22.3% 和 23.5%。其中，工业废水排放量最大的是河南省、工业化学需氧量和氨氮排放量最大的是宁夏回族自治区，分别占该流域工业排放总量的 24.8%、28.1% 和 23.5%。农业化学需氧量排放量最大的是内蒙古自治区，农业氨氮排放量最大的是河南省，分别占该流域农业排放总量的 21.8% 和 26.1%。城镇生活废水、化学需氧量和氨氮排放量最大的均是陕西省，分别占该流域生活排放总量的 26.0%、26.8% 和 26.0%。

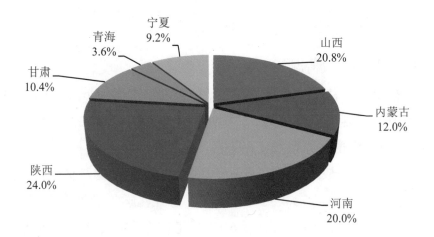

图 2-41 黄河中上游流域废水排放区域构成

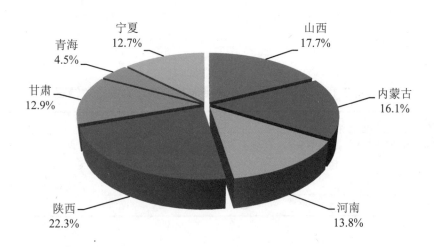

图 2-42 黄河中上游流域化学需氧量排放区域构成

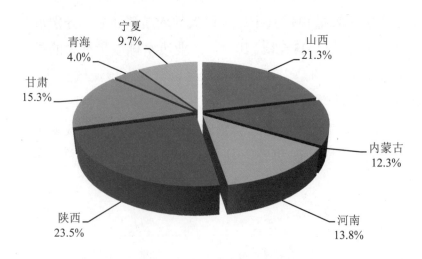

图 2-43 黄河中上游流域氨氮排放区域构成

（3）废水及主要污染物在行业的分布

2012 年，在调查统计的 41 个工业行业中，黄河中上游流域废水排放量位于前 5 位的行业依次为造纸和纸制品业，化学原料和化学制品制造业，煤炭开采和洗选业，黑色金属冶炼和压延加工业，石油加工、炼焦和核燃料加工业；5 个行业的废水排放量为 7.7 亿吨，占重点调查工业企业废水排放总量的 63.0%。

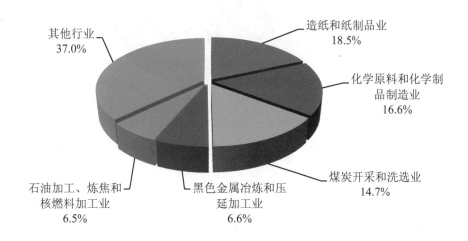

图 2-44 黄河中上游流域工业废水排放量行业构成

2012 年，在调查统计的 41 个工业行业中，黄河中上游流域化学需氧量位于前 5 位的行业依次为造纸和纸制品业，农副食品加工业，化学原料和化学制品制造业，食品制造业，酒、饮料和精制茶制造业；5 个行业的化学需氧量排放量为 23.4 万吨，占重点调查工业企业化学需氧量排放总量的 70.0%。

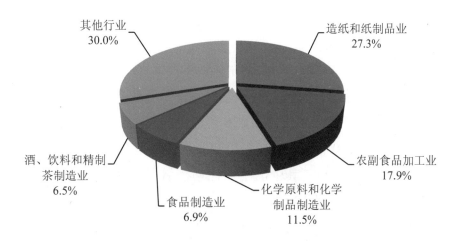

图 2-45 黄河中上游流域工业化学需氧量排放量行业构成

2012 年，在调查统计的 41 个工业行业中，黄河中上游流域氨氮排放量位于前 4 位的行业依次为化学原料和化学制品制造业，有色金属冶炼和压延加工业，石油加工、炼焦和核燃料加工业，食品制造业；4 个行业的氨氮排放量为 2.3 万吨，占重点调查工业企业氨氮排放总量的 71.7%。

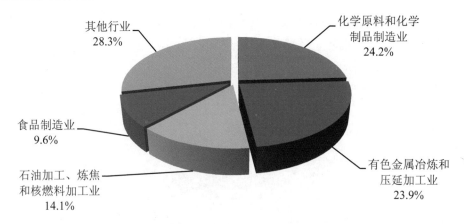

其他行业
28.3%

化学原料和化学
制品制造业
24.2%

食品制造业
9.6%

石油加工、炼焦
和核燃料加工业
14.1%

有色金属冶炼和
压延加工业
23.9%

图 2-46　黄河中上游流域工业氨氮排放量行业构成

（4）废水及污染物治理情况

2012 年，黄河中上游流域纳入统计的污水处理厂 388 座，形成了 918 万吨/日的处理能力，年运行费用达 19.9 亿元，共处理污水 22.2 亿吨，其中生活污水 20.1 亿吨。去除化学需氧量 66.3 万吨、氨氮 6.1 万吨、油类 0.5 万吨、总氮 6.7 万吨、总磷 0.7 万吨。

黄河中上游流域重点调查工业企业共有废水治理设施 6 551 套，形成了 2 024 万吨/日的废水处理能力，年运行费用达 49.6 亿元，处理了 40.1 亿吨工业废水。去除工业化学需氧量 147.1 万吨、氨氮 19.6 万吨、石油类 5.3 万吨、挥发酚 1.2 万吨、氰化物 506.9 吨。

2.4.5　淮河流域

（1）废水及污染物排放情况

《重点流域水污染防治"十二五"规划》中淮河流域含江苏、安徽、山东、河南 4 个省份的 35 个地市和 218 个区县。2012 年，重点调查了工业企业 12 428 家，规模化畜禽养殖场 25 129 家，规模化畜禽养殖小区 1 026 家。

淮河流域共排放废水 66.0 亿吨，其中，工业废水 21.6 亿吨，城镇生活污水 44.4 亿吨。化学需氧量排放量为 262.9 万吨，其中，工业化学需氧量为 25.9 万吨，农业化学需氧量 136.1 万吨，城镇生活化学需氧量 99.2 万吨。氨氮排放量为 28.7 万吨，其中，工业氨氮为 2.2 万吨，农业氨氮 11.3 万吨，城镇生活氨氮 15.1 万吨。

淮河流域工业石油类排放量为 1 752.8 吨，工业挥发酚排放量为 153.6 吨，工业氰化物排放量为 19.0 吨，工业废水中 6 种重金属（包括铅、镉、汞、六价铬、总铬及砷）排放总量为 20.2 吨。

表 2-12　淮河流域废水及主要污染物排放情况

年份 \ 污染物	废水/亿吨			化学需氧量/万吨				氨氮/万吨			
	工业源	生活源	集中式	工业源	农业源	生活源	集中式	工业源	农业源	生活源	集中式
2011	22.16	41.76	0.03	27.44	140.94	100.59	1.83	2.27	11.60	15.24	0.20
2012	21.60	44.37	0.03	25.87	136.08	99.16	1.82	2.15	11.28	15.09	0.20
变化率/%	-2.5	6.3	—	-5.7	-3.4	-1.4	—	-5.3	-2.7	-1.0	—

（2）废水及主要污染物在各地区的分布

2012年，淮河流域废水、化学需氧量和氨氮排放量在各地区的分布相对比较平均。废水、化学需氧量和氨氮排放量最大的均是河南省，分别占该流域各类污染物排放量的31.9%、31.0%和30.5%。其中，工业废水、化学需氧量和氨氮排放量最大的均是江苏省，分别占该流域工业排放总量的32.3%、40.1%和36.7%。农业化学需氧量和氨氮排放量最大的均是河南省，分别占该流域农业排放总量的35.1%和34.8%。城镇生活废水排放量最大的是河南省，生活化学需氧量和氨氮排放量最大的是江苏省，分别占该流域生活排放总量的32.4%、35.1%和32.9%。

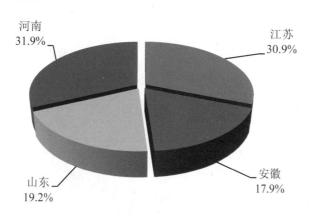

图 2-47　淮河流域废水排放区域构成

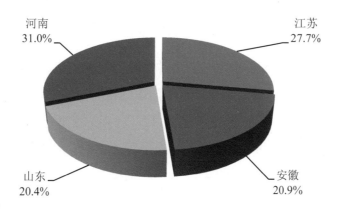

图 2-48　淮河流域化学需氧量排放区域构成

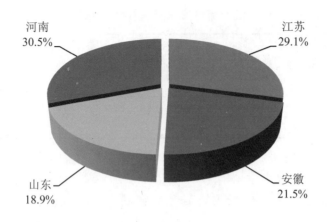

图 2-49   淮河流域氨氮排放区域构成

（3）废水及主要污染物在行业的分布

2012 年，在调查统计的 41 个工业行业中，淮河流域废水排放量位于前 4 位的行业依次为煤炭开采和洗选业，造纸和纸制品业，化学原料和化学制品制造业，农副食品加工业；4 个行业的废水排放量为 12.4 亿吨，占重点调查工业企业废水排放总量的 62.2%。

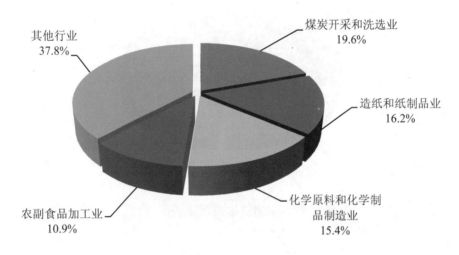

图 2-50   淮河流域工业废水排放量行业构成

2012 年，在调查统计的 41 个工业行业中，淮河流域化学需氧量位于前 4 位的行业依次为造纸和纸制品业，农副食品加工业，化学原料和化学制品制造业，酒、饮料和精制茶制造业；4 个行业的化学需氧量排放量为 14.4 万吨，占重点调查工业企业化学需氧量排放总量的 61.4%。

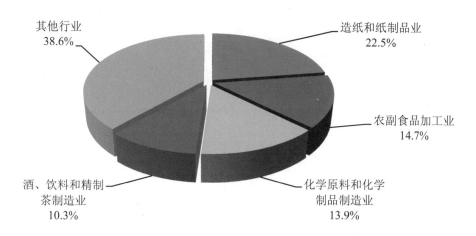

图 2-51 淮河流域工业化学需氧量排放量行业构成

2012 年，在调查统计的 41 个工业行业中，淮河流域氨氮排放量位于前 4 位的行业依次为化学原料和化学制品制造业，造纸和纸制品业，农副食品加工业，酒、饮料和精制茶制造业；4 个行业的氨氮排放量为 1.4 万吨，占重点调查工业企业氨氮排放总量的 68.4%。

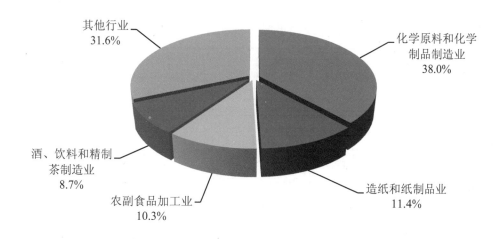

图 2-52 淮河流域工业氨氮排放量行业构成

（4）废水及污染物治理情况

2012 年，淮河流域纳入统计的污水处理厂 405 座，形成了 1 075 万吨/日的处理能力，年运行费用达 20.1 亿元，共处理污水 31.2 亿吨，其中生活污水 28.4 亿吨。去除化学需氧量 66.9 万吨、氨氮 6.4 万吨、油类 0.2 万吨、总氮 5.4 万吨、总磷 0.7 万吨。

淮河流域重点调查工业企业共有废水治理设施 5 229 套，形成了 1 696 万吨/日的废水处理能力，年运行费用达 41.2 亿元，处理了 27.5 亿吨工业废水。去除工业化学需氧量 229.6 万吨、氨氮 7.0 万吨、石油类 1.0 万吨、挥发酚 2 814.2 吨、氰化物 124.7 吨。

2.4.6 长江中下游流域

（1）废水及污染物排放情况

《重点流域水污染防治"十二五"规划》中长江中下游流域含上海、江苏、安徽、江西、河南、湖北、湖南、广西 8 个省份的 55 个地市和 408 个区县，包括长江干流控制区 109 个区县、长江口控制区 45 个区县、汉江中下游控制区 35 个区县、洞庭湖控制区 130 个区县和鄱阳湖控制区 89 个区县。2012 年，重点调查了工业企业 22 911 家，规模化畜禽养殖场 31 631 家，规模化畜禽养殖小区 645 家。

长江中下游流域共排放废水 124.2 亿吨，其中，工业废水 40.2 亿吨，城镇生活污水 83.8 亿吨。化学需氧量排放量为 381.1 万吨，其中，工业化学需氧量为 49.9 万吨，农业化学需氧量 145.3 万吨，城镇生活化学需氧量 180.4 万吨。氨氮排放量为 48.8 万吨，其中，工业氨氮为 5.8 万吨，农业氨氮 15.7 万吨，城镇生活氨氮 26.8 万吨。

长江中下游流域工业石油类排放量为 4 101.4 吨，工业挥发酚排放量为 62.9 吨，工业氰化物排放量为 37.7 吨，工业废水中 6 种重金属（包括铅、镉、汞、六价铬、总铬及砷）排放总量为 221.7 吨。

表 2-13 长江中下游流域废水及主要污染物排放情况

年份	污染物	废水/亿吨			化学需氧量/万吨				氨氮/万吨			
		工业源	生活源	集中式	工业源	农业源	生活源	集中式	工业源	农业源	生活源	集中式
2011	长江中下游	41.90	79.03	0.14	55.01	147.50	181.37	5.82	6.42	16.11	26.90	0.52
	其中：洞庭湖	9.77	18.58	0.04	17.37	59.44	54.92	1.68	2.80	6.59	7.36	0.15
	鄱阳湖	6.81	11.73	0.02	10.91	23.60	36.40	1.07	0.99	2.94	4.59	0.09
2012	长江中下游	40.25	83.77	0.17	49.89	145.35	180.37	5.46	5.79	15.73	26.81	0.48
	其中：洞庭湖	9.78	21.28	0.04	15.28	57.09	54.94	1.61	2.60	6.38	7.39	0.15
	鄱阳湖	6.44	12.53	0.02	9.41	22.63	37.24	1.00	0.90	2.80	4.72	0.08
变化率/%	长江中下游	-3.9	6.0	—	-9.3	-1.5	-0.6	—	-9.7	-2.4	-0.3	—
	其中：洞庭湖	0.1	14.6	—	-12.0	-4.0	0.0	—	-7.3	-3.1	0.4	—
	鄱阳湖	-5.4	6.8	—	-13.7	-4.1	2.3	—	-9.3	-4.7	2.7	—

表 2-14 洞庭湖及鄱阳湖其他污染物排放情况

年份	污染物	工业石油类/吨	工业挥发酚/吨	工业氰化物/吨	工业重金属/吨
2011	长江中下游	4 926.07	229.55	46.22	272.39
	其中：洞庭湖	907.44	103.22	12.83	148.52
	鄱阳湖	663.90	74.64	10.20	62.23
2012	长江中下游	4 101.43	62.89	37.69	221.69
	其中：洞庭湖	810.45	16.80	11.30	125.78
	鄱阳湖	536.51	7.73	7.69	51.76
变化率/%	长江中下游	-16.7	-72.6	-18.5	-18.6
	其中：洞庭湖	-10.7	-83.7	-11.9	-15.3
	鄱阳湖	-19.2	-89.6	-24.6	-16.8

（2）废水及主要污染物在各地区的分布

2012 年，长江中下游流域废水、化学需氧量和氨氮排放量主要集中在湖南、湖北和江西 3 个省。废水、化学需氧量和氨氮排放量最大的均是湖南省，分别占该流域各类污染物排放量的 23.9%、33.0%和 32.4%。其中，工业废水、化学需氧量和氨氮排放量最大的均是湖南省，分别占该流域工业排放总量的 23.2%、28.6%和 44.1%。农业化学需氧量和氨氮排放量最大的均是湖南省，分别占该流域农业排放总量的 38.2%和 39.0%。城镇生活废水、化学需氧量和氨氮排放量最大的均是湖南省，分别占该流域生活排放总量的24.2%、28.7%和 26.1%。

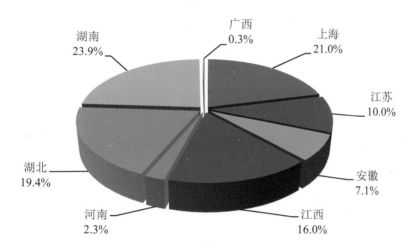

图 2-53　长江中下游流域废水排放区域构成

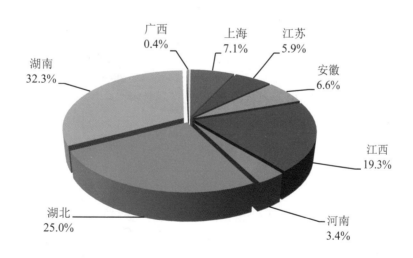

图 2-54　长江中下游流域化学需氧量排放区域构成

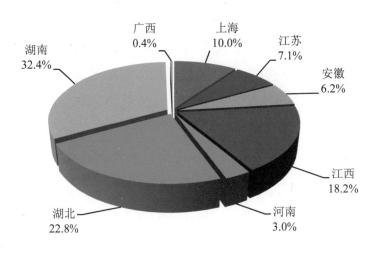

图 2-55 长江中下游流域氨氮排放区域构成

（3）废水及主要污染物在行业的分布

2012 年，在调查统计的 41 个工业行业中，长江流域废水排放量位于前 4 位的行业依次为化学原料和化学制品制造业，造纸和纸制品业，黑色金属冶炼和压延加工业，石油加工、炼焦和核燃料加工业；4 个行业的废水排放量为 19.0 亿吨，占重点调查工业企业废水排放总量的 51.3%。

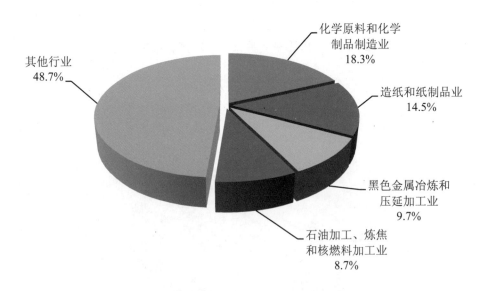

图 2-56 长江流域工业废水排放量行业构成

2012 年，在调查统计的 41 个工业行业中，长江流域化学需氧量位于前 4 位的行业依次为造纸和纸制品业，化学原料和化学制品制造业，农副食品加工业，石油加工、炼焦和核燃料加工业；4 个行业的化学需氧量排放量为 23.4 万吨，占重点调查工业企业化学需氧量排放总量的 51.9%。

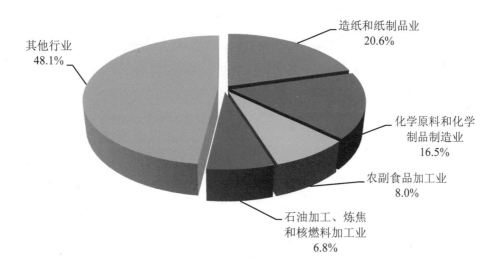

图 2-57   长江流域工业化学需氧量排放量行业构成

2012 年，在调查统计的 41 个工业行业中，长江流域氨氮排放量位于前 4 位的行业依次为化学原料和化学制品制造业，造纸和纸制品业，有色金属冶炼和压延加工业，石油加工、炼焦和核燃料加工业；4 个行业的氨氮排放量为 3.6 万吨，占重点调查工业企业氨氮排放总量的 68.4%。

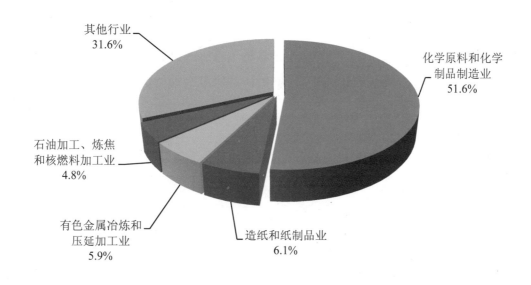

图 2-58   长江流域工业氨氮排放量行业构成

（4）废水及污染物治理情况

2012 年，长江中下游流域纳入统计的污水处理厂 673 座，形成了 3 199 万吨/日的处理能力，年运行费用达 67.0 亿元，共处理污水 93.6 亿吨，其中生活污水 80.1 亿吨。去除化学需氧量 198.6 万吨、氨氮 16.1 万吨、油类 0.4 万吨、总氮 13.4 万吨、总磷 2.0 万吨。

长江中下游流域重点调查工业企业共有废水治理设施 13 766 套，形成了 4 562 万吨/日的废水处理能力，年运行费用达 144.0 亿元，处理了 115.0 亿吨工业废水。共去除工业化学需氧量 253.5 万吨、氨氮 12.5 万吨、石油类 3.7 万吨、挥发酚 1.1 万吨、氰化物 1 340.0 吨。

2.4.7 太湖、巢湖及滇池

（1）废水及污染物排放情况

《重点流域水污染防治"十二五"规划》中太湖流域含上海、江苏、浙江 3 个省份的 8 个地市和 51 个区县。巢湖流域含安徽省的 4 个地市和 13 个区县。滇池流域含云南省的 1 个地市和 6 个区县。

2012 年，重点调查了太湖流域工业企业 7 739 家，规模化畜禽养殖场 2 401 家，规模化畜禽养殖小区 83 家。巢湖流域工业企业 1 290 家，规模化畜禽养殖场 608 家，规模化畜禽养殖小区 23 家。滇池流域工业企业 168 家，规模化畜禽养殖场 72 家，规模化畜禽养殖小区 16 家。

太湖、巢湖、滇池流域共排放废水 42.8 亿吨。其中，工业废水 15.7 亿吨，城镇生活污水 27.0 亿吨。化学需氧量排放总量为 48.0 万吨。其中，工业化学需氧量为 11.4 万吨，农业化学需氧量 16.7 万吨，城镇生活化学需氧量 19.6 万吨。氨氮排放总量为 7.2 万吨。其中，工业氨氮为 0.8 万吨，农业氨氮 1.8 万吨，城镇生活氨氮 4.5 万吨。

太湖、巢湖、滇池流域工业石油类排放总量为 321.3 吨，工业挥发酚排放总量为 9.4 吨，工业氰化物排放总量为 6.8 吨，工业废水中 6 种重金属（包括铅、镉、汞、六价铬、总铬及砷）排放总量为 13.1 吨。

表 2-15　太湖、巢湖及滇池流域废水及主要污染物排放情况

年份	污染物	废水/亿吨			化学需氧量/万吨				氨氮/万吨			
		工业源	生活源	集中式	工业源	农业源	生活源	集中式	工业源	农业源	生活源	集中式
2011	太湖	15.72	17.15	0.02	10.71	11.50	14.36	0.10	0.81	1.34	3.45	0.02
	巢湖	0.67	3.67	0.00	1.23	4.69	6.83	0.14	0.07	0.37	0.97	0.01
	滇池	0.09	2.86	0.00	0.12	0.36	0.47	0.08	0.01	0.05	0.26	0.01
2012	太湖	14.99	19.52	0.02	10.11	11.28	13.39	0.13	0.78	1.31	3.33	0.02
	巢湖	0.62	3.69	0.00	1.15	5.03	6.02	0.14	0.06	0.44	0.88	0.01
	滇池	0.08	3.82	0.00	0.13	0.42	0.15	0.08	0.01	0.06	0.25	0.01
变化率/%	太湖	−4.6	13.8	—	−5.7	−1.9	−6.7	—	−3.9	−2.1	−3.7	—
	巢湖	−7.6	0.6	—	−6.8	7.3	−11.9	—	−18.5	20.8	−9.1	—
	滇池	−6.3	33.7	—	3.0	15.7	−67.9	—	5.3	7.1	−4.6	—

（2）废水及主要污染物在各地区的分布

2012 年，太湖流域废水、化学需氧量和氨氮排放量最大的均是江苏省，分别占该流域各类污染物排放量的 67.7%、52.9% 和 49.7%。其中，工业废水、化学需氧量和氨氮排放量最大的均是江苏省，分别占该流域工业排放总量的 72.1%、65.1% 和 55.3%；农业化学需氧量和氨氮排放量最大的均是浙江省，分别占该流域农业排放总量的 54.1% 和 57.9%；城镇生活废水、化学需氧量和氨氮排放量最大的均是江苏省，分别占该流域生活排放总量的 64.4%、51.2% 和 51.9%。

巢湖流域均分布在安徽省，滇池流域均分布在云南省。

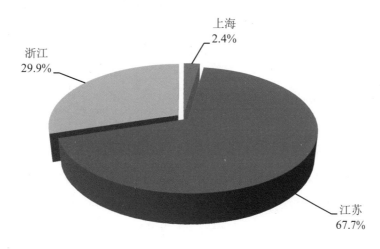

图 2-59　太湖流域废水排放区域构成

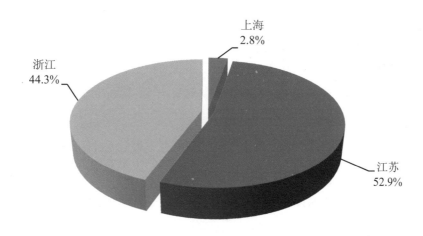

图 2-60　太湖流域化学需氧量排放区域构成

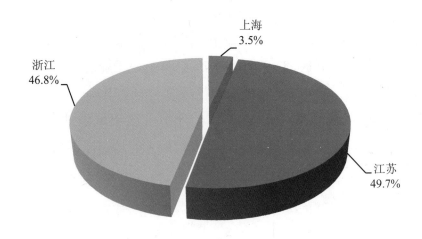

图 2-61 太湖流域氨氮排放区域构成

（3）废水及主要污染物在行业的分布

1）太湖流域

2012年，在调查统计的41个工业行业中，太湖流域废水排放量位于前4位的行业依次为纺织业，化学原料和化学制品制造业，造纸和纸制品业，计算机、通信和其他电子设备制造业；4个行业的废水排放量为10.2亿吨，占重点调查工业企业废水排放总量的75.5%。

太湖流域化学需氧量排放量位于前4位的行业依次为纺织业，化学原料和化学制品制造业，造纸和纸制品业，计算机、通信和其他电子设备制造业；4个行业的化学需氧量排放量为7.1万吨，占重点调查工业企业化学需氧量排放总量的77.2%。

太湖流域氨氮排放量位于前4位的行业依次为纺织业，化学原料和化学制品制造业，造纸和纸制品业，计算机、通信和其他电子设备制造业；4个行业的氨氮排放量为0.6万吨，占重点调查工业企业氨氮排放总量的81.6%。

2）巢湖流域

2012年，在调查统计的41个工业行业中，巢湖流域废水排放量位于前4位的行业依次为化学纤维制造业，黑色金属冶炼和压延加工业，化学原料和化学制品制造业，造纸和纸制品业；4个行业的废水排放量为0.33亿吨，占重点调查工业企业废水排放总量的61.3%。

巢湖流域化学需氧量排放量位于前4位的行业依次为黑色金属冶炼和压延加工业，造纸和纸制品业，化学纤维制造业，化学原料和化学制品制造业；4个行业的化学需氧量排放量为0.76万吨，占重点调查工业企业化学需氧量排放总量的73.0%。

巢湖流域氨氮排放量位于前4位的行业依次为黑色金属冶炼和压延加工业，化学纤维制造业，酒、饮料和精制茶制造业，化学原料和化学制品制造业；4个行业的氨氮排放量为 0.04 万吨，占重点调查工业企业氨氮排放总量的 73.1%。

3）滇池流域

2012 年，在调查统计的 41 个工业行业中，滇池流域废水排放量位于前 4 位的行业依次为化学原料和化学制品制造业，农副食品加工业，酒、饮料和精制茶制造业，医药制造业；4 个行业的废水排放量为 0.05 亿吨，占重点调查工业企业废水排放总量的 64.1%。

滇池流域化学需氧量排放量位于前 4 位的行业依次为农副食品加工业，化学原料和化学制品制造业，烟草制品业，酒、饮料和精制茶制造业；4 个行业的化学需氧量排放量为 0.07 万吨，占重点调查工业企业化学需氧量排放总量的 65.3%。

滇池流域氨氮排放量位于前 4 位的行业依次为化学原料和化学制品制造业，酒、饮料和精制茶制造业，农副食品加工业，造纸和纸制品业；4 个行业的氨氮排放量为 55 吨，占重点调查工业企业氨氮排放总量的 89.7%。

（4）废水及污染物治理情况

2012 年，太湖、巢湖、滇池流域纳入统计的污水处理厂分别为 332、19、10 座，形成了 1 056、107、84 万吨/日的处理能力，年运行费用达 31.2、2.4、2.4 亿元，共处理污水 26.9、3.6、2.8 亿吨，其中生活污水 17.4、3.6、2.8 亿吨。分别去除化学需氧量 77.8、5.7、7.6 万吨、氨氮 6.0、0.6、0.6 万吨、油类 0.3、0.04、0.04 万吨、总氮 4.3、0.5、0.8 万吨、总磷 0.7、0.2、0.1 万吨。

太湖、巢湖、滇池流域重点调查工业企业分别有废水治理设施 5 534、290、182 套，形成了 1 188、156、18 万吨/日的废水处理能力，年运行费用达 69.1、1.9、0.8 亿元，处理了 21.3、2.6、0.2 亿吨工业废水。分别去除工业化学需氧量 106.2、7.2、0.8 万吨、氨氮 2.5、0.5、0.1 万吨、石油类 0.4、0.1、0.002 万吨、挥发酚 366.6、0.1、0.8 吨、氰化物 122.2、3.3、0.7 吨。

2.4.8　三峡库区及其上游

（1）废水及污染物排放情况

《重点流域水污染防治"十二五"规划》中三峡库区及其上游流域含湖北、重庆、四川、贵州、云南 5 个省份的 42 个地市，含库区、影响区及上游区共 320 个区县。2012年，重点调查了工业企业 14 113 家，规模化畜禽养殖场 11 036 家，规模化畜禽养殖小区 802 家。

三峡库区及其上游流域共排放废水 55.5 亿吨，其中，工业废水 13.0 亿吨，城镇生活污水 42.4 亿吨。化学需氧量排放量为 202.6 万吨，其中，工业化学需氧量为 22.9 万吨，农业化学需氧量 73.9 万吨，城镇生活化学需氧量 104.3 万吨。氨氮排放量为 24.1 万吨。

其中，工业氨氮为 1.2 万吨，农业氨氮 8.1 万吨，城镇生活氨氮 14.6 万吨。

三峡库区及其上游流域工业石油类排放量为 1 271 吨，工业挥发酚排放量为 11.6 吨，工业氰化物排放量为 3.4 吨，工业废水中 6 种重金属（包括铅、镉、汞、六价铬、总铬及砷）排放总量为 23.0 吨。

表 2-16　三峡库区及其上游流域主要污染物排放情况

污染物 年份	废水/亿吨			化学需氧量/万吨				氨氮/万吨			
	工业源	生活源	集中式	工业源	农业源	生活源	集中式	工业源	农业源	生活源	集中式
2011	14.4	38.4	0.04	25.0	75.0	106.3	2.3	1.2	8.3	14.9	0.2
2012	13.0	42.4	0.05	22.9	73.9	104.3	1.5	1.2	8.1	14.6	0.2
变化率/%	−9.7	10.4	—	−8.4	−1.5	−1.9	—	0.5	−2.4	−2.0	—

（2）废水及主要污染物在各地区的分布

2012 年，三峡库区及其上游流域废水、化学需氧量和氨氮排放量最大的均是四川省，分别占该流域各类污染物排放量的 53.0%、62.6% 和 58.4%。其中，工业废水、化学需氧量和氨氮排放量最大的均是四川省，分别占该流域工业排放总量的 53.6%、51.6% 和 45.5%。农业化学需氧量和氨氮排放量最大的均是四川省，分别占该流域农业排放总量的 73.1% 和 70.9%。城镇生活废水、化学需氧量和氨氮排放量最大的均是四川省，分别占该流域生活排放总量的 50.3%、58.0% 和 52.8%。

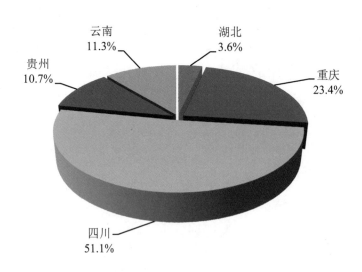

图 2-62　三峡库区及其上游流域废水排放区域构成

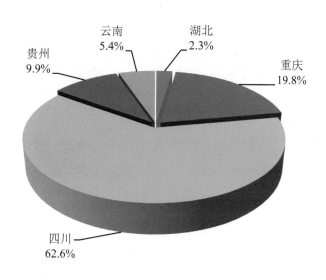

图 2-63　三峡库区及其上游流域化学需氧量排放区域构成

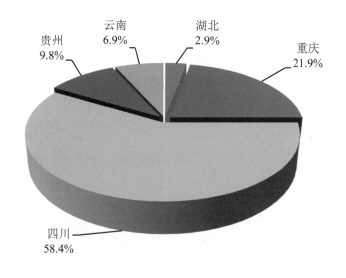

图 2-64　三峡库区及其上游流域氨氮排放区域构成

（3）废水及主要污染物在行业的分布

2012 年，在调查统计的 41 个工业行业中，三峡库区及其上游流域废水排放量位于前 4 位的行业依次为化学原料和化学制品制造业，煤炭开采和洗选业，造纸和纸制品业，酒、饮料和精制茶制造业；4 个行业的废水排放量为 7.2 亿吨，占重点调查工业企业废水排放总量的 60.7%。

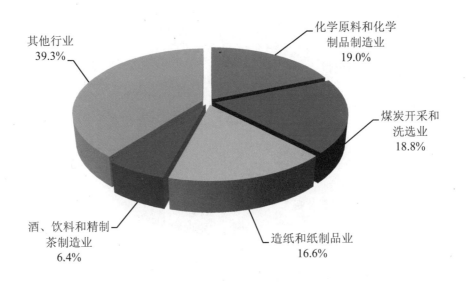

图 2-65　三峡库区及其上游流域工业废水排放量行业构成

2012 年，在调查统计的 41 个工业行业中，三峡库区及其上游流域化学需氧量位于前 5 位的行业依次为酒、饮料和精制茶制造业，造纸和纸制品业，农副食品加工业，煤炭开采和洗选业，化学原料和化学制品制造业；5 个行业的化学需氧量排放量为 15.4 万吨，占重点调查工业企业化学需氧量排放总量的 75.5%。

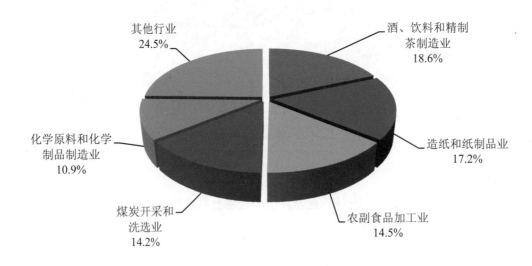

图 2-66　三峡库区及其上游流域工业化学需氧量排放量行业构成

2012 年，在调查统计的 41 个工业行业中，三峡库区及其上游流域氨氮排放量位于前 4 位的行业依次为化学原料和化学制品制造业，农副食品加工业，酒、饮料和精制茶制造业，造纸和纸制品业；4 个行业的氨氮排放量为 0.7 万吨，占重点调查工业企业氨氮排放总量的 65.7%。

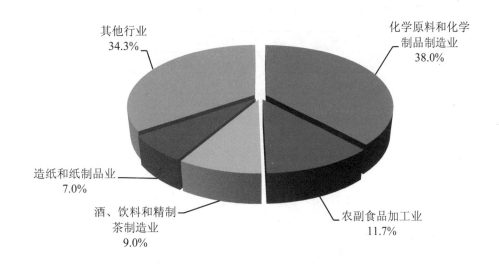

图 2-67　三峡库区及其上游流域工业氨氮排放量行业构成

（4）废水及污染物治理情况

2012 年，三峡库区及其上游流域纳入统计的污水处理厂 533 座，形成了 1 058 万吨/日的处理能力，年运行费用达 28.6 亿元，共处理污水 31.6 亿吨，其中生活污水 31.1 亿吨。去除化学需氧量 67.6 万吨、氨氮 6.6 万吨、油类 0.4 万吨、总氮 7.4 万吨、总磷 0.7 万吨。

三峡库区及其上游流域重点调查工业企业共有废水治理设施 7 701 套，形成了 1 756 万吨/日的废水处理能力，年运行费用达 43.9 亿元，处理了 33.3 亿吨工业废水。去除工业化学需氧量 94.5 万吨、氨氮 11.0 万吨、石油类 0.8 万吨、挥发酚 2 688.7 吨、氰化物 219.2 吨。

2.4.9　丹江口库区及其上游

（1）废水及污染物排放情况

《重点流域水污染防治"十二五"规划》中丹江口库区及其上游流域含河南、湖北、陕西 3 个省份的 8 个地市和 43 个区县。2012 年，重点调查了工业企业 1 686 家，规模化畜禽养殖场 1 291 家，规模化畜禽养殖小区 57 家。

丹江口库区及其上游流域共排放废水 4.5 亿吨，其中，工业废水 1.1 亿吨，城镇生活污水 3.4 亿吨。化学需氧量排放量为 21.9 万吨，其中，工业化学需氧量为 2.8 万吨，农业化学需氧量 8.8 万吨，城镇生活化学需氧量 9.8 万吨。氨氮排放量为 2.9 万吨，其中，工业氨氮为 0.3 万吨，农业氨氮 1.1 万吨，城镇生活氨氮 1.5 万吨。

丹江口库区及其上游流域工业石油类排放量为 152.7 吨，工业挥发酚排放量为 1.7 吨，工业氰化物排放量为 0.66 吨，工业废水中 6 种重金属（包括铅、镉、汞、六价铬、总铬及砷）排放总量为 17.0 吨。

表 2-17　丹江口库区及其上游流域主要污染物排放情况

| 年份 | 废水/亿吨 | | | 化学需氧量/万吨 | | | | 氨氮/万吨 | | | |
	工业源	生活源	集中式	工业源	农业源	生活源	集中式	工业源	农业源	生活源	集中式
2011	1.13	3.21	0.00	2.77	8.36	9.80	0.43	0.35	1.03	1.47	0.04
2012	1.08	3.44	0.01	2.79	8.82	9.83	0.42	0.34	1.05	1.48	0.04
变化率/%	−4.4	7.2	—	0.7	5.5	0.3	—	−2.9	1.9	0.7	—

（2）废水及主要污染物在各地区的分布

2012 年，丹江口库区及其上游流域废水、化学需氧量和氨氮排放量最大的均是陕西省，分别占该流域各类污染物排放量的 45.5%、55.5%和 59.3%。其中，工业废水、化学需氧量和氨氮排放量最大的均是陕西省，分别占该流域工业排放总量的 44.4%、49.9%和 68.1%。农业化学需氧量和氨氮排放量最大的均是陕西省，分别占该流域农业排放总量的 54.0%和 63.3%。城镇生活废水、化学需氧量和氨氮排放量最大的均是陕西省，分别占该流域生活排放总量的 45.8%、57.9%和 54.2%。

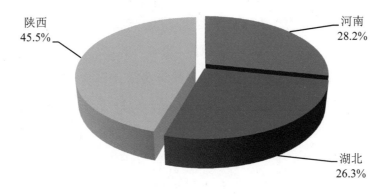

图 2-68　丹江口库区及其上游流域废水排放区域构成

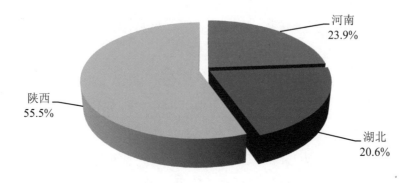

图 2-69　丹江口库区及其上游流域化学需氧量排放区域构成

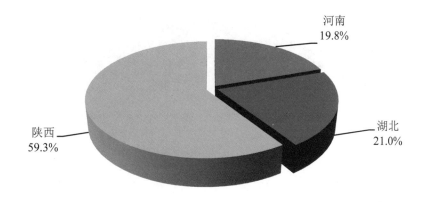

河南
19.8%

湖北
21.0%

陕西
59.3%

图 2-70 丹江口库区及其上游流域氨氮排放区域构成

（3）废水及主要污染物在行业的分布

2012 年，在调查统计的 41 个工业行业中，丹江口库区及其上游流域废水排放量位于前 3 位的行业依次为有色金属矿采选业、造纸和纸制品业、化学原料和化学制品制造业；3 个行业的废水排放量为 0.6 亿吨，占重点调查工业企业废水排放总量的 59.2%。

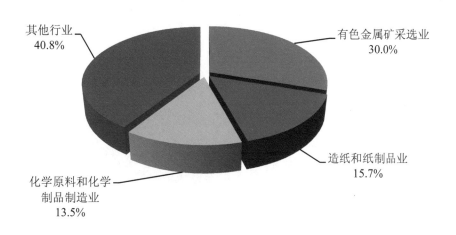

其他行业
40.8%

有色金属矿采选业
30.0%

化学原料和化学
制品制造业
13.5%

造纸和纸制品业
15.7%

图 2-71 丹江口库区及其上游流域工业废水排放量行业构成

2012 年，在调查统计的 41 个工业行业中，丹江口库区及其上游流域化学需氧量位于前 5 位的行业依次为医药制造业，农副食品加工业，化学原料和化学制品制造业，造纸和纸制品业，有色金属矿采选业；5 个行业的化学需氧量排放量为 1.9 万吨，占重点调查工业企业化学需氧量排放总量的 77.1%。

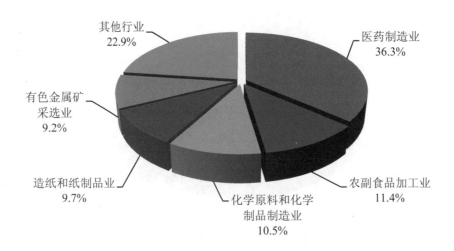

图 2-72　丹江口库区及其上游流域工业化学需氧量排放量行业构成

2012 年，在调查统计的 41 个工业行业中，丹江口库区及其上游流域氨氮排放量位于前 4 位的行业依次为化学原料和化学制品制造业，医药制造业，有色金属矿采选业，有色金属冶炼和压延加工业；4 个行业的氨氮排放量为 0.26 万吨，占重点调查工业企业氨氮排放总量的 81.0%。

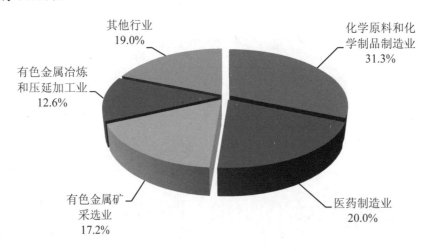

图 2-73　丹江口库区及其上游流域工业氨氮排放量行业构成

（4）废水及污染物治理情况

2012 年，丹江口库区及其上游流域纳入统计的污水处理厂 36 座，形成了 90 万吨/日的处理能力，年运行费用达 1.7 亿元，共处理污水 2.1 亿吨，其中生活污水 2.0 亿吨。去除化学需氧量 4.5 万吨、氨氮 0.3 万吨、总氮 0.2 万吨、总磷 0.03 万吨。

丹江口库区及其上游流域重点调查工业企业共有废水治理设施 1 185 套，形成了 171 万吨/日的废水处理能力，年运行费用达 3.7 亿元，处理了 2.9 亿吨工业废水。去除工业化学需氧量 6.9 万吨、氨氮 0.6 万吨、石油类 0.1 万吨、挥发酚 5.0 吨、氰化物 2.8 吨。

2.5 沿海地区废水及主要污染物排放情况

2.5.1 废水及污染物排放情况

2012 年，沿海地区的统计范围为天津、河北、辽宁、上海、江苏、浙江、福建、山东、广东、广西、海南等 11 个沿海省份的 56 个地市和 212 个区县。我国沿海地区重点调查工业企业数为 22 559 家，规模化畜禽养殖场 13 163 家，规模化畜禽养殖小区 427 家。

表 2-18　沿海地区废水及主要污染物接纳情况

年份 \ 污染物	废水/亿吨			化学需氧量/万吨				氨氮/万吨			
	工业源	生活源	集中式	工业源	农业源	生活源	集中式	工业源	农业源	生活源	集中式
2011	45.8	59.8	0.1	42.5	110.6	109.7	2.1	3.2	9.6	18.5	0.2
2012	38.6	63.9	0.1	42.5	108.1	106.1	2.0	3.1	9.5	18.1	0.2
变化率/%	−15.7	6.9	—	0	−2.3	−3.3	—	−3.1	−1.0	−2.2	—

表 2-19　沿海地区工业其他污染物排放情况

年份 \ 污染物	工业石油类/吨	工业挥发酚/吨	工业氰化物/吨	工业重金属/吨
2011	2 481.6	66.2	17.0	60.2
2012	2 492.2	62.8	13.0	48.3
变化率/%	0.4	−5.1	−23.5	−19.8

沿海地区废水排放总量为 102.5 亿吨。其中，工业废水排放量为 38.6 亿吨，城镇生活污水排放量为 63.9 亿吨。其中，东海沿海地区工业和生活的废水排放量均居四大海域之首，分别占整个沿海地区的 45.9%和 39.1%。

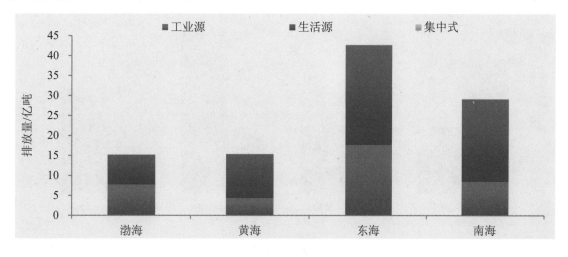

图 2-74　四大海域废水排放情况

沿海地区化学需氧量排放总量为 258.6 万吨。其中，工业废水化学需氧量为 42.5 万吨，农业源化学需氧量 108.1 万吨，城镇生活污水化学需氧量 106.1 万吨。其中，工业化学需氧量排放量最大的是东海，占整个沿海地区的 36.0%；农业源化学需氧量较大的是渤海和黄海，分别占沿海地区的 30.2% 和 29.7%；生活化学需氧量最大的是东海，占沿海地区的 41.1%。

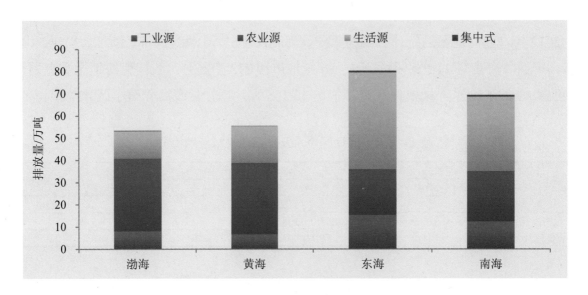

图 2-75　四大海域化学需氧量排放情况

沿海地区氨氮排放总量为 30.9 万吨。其中，工业废水氨氮为 3.1 万吨，农业源氨氮为 9.5 万吨，城镇生活污水氨氮为 18.1 万吨。其中，工业、农业、生活氨氮排放量最大的均为东海沿海，分别占整个沿海地区的 33.3%、34.8% 和 43.3%。

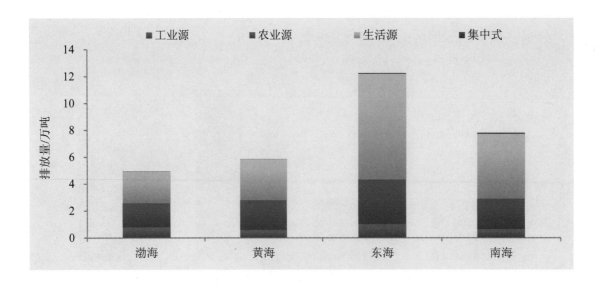

图 2-76　四大海域氨氮排放情况

沿海地区工业石油类排放量为 2 492.2 吨，工业挥发酚排放量为 62.8 吨，工业氰化物排放量为 13.0 吨，工业废水中 6 种重金属（包括铅、镉、汞、六价铬、总铬及砷）排放总量为 48.3 吨。

2.5.2 废水及主要污染物在各地区的分布

2012 年，沿海地区废水、化学需氧量和氨氮排放量主要来自于广东、山东、福建和浙江 4 个省。废水、化学需氧量和氨氮排放量最大的均是广东省，分别占该流域各类污染物排放量的 26.1%、20.9% 和 21.1%。其中，工业废水和化学需氧量排放量最大的是广东省，工业氨氮排放量最大的是浙江省，分别占该流域工业排放总量的 20.1%、24.3% 和 19.4%。农业化学需氧量排放量最大的是山东省，农业氨氮排放量最大的是福建省，分别占该流域农业排放总量的 25.2% 和 19.2%。城镇生活废水、化学需氧量和氨氮排放量最大的均是广东省，分别占该流域生活排放总量的 29.7%、25.8% 和 23.0%。

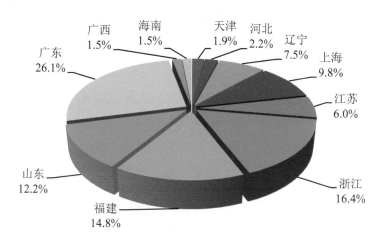

图 2-77 沿海地区废水排放区域构成

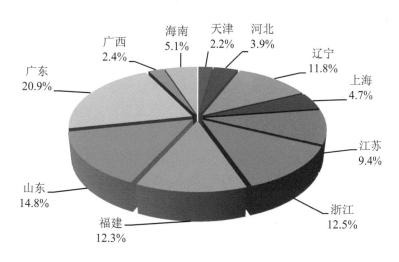

图 2-78 沿海地区化学需氧量排放区域构成

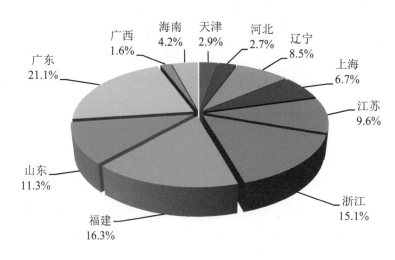

图 2-79　沿海地区氨氮排放区域构成

2.5.3　主要污染物在行业的分布

2012 年，在调查统计的 41 个工业行业中，沿海地区废水排放量位于前 4 位的行业依次为纺织业，造纸和纸制品业，电力、热力生产和供应业，石油加工、炼焦和核燃料加工业；4 个行业的废水排放量为 21.2 亿吨，占重点调查工业企业废水排放总量的 60.2%。

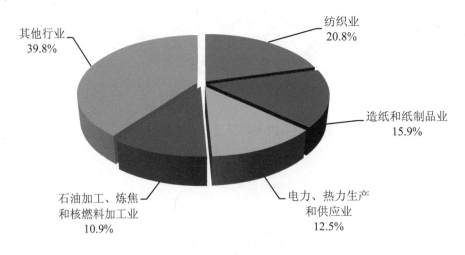

图 2-80　沿海地区工业废水排放量行业构成

2012 年，在调查统计的 41 个工业行业中，沿海地区化学需氧量位于前 4 位的行业依次为纺织业，造纸和纸制品业，农副食品加工业，化学原料和化学制品制造业；4 个行业的化学需氧量排放量为 24.4 万吨，占重点调查工业企业化学需氧量排放总量的 66.1%。

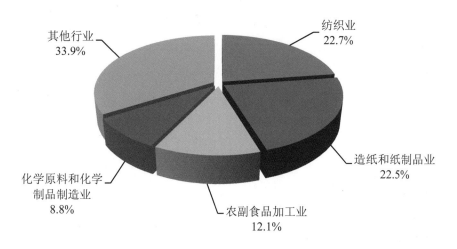

图 2-81 沿海地区工业化学需氧量排放量行业构成

2012 年，在调查统计的 41 个工业行业中，沿海地区氨氮排放量位于前 5 位的行业依次为纺织业，化学原料和化学制品制造业，造纸和纸制品业，石油加工、炼焦和核燃料加工业，农副食品加工业；5 个行业的氨氮排放量为 2.0 万吨，占重点调查工业企业氨氮排放总量的 72.1%。

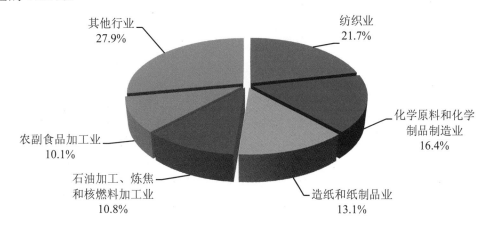

图 2-82 沿海地区工业氨氮排放量行业构成

2.5.4 废水及主要污染物治理情况

2012 年，沿海地区纳入统计的集中式污水处理厂 666 座，形成了 3 011 万吨/日的处理能力，年运行费用达 71.0 亿元，共处理污水 83.4 亿吨，其中生活污水 68.3 亿吨。去除化学需氧量 230.3 万吨、氨氮 18.3 万吨、油类 0.7 万吨、总氮 14.7 万吨、总磷 2.3 万吨。

沿海地区重点调查工业企业共有废水治理设施 14 702 套，形成了 3 187 万吨/日的处理能力，年运行费用达 110.3 亿元，处理了 63.1 亿吨工业废水。去除工业化学需氧量 330.4 万吨，氨氮 13.1 万吨，石油类 6.8 万吨，挥发酚 3 234.0 吨，氰化物 1 258.4 吨。

3

废　气

ANNUAL STATISTIC REPORT ON ENVIRONMENT IN CHINA
2012

3.1 废气及废气中主要污染物排放情况

3.1.1 二氧化硫排放情况

2012 年，全国工业废气排放量 635 519 亿立方米（标态），比 2011 年减少 5.8%；集中式废气排放量 36 832.3 亿立方米（标态）。

全国二氧化硫排放量 2 117.6 万吨，比 2011 年减少 4.52%。

工业二氧化硫排放量 1 911.7 万吨，比 2011 年减少 5.2%，占全国二氧化硫排放总量的 90.3%。

城镇生活二氧化硫排放量 205.7 万吨，比 2011 年增加 2.6%，占全国二氧化硫排放总量的 9.7%。

集中式污染治理设施二氧化硫排放量 0.3 万吨。

表 3-1　全国二氧化硫排放量　　　　　　　　　　　　　　单位：万吨

排放源 年份	合计	工业源	城镇生活源	集中式
2011	2 217.9	2 017.2	200.4	0.3
2012	2 117.6	1 911.7	205.7	0.3
变化率/%	−4.5	−5.2	2.6	—

注：集中式污染治理设施包括生活垃圾处理厂（场）和危险废物（医疗废物）集中处理（置）厂焚烧废气中排放的污染物。（下同）

3.1.2 氮氧化物排放情况

2012 年，全国氮氧化物排放量 2 337.8 万吨，比 2011 年减少 2.77%。

工业氮氧化物排放量 1 658.1 万吨，比 2011 年减少 4.1%，占全国氮氧化物排放总量的 70.9%。

城镇生活氮氧化物排放量 39.3 万吨，比 2011 年增加 7.4%，占全国氮氧化物排放总量的 1.7%。

机动车氮氧化物排放量 640.0 万吨，比 2011 年增加 0.4%，占全国氮氧化物排放总量的 27.4%。

集中式污染治理设施氮氧化物排放量 0.4 万吨。

表 3-2　全国氮氧化物排放量

単位：万吨

年份 \ 排放源	合计	工业源	城镇生活源	机动车	集中式
2011	2 404.3	1 729.7	36.6	637.6	0.3
2012	2 337.8	1 658.1	39.3	640.0	0.4
变化率/%	-2.8	-4.1	7.4	0.4	—

注：自 2011 年起机动车排气污染物排放情况与生活源分开单独统计。

3.1.3　烟（粉）尘排放情况

2012 年，全国烟（粉）尘排放量 1 234.3 万吨，比 2011 年减少 3.5%。

工业烟（粉）尘排放量 1 029.3 万吨，比 2011 年减少 6.5%，占全国烟（粉）尘排放总量的 83.4%。

城镇生活烟尘排放量 142.7 万吨，比 2011 年增加 24.3%，占全国烟（粉）尘排放总量的 11.6%。

机动车颗粒物排放量 62.1 吨，比 2011 年减少 1.3%，占全国烟（粉）尘排放总量的 5.0%。

集中式污染治理设施烟（粉）尘排放量 0.2 万吨。

表 3-3　全国烟（粉）尘排放量　　　　　　　　　　　　　单位：万吨

年份 \ 排放源	合计	工业源	城镇生活源	机动车	集中式
2011	1 278.8	1 100.9	114.8	62.9	0.2
2012	1 234.3	1 029.3	142.7	62.1	0.2
变化率/%	-3.5	-6.5	24.3	-1.3	—

注：1. 自 2011 年起不再单独统计烟尘和粉尘，统一以烟（粉）尘进行统计；2. 机动车的烟（粉）尘排放量指机动车的颗粒物排放量。

3.2　各地区废气中主要污染物排放情况

3.2.1　二氧化硫排放情况

2012 年，二氧化硫排放量超过 100 万吨的省份依次为山东、内蒙古、河北、山西、河南、辽宁和贵州，7 个省份的二氧化硫排放量占全国排放量的 43.2%。各地区中，工业和城镇生活二氧化硫排放量最大的均是山东，集中式污染治理设施二氧化硫排放量最大的是广东。

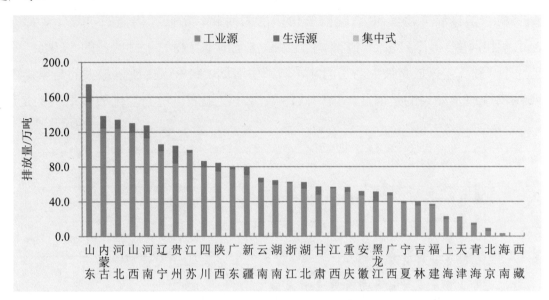

图 3-1　各地区二氧化硫排放情况

3.2.2　氮氧化物排放情况

氮氧化物排放量超过 100 万吨的省份依次为河北、山东、河南、江苏、内蒙古、广东、山西和辽宁，8 个省份氮氧化物排放量占全国氮氧化物排放量的 49.7%。工业氮氧化物排放量最大的是山东，城镇生活氮氧化物排放量最大的是黑龙江，机动车氮氧化物排放量最大的是河北，集中式污染治理设施氮氧化物排放量最大的是广东。

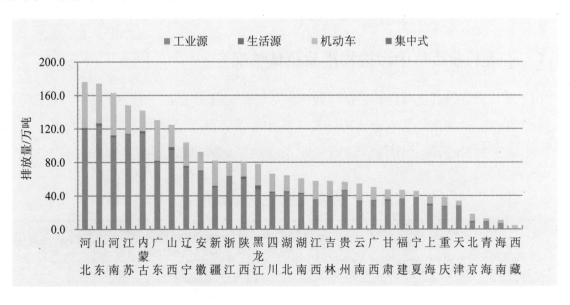

图 3-2　各地区氮氧化物排放情况

3.2.3 烟（粉）尘排放情况

烟（粉）尘排放量超过 50 万吨的省份依次为河北、山西、内蒙古、辽宁、黑龙江、新疆、山东和河南，8 个省份烟（粉）尘排放量占全国烟（粉）尘排放量的 53.0%。各地区中，工业烟（粉）尘排放量最大的是河北，城镇生活烟（粉）尘排放量最大的是黑龙江，机动车颗粒物排放量最大的是河北，集中式污染治理设施烟（粉）尘排放量最大的是广东。

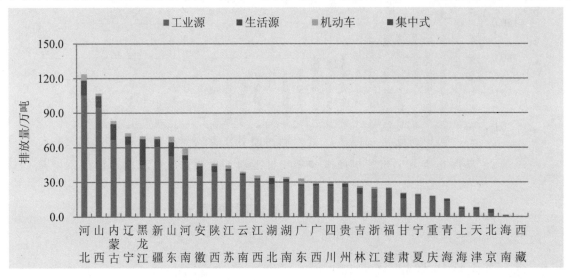

图 3-3　各地区烟（粉）尘排放情况

3.3　工业行业废气中主要污染物排放情况

3.3.1　工业行业废气中污染物排放总体情况

2012 年，调查统计的 41 个工业行业中，二氧化硫排放量位于前 3 位的行业依次为电力、热力生产和供应业，黑色金属冶炼及压延加工业，非金属矿物制品业，3 个行业共排放二氧化硫 1 237.4 万吨，占重点调查工业企业二氧化硫排放总量的 69.7%。

表 3-4　重点行业二氧化硫排放情况 单位：万吨

行业 年份	合计	电力、热力生产和供应业	黑色金属冶炼及压延加工业	非金属矿物制品业
2011	1 354.3	901.2	251.4	201.7
2012	1 237.4	797.0	240.6	199.8
变化率/%	−8.6	−11.6	−4.3	−0.9

2012 年，调查统计的工业行业中，氮氧化物排放量位于前 3 位的行业依次为电力、热力生产和供应业，非金属矿物制品业，黑色金属冶炼及压延加工业，3 个行业共排放氮氧化物 1 390.1 万吨，占重点调查工业企业氮氧化物排放总量的 87.9%。

表 3-5 重点工业行业氮氧化物排放情况   单位：万吨

年份 \ 行业	合计	电力、热力生产和供应业	非金属矿物制品业	黑色金属冶炼及压延加工业
2011	1 471.3	1 106.8	269.4	95.1
2012	1 390.1	1 018.7	274.2	97.2
变化率/%	−5.5	−8.0	1.8	2.1

2012 年，调查统计工业行业中，烟（粉）尘排放量位于前 3 位的行业依次为非金属矿物制品业，电力、热力生产和供应业，黑色金属冶炼及压延加工业；3 个行业共排放烟（粉）尘 659.3 万吨，占重点调查统计工业企业烟（粉）尘排放量的 68.9%。

表 3-6 重点工业行业烟（粉）尘排放情况   单位：万吨

年份 \ 行业	合计	非金属矿物制品业	电力、热力生产和供应业	黑色金属冶炼及压延加工业
2011	700.9	279.1	215.6	206.2
2012	659.3	255.2	222.8	181.3
变化率/%	−5.9	−8.6	3.3	−12.1

3.3.2 电力、热力生产和供应业废气污染物排放及处理情况

（1）电力、热力生产和供应业总体情况

2012 年，电力、热力生产和供应业重点调查工业企业 5 525 家，占重点调查工业企业的 3.7%；全年工业废气排放量为 203 436.4 亿立方米（标态），占重点调查工业企业废气排放量的 32.0%；二氧化硫排放量为 797.0 万吨，占重点调查工业企业的 44.9%；氮氧化物排放量为 1 018.7 万吨，占重点调查工业企业的 64.4%；烟（粉）尘排放量为 222.8 万吨，占重点调查工业企业的 23.3%。

重点调查工业企业拥有废气治理设施 22 642 套，占重点调查工业企业废气治理设施总数的 10.0%，其中，脱硫设施 5 677 套，占重点调查工业企业脱硫设施总数的 27.1%；脱硝设施 480 套，占重点调查工业企业脱硝设施总数的 57.7%；除尘设施 16 162 套，占重点调查工业企业除尘设施总数的 9.1%。二氧化硫去除量为 2 439.4 万吨，去除率为 75.4%，较重点调查工业企业平均水平（68.9%）高出 6.5 个百分点；氮氧化物去除量为

115.2 万吨，去除率为 10.2%，较重点调查工业企业平均水平（7.5%）高出 2.7 个百分点；烟（粉）尘去除量为 38 818.2 万吨，去除率为 99.4%，较重点调查工业企业平均水平（98.7%）高出 0.7 个百分点。

电力、热力生产和供应业二氧化硫排放量居全国前 4 位的省份依次为内蒙古、山东、山西和贵州，其二氧化硫排放量占电力、热力生产和供应业总排放量的 34.0%；氮氧化物排放量居前 4 位的省份依次为内蒙古、江苏、山东和河南，其氮氧化物排放量占电力、热力生产和供应业总排放量的 30.5%；烟（粉）尘排放量居前 4 位的省份依次为内蒙古、黑龙江、辽宁和山西，其烟（粉）尘排放量占电力、热力生产和供应业总排放量的 45.5%。

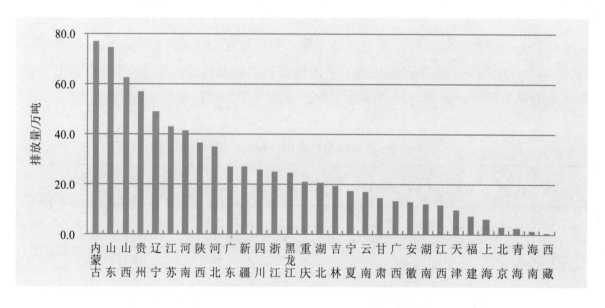

图 3-4 各地区电力、热力生产和供应业二氧化硫排放情况

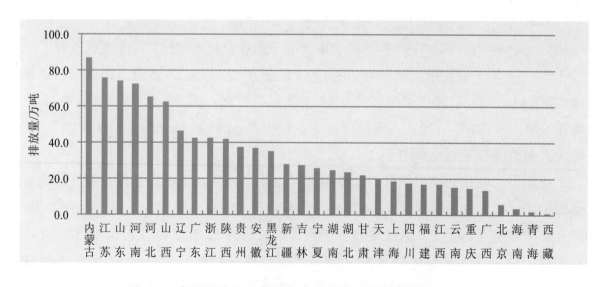

图 3-5 各地区电力、热力生产和供应业氮氧化物排放情况

76

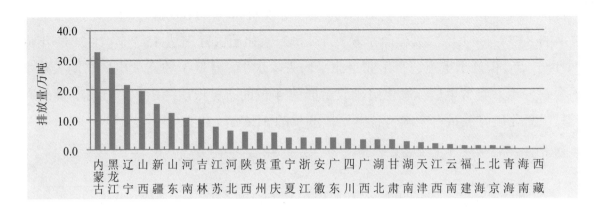

图 3-6　各地区电力、热力生产和供应业烟（粉）尘排放情况

（2）火电厂废气污染物排放及治理情况

2012 年，纳入重点调查统计范围的火电厂共 3 127 家，占重点调查工业企业数量的 2.1%。其中，独立火电厂 1 824 家，拥有 4 625 台机组，共有脱硫设施 3 465 套，脱硝设施 438 套，除尘设施 5 171 套；自备电厂 1 303 家，拥有 2 517 台机组，其中 1 011 台机组有脱硫设施的，71 台有脱硝设施，1 539 台有除尘设施。

独立火电厂二氧化硫排放量为 706.3 万吨，比 2011 年减少 13.6%，占重点调查工业企业排放量的 39.8%；共去除二氧化硫 2 396.3 万吨，二氧化硫去除率达 77.2%，比 2011 年提高 2.7 个百分点，比重点调查工业企业平均去除水平高 8.3 个百分点。

独立火电厂二氧化硫排放量居前 4 位的省份依次为山东、内蒙古、山西和贵州，其二氧化硫排放量占全国独立火电厂排放量的 35.1%。独立火电厂单位发电量二氧化硫平均排放强度为 1.90 克/千瓦·时，其中重庆、贵州、四川和云南 4 个省份的排放强度大于 4.0 克/千瓦·时。

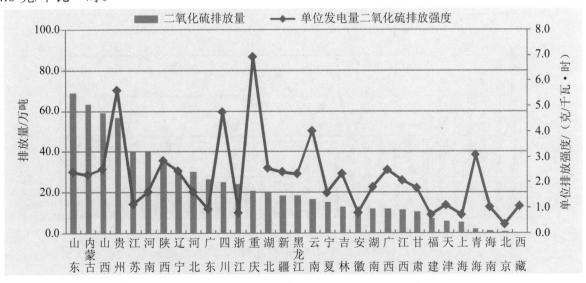

图 3-7　各地区独立火电厂二氧化硫排放情况

自备电厂二氧化硫排放量为 152.6 万吨，比 2011 年增加 0.8%，占重点调查工业企业排放量的 8.6%；共去除二氧化硫 126.5 万吨，二氧化硫去除率达 45.3%，比 2011 年提高 0.2 个百分点，但低于重点调查工业企业平均去除水平，且低于独立火电厂平均去除水平 31.9 个百分点。自备电厂二氧化硫排放量居前 4 位的省份依次为山东、江苏、新疆和宁夏，其二氧化硫排放量占全国自备电厂排放量的 31.6%。

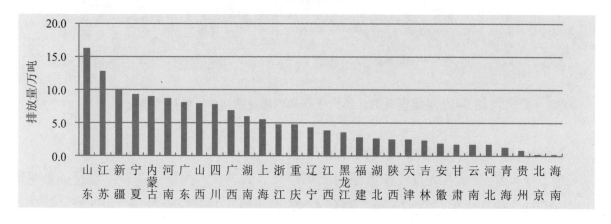

图 3-8 各地区自备电厂二氧化硫排放情况

独立火电厂氮氧化物排放量为 981.6 万吨，占重点调查工业企业排放量的 62.1%。共去除氮氧化物 111.6 万吨，氮氧化物去除率为 10.2%，比 2011 年提高 4.1 个百分点，比重点调查工业企业平均去除水平高出 2.7 个百分点。

独立火电厂氮氧化物排放量居前 4 位的省份依次为内蒙古、江苏、山东和河南，其氮氧化物排放量占独立火电厂排放量的 30.7%。独立火电厂单位发电量氮氧化物平均排放强度为 2.6 克/千瓦·时，其中重庆、吉林和黑龙江 3 个省份的排放强度大于 4.0 克/千瓦·时。

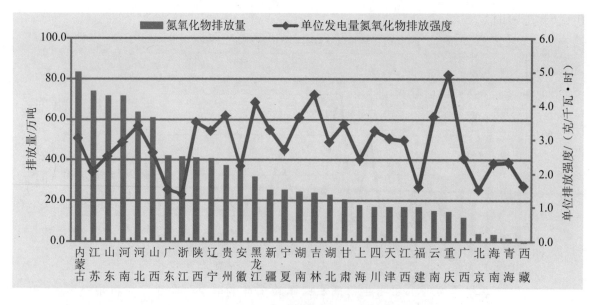

图 3-9 各地区独立火电厂氮氧化物排放情况

自备电厂氮氧化物排放量为 103.1 万吨，占重点调查工业企业排放量的 6.5%。共去除氮氧化物 3.6 万吨，氮氧化物去除率为 3.4%，比 2011 年提高个 2.1 百分点，但低于重点调查工业企业平均去除水平，且低于独立火电厂平均去除水平 6.8 个百分点。自备电厂氮氧化物排放量居前 4 位的省份依次为江苏、山东、内蒙古和新疆，其氮氧化物排放量占自备电厂排放量的 35.3%。

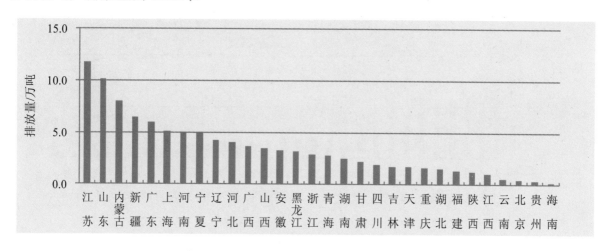

图 3-10　各地区自备电厂氮氧化物排放情况

独立火电厂烟（粉）尘排放量为 144.2 万吨，占重点调查工业企业排放量的 15.1%。共去除烟（粉）尘 3.6 亿吨，烟（粉）尘去除率达 99.6%，与 2011 年持平，比重点调查工业企业平均去除水平高出 0.9 个百分点。

独立火电厂烟（粉）尘排放量居前 4 位的省份依次为内蒙古、山西、黑龙江和山东，其烟（粉）尘排放量占独立火电厂排放量的 43.2%。独立火电厂单位发电量烟粉尘平均排放强度为 0.4 克/千瓦·时，其中黑龙江和重庆 2 个省份的排放强度大于 1 克/千瓦·时。

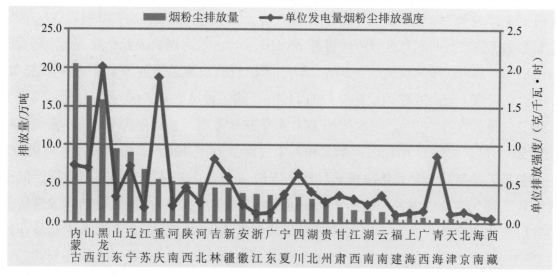

图 3-11　各地区独立火电厂烟（粉）尘排放情况

自备电厂烟（粉）尘排放量为 34.1 万吨，占重点调查工业企业排放量的 3.6%。重点调查工业企业共去除烟（粉）尘 2 777.0 万吨，烟（粉）尘去除率达 98.8%，比 2011 年提高 0.1 个百分点，比重点调查工业企业平均去除水平高 0.1 个百分点，但低于独立火电厂平均去除水平 0.8 个百分点。自备电厂烟（粉）尘排放量居前 4 位的省份依次为内蒙古、广西、山东和辽宁，其烟（粉）尘排放量占自备电厂排放量的 34.3%。

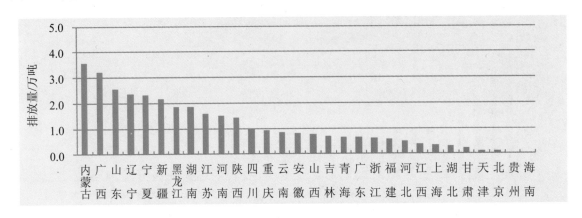

图 3-12　各地区自备电厂烟（粉）尘排放情况

3.3.3　非金属矿物制品业废气污染物排放及处理情况

（1）非金属矿物制品业总体情况

2012 年，非金属矿物制品业重点调查工业企业 33 164 家，占重点调查工业企业总数的 22.4%；全年工业废气排放量为 123 285.2 亿立方米（标态），占重点调查统计工业企业废气排放量的 19.4%；二氧化硫排放量为 199.8 万吨，占重点调查统计工业企业排放量的 11.3%；氮氧化物排放量为 274.2 万吨，占全重点调查统计工业企业排放量的 17.3%；烟（粉）尘排放量为 255.2 万吨，占重点调查统计工业企业排放量的 26.7%。

非金属矿物制品业拥有废气治理设施 65 810 套，占重点调查工业企业废气治理设施总数的 29.1%，其中，脱硫设施数 1 976 套，占全国脱硫设施总数的 9.4%，脱硝设施数 95 套，占重点调查工业企业脱硝设施总数的 11.4%，除尘设施数 61 700 套，占重点调查工业企业除尘设施总数的 34.7%。二氧化硫去除量为 28.9 万吨，去除率为 12.6%；氮氧化物去除量为 5.0 万吨，去除率为 1.8%；烟（粉）尘去除量为 22 126.6 万吨，去除率为 98.9%。

非金属矿物制品业二氧化硫排放量居全国前 4 位的省份依次为山东、安徽、河南和江西，其二氧化硫排放量占非金属矿物制品业排放量的 31.0%；氮氧化物排放量居前 4 位的省份依次为安徽、山东、广东和河南，其氮氧化物排放量占非金属矿物制品业排放量的 28.5%；烟（粉）尘排放量居前 4 位的省份依次为河北、安徽、云南和河南，其烟（粉）尘排放量占非金属矿物制品业排放量的 28.2%。

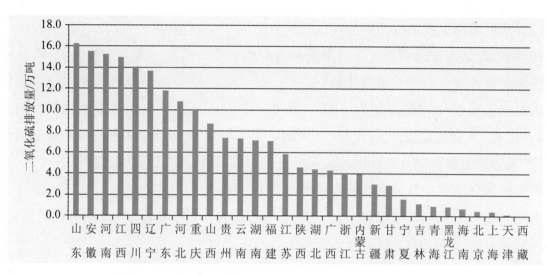

图 3-13　各地区非金属矿物制品业二氧化硫排放情况

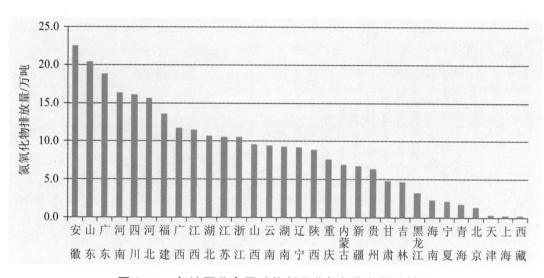

图 3-14　各地区非金属矿物制品业氮氧化物排放情况

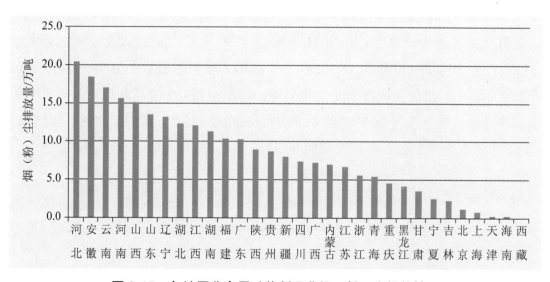

图 3-15　各地区非金属矿物制品业烟（粉）尘排放情况

（2）水泥制造企业废气污染物排放及处理情况

2012 年，纳入重点调查统计范围的水泥制造企业（简称"水泥企业"，下同）3 288 家，占重点调查工业企业数量的 2.2%。

水泥企业氮氧化物排放量为 197.9 万吨，占重点调查工业企业氮氧化物排放量的 12.5%。拥有 84 套脱硝设施，比 2011 年增加 77 套，占重点调查工业企业脱硝设施数的 10.1%；共去除氮氧化物 3.8 万吨，氮氧化物去除率 1.9%，比 2011 年提高 0.9 个百分点，但低于重点调查工业企业平均去除水平低 5.6 个百分点。

水泥企业氮氧化物排放量居前 4 位的省份依次为安徽、山东、河南和四川，其氮氧化物排放量占全国水泥企业排放量的 27.3%。

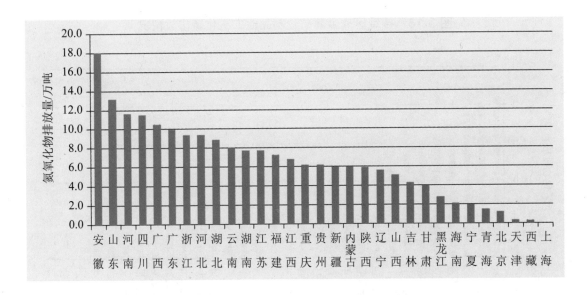

图 3-16　各地区水泥企业氮氧化物排放及处理情况

水泥企业烟（粉）尘排放量为 67.1 万吨，占重点调查工业企业烟（粉）尘排放量的 7.0%。拥有除尘设施 48 339 套，比 2011 年增加 3 597 套，占重点调查工业企业除尘设施数的 27.2%；共去除烟（粉）尘 18 182.6 万吨，烟（粉）尘去除率为 99.6%，比 2011 年增加 0.2 个百分点，比重点调查工业企业平均去除水平高 0.9 个百分点。

水泥企业烟（粉）尘排放量居前 4 位的省份依次为辽宁、湖南、安徽和山东，其烟（粉）尘排放量占全国水泥企业排放量的 30.1%。

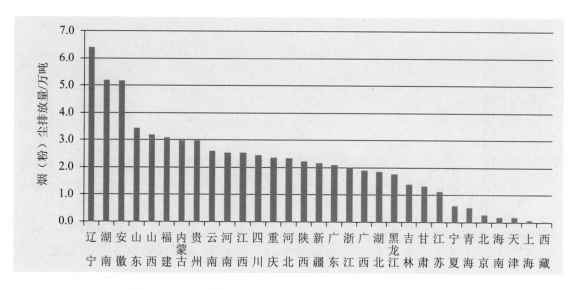

图 3-17　各地区水泥企业烟（粉）尘排放及处理情况

3.3.4　黑色金属冶炼及压延加工业废气污染物排放及处理情况

（1）黑色金属冶炼及压延加工业总体情况

2012 年，黑色金属冶炼及压延加工业重点调查工业企业 4 599 家，占重点调查工业企业总数的 3.1%；全年工业废气排放量为 160 874.9 亿立方米（标态），占重点调查工业企业废气排放量的 25.3%；二氧化硫排放量为 240.6 万吨，占重点调查工业企业排放量的 13.5%；氮氧化物排放量为 97.2 万吨，占重点调查工业企业排放量的 6.1%；烟（粉）尘排放量为 181.3 万吨，占重点调查工业企业排放量的 18.9%。

黑色金属冶炼及压延加工业拥有废气治理设施 17 009 套，占重点调查工业企业废气治理设施总数的 7.5%，其中，脱硫设施 800 套，占重点调查工业企业脱硫设施总数的 3.8%，脱硝设施数 47 套，占重点调查工业企业脱硝设施总数的 5.6%；除尘设施 15 490 套，占重点调查工业企业除尘设施总数的 8.7%。二氧化硫去除量为 84.9 万吨，去除率为 26.1%；氮氧化物去除量为 1.1 万吨，去除率为 1.1%；烟（粉）尘去除量为 7 575.0 万吨，去除率为 97.7%。

黑色金属冶炼及压延加工业二氧化硫排放量居全国前 4 位的省份依次为河北、四川、山西和山东，其二氧化硫排放量占黑色金属冶炼及压延加工业排放量的 39.4%；氮氧化物排放量居前 4 位的省份依次为河北、山东、江苏和辽宁，其氮氧化物排放量占黑色金属冶炼及压延加工业排放量的 46.5%；烟（粉）尘排放量居前 4 位的省份依次为河北、山西、山东和辽宁，其烟（粉）尘排放量占黑色金属冶炼及压延加工业排放量的 54.9%。

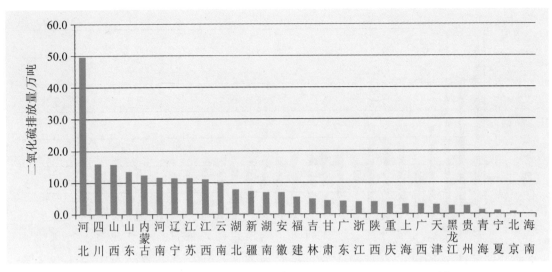

图 3-18　各地区黑色金属冶炼及压延加工业二氧化硫排放情况

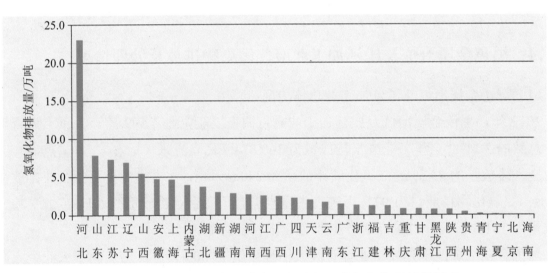

图 3-19　各地区黑色金属冶炼及压延加工业氮氧化物排放情况

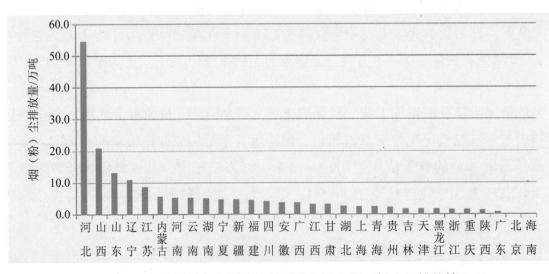

图 3-20　各地区黑色金属冶炼及压延加工业烟（粉尘）排放情况

（2）钢铁冶炼企业废气污染物排放及处理情况

2012 年，纳入重点调查统计范围的、有烧结机或球团设备的钢铁冶炼企业 1 180 家，占重点调查工业企业总数的 0.8%。共拥有烧结机数 1 127 台，其中有 333 台有脱硫设施，749 台有除尘设施；有球团设备数 657 套，其中有 45 套有脱硫设施，386 套有除尘设施。

钢铁冶炼企业二氧化硫排放量为 200.0 万吨，占重点调查工业企业排放量的 11.3%。二氧化硫去除量为 51.3 万吨，二氧化硫去除率为 20.4%，比重点调查工业企业平均去除水平低 48.5 个百分点。钢铁冶炼企业二氧化硫排放量居前 4 位的省份依次为河北、四川、山西和山东，占全国钢铁企业排放量的 40.2%。

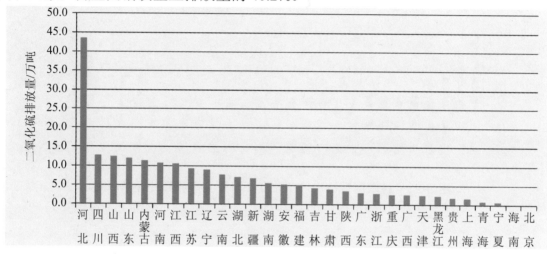

图 3-21　各地区钢铁冶炼企业二氧化硫排放及处理情况

钢铁冶炼企业氮氧化物排放量为 54.3 万吨，占重点调查工业企业排放量的 3.4%。氮氧化物去除量 0.6 万吨，氮氧化物去除率为 1.2%，比重点调查工业企业平均去除水平低 6.3 个百分点。钢铁冶炼企业氧化物排放量居前 4 位的省份依次为河北、山东、江苏和辽宁，占全国钢铁企业排放量的 53.6%。

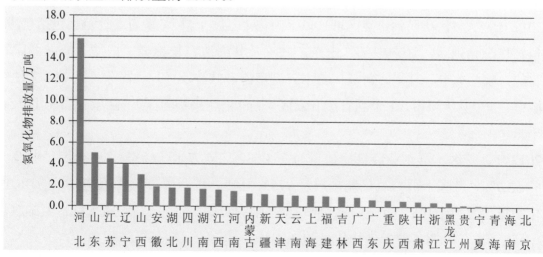

图 3-22　各地区钢铁冶炼企业氮氧化物排放及处理情况

钢铁冶炼企业烟（粉）尘排放量为 60.1 万吨，占重点调查工业企业排放量的 6.3%。烟（粉）尘去除量 2 206.4 万吨，烟（粉）尘去除率为 97.3%，比 2011 年提高 0.4 个百分点，比重点调查工业平均去除水平低 1.4 个百分点。钢铁冶炼企业烟（粉）尘放量居前 4 位的省份依次为河北、山西、辽宁和山东，占全国钢铁企业排放量的 62.1%。

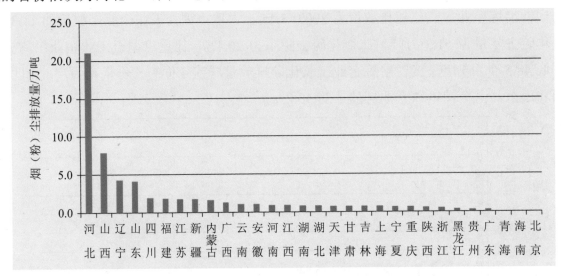

图 3-23　各地区钢铁冶炼企业烟（粉）尘排放及处理情况

3.4　大气污染防治重点区域（三区十群）废气污染物排放及处理情况

3.4.1　三区十群总体情况

2012 年，大气污染防治重点区域（以下简称"三区十群"）调查统计范围包括京津冀、长三角、珠三角地区，辽宁中部、山东、武汉及其周边、长株潭、成渝、海峡西岸、山西中北部、陕西关中、甘宁、新疆乌鲁木齐城市群，共涉及 19 个省份（含 114 个地级及以上城市、3 个县级城市、1 个副省级开发区、1 个正厅级实验区），区域总面积 132.56 万平方千米。

2012 年，三区十群工业废气排放量为 341 532 亿立方米（标态），占全国工业废气排放量的 53.7%。三区十群二氧化硫排放量 979.6 万吨，占全国二氧化硫排放量的 46.3%。氮氧化物排放量为 1 148.7 万吨，占全国氮氧化物排放量的 49.1%。烟（粉）尘排放量为 491.9 万吨，占全国烟（粉）尘排放量的 39.9%。

3.4.2 三区十群二氧化硫排放及处理情况

2012 年，三区十群二氧化硫排放量为 979.6 万吨，其中，工业二氧化硫排放量为 899.2 万吨，占三区十群二氧化硫排放总量的 91.8%；城镇生活二氧化硫排放量为 80.3 万吨，占三区十群二氧化硫排放总量的 8.2%；集中式二氧化硫排放量为 0.1 万吨。

表 3-7 三区十群二氧化硫排放情况　　　　　　　　　　　单位：万吨

排放源 区域	合计	工业源	城镇生活源	单位面积排放强度/ （吨/千米^2）
京津冀	166.0	151.4	14.6	7.6
长三角	184.6	176.3	8.2	8.8
珠三角	46.5	45.6	0.9	8.5
辽宁中部城市群	53.1	48.8	4.2	8.2
山东城市群	174.9	154.4	20.5	11.2
武汉及其周边城市群	35.5	32.1	3.4	6.0
长株潭城市群	10.9	10.4	0.5	3.9
成渝城市群	124.7	113.6	11.1	5.6
海峡西岸城市群	37.1	35.2	1.9	3.0
山西中北部城市群	45.4	39.6	5.8	8.0
陕西关中城市群	52.2	46.3	5.9	9.5
甘宁城市群	30.1	28.1	2.0	7.0
新疆乌鲁木齐城市群	18.7	17.3	1.4	5.9
总　计	979.6	899.2	80.3	7.4

三区十群共有工业脱硫设施 13 880 套，占全国工业脱硫设施数的 66.2%，二氧化硫去除率为 67.7%。全年工业二氧化硫去除量为 1 883.8 万吨，占全国工业二氧化硫去除量的 46.8%。

三区十群中，二氧化硫排放量居前 4 位的区域依次为长三角地区、山东城市群、京津冀地区和成渝城市群，其二氧化硫排放量占三区十群排放量的 66.4%。

三区十群单位面积二氧化硫排放强度为 7.4 吨/平方千米，是全国平均排放强度（2.2 吨/平方千米）的 3.4 倍。三区十群中各区域的单位面积二氧化硫排放强度均高于全国平均水平。其中，单位面积二氧化硫排放强度较大的区域为山东城市群和陕西关中城市群，其污染物排放强度分别为 11.2 吨/平方千米和 9.5 吨/平方千米。二氧化硫排放量和单位面积二氧化硫排放强度均较大的区域为长三角地区、山东城市群和京津冀地区。

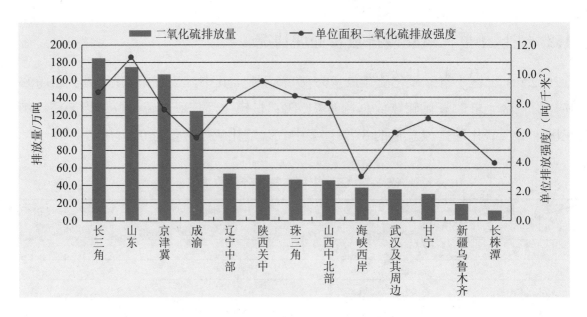

图 3-24　三区十群二氧化硫排放情况

3.4.3　三区十群氮氧化物排放及处理情况

2012 年，氮氧化物排放量 1 148.7 万吨。其中，工业氮氧化物排放量 827.6 万吨，占区域氮氧化物排放量的 72.0%；生活氮氧化物排放量 17.0 万吨，占区域氮氧化物排放量的 1.5%；机动车氮氧化物排放量 303.9 万吨，占区域氮氧化物排放量的 26.5%；集中式氮氧化物排放量 0.2 万吨。

表 3-8　三区十群氮氧化物排放量　　　　　　　　　　　　　　单位：万吨

排放源 区域	合计	工业源	城镇生活源	机动车	单位面积排放强度/ （吨/千米²）
京津冀	227.3	155.6	3.5	68.2	10.4
长三角	269.0	205.3	2.9	60.7	12.8
珠三角	81.4	48.2	0.4	32.8	14.9
辽宁中部城市群	56.6	41.8	1.0	13.8	8.7
山东城市群	173.9	123.8	3.2	46.9	11.1
武汉及其周边城市群	38.8	28.6	0.7	9.5	6.5
长株潭城市群	13.7	9.3	0.1	4.3	4.9
成渝城市群	92.8	64.6	1.3	26.8	4.2
海峡西岸城市群	46.7	36.1	0.2	10.4	3.8
山西中北部城市群	45.1	35.7	1.3	8.1	7.9
陕西关中城市群	51.5	39.9	1.8	9.8	9.4
甘宁城市群	29.8	22.6	0.4	6.8	6.9
新疆乌鲁木齐城市群	22.1	16.1	0.3	5.7	7.0
总计	1 148.7	827.6	17.0	303.9	8.7

三区十群共有工业脱硝设施 595 套，占全国工业脱硝设施数的 71.5%，氮氧化物去除率为 9.2%。全年工业氮氧化物去除量为 84.3 万吨，占全国工业氮氧化物去除量的 62.4%。

三区十群中，氮氧化物排放量居前 4 位的区域依次为长三角地区、京津冀地区、山东城市群和成渝城市群，其氮氧化物排放量占三区十群排放量的 66.4%。

三区十群单位面积氮氧化物排放强度为 8.7 吨/平方千米，是全国平均排放强度（2.4 吨/平方千米）的 3.6 倍。三区十群中各区域单位面积氮氧化物排放强度均高于全国平均水平。其中，单位面积氮氧化物排放强度较大的区域为珠三角地区、长三角地区、山东城市群和京津冀地区，其污染物排放强度分别为 14.9、12.8、11.1 和 10.4 吨/平方千米。氮氧化物排放量和单位面积氮氧化物排放强度均较大的区域为长三角地区、京津冀地区和山东城市群。

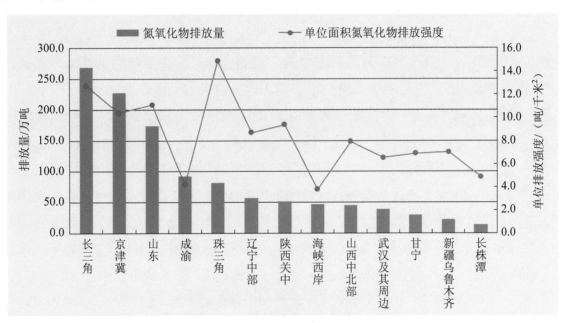

图 3-25 三区十群氮氧化物排放情况

3.4.4 三区十群烟（粉）尘排放及处理情况

2012 年，三区十群烟（粉）尘排放量为 491.9 万吨。其中，工业烟（粉）尘排放量 416.9 万吨，占区域烟（粉）尘排放量的 84.8%；生活烟（粉）尘排放量 47.1 万吨，占区域烟（粉）尘排放量的 9.6%；机动车烟（粉）尘排放量 27.8 万吨，占区域烟（粉）尘排放量的 5.7%；集中式烟（粉）尘排放量 0.1 万吨。

表 3-9 三区十群烟（粉）尘排放情况   单位：万吨

排放源 区域	合计	工业源	城镇生活源	机动车	单位面积排放强度/ （吨/千米²）
京津冀	132.6	114.6	11.6	6.4	6.1
长三角	78.4	69.3	3.8	5.3	3.7
珠三角	18.5	14.8	0.4	3.4	3.4
辽宁中部城市群	44.2	39.7	3.2	1.4	6.8
山东城市群	69.5	52.6	12.0	5.0	4.4
武汉及其周边城市群	16.8	13.3	2.7	0.9	2.8
长株潭城市群	5.2	4.4	0.5	0.3	1.9
成渝城市群	38.5	34.9	1.7	1.9	1.7
海峡西岸城市群	24.8	23.4	0.5	0.9	2.0
山西中北部城市群	29.5	23.2	5.5	0.8	5.2
陕西关中城市群	16.5	12.5	3.5	0.5	3.0
甘宁城市群	8.5	7.2	0.7	0.6	2.0
新疆乌鲁木齐城市群	8.6	7.1	1.0	0.5	2.7
总计	491.6	416.9	47.1	27.8	3.7

三区十群共有工业除尘设施 98 027 套，占全国工业除尘设施数的 55.1%。全年工业烟（粉）尘去除量为 36 876.0 万吨，占全国工业烟（粉）尘去除量的 48.0%；氮氧化物去除率为 98.9%，比全国工业烟（粉）尘平均去除率高 0.1 个百分点。

三区十群中，烟（粉）尘排放量居前 4 位的区域依次为京津冀地区、长三角地区、山东城市群和辽宁中部城市群，其烟（粉）尘排放量占三区十群排放量的 66.0%。

三区十群单位面积烟（粉）尘排放强度为 3.7 吨/平方千米，是全国平均排放强度（1.3 吨/平方千米）的 2.8 倍。三区十群中各区域的单位面积烟（粉）尘排放强度均高于全国平均水平。其中，单位面积烟（粉）尘排放强度较大的区域为辽宁中部城市群、京津冀地区和山西中北部城市群，其污染物排放强度分别为 6.8、6.1 和 5.2 吨/平方千米。烟（粉）尘排放量和单位面积烟（粉）尘排放强度均较大的区域为京津冀地区、山东城市群和辽宁中部城市群。

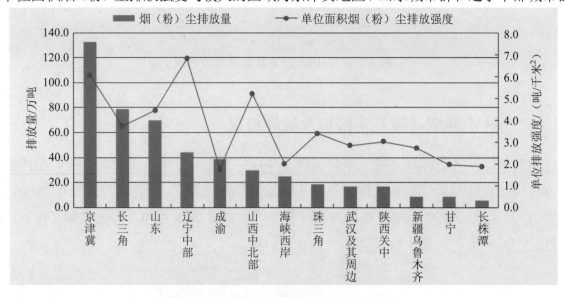

图 3-26 三区十群氮氧化物排放情况

3.5 机动车排气中主要污染物排放情况

2012 年，全国机动车排放一氧化碳 3 471.7 万吨，碳氢化合物 438.2 万吨，氮氧化物 640.0 万吨，颗粒物 62.1 万吨。机动车污染物排放量较大的是汽车，其一氧化碳、碳氢化合物、氮氧化物和颗粒物排放量分别占机动车排放量的 82.5%、78.7%、91.1%和 95.3%。

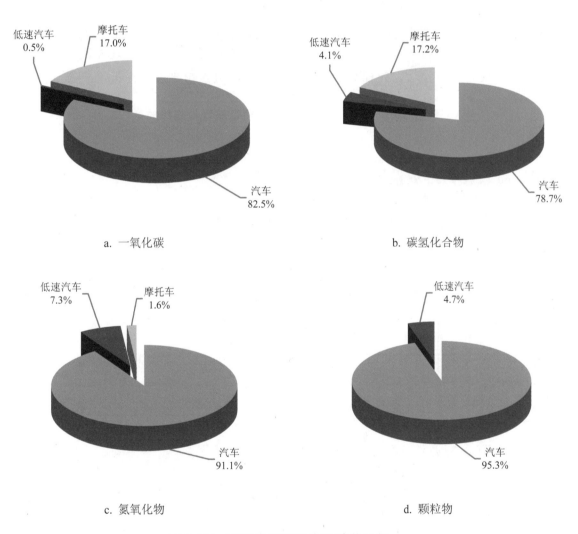

a. 一氧化碳
b. 碳氢化合物
c. 氮氧化物
d. 颗粒物

图 3-27　不同类型机动车污染物比例

3.5.1　颗粒物排放情况

2012 年全国机动车颗粒物排放量为 62.1 万吨。其中，汽车排放 59.1 万吨，占 95.3%；

低速汽车排放 3.0 万吨，占 4.7%。

2012 年全国机动车颗粒物排放量前 5 位的省份依次为河北、河南、山东、广东、辽宁。

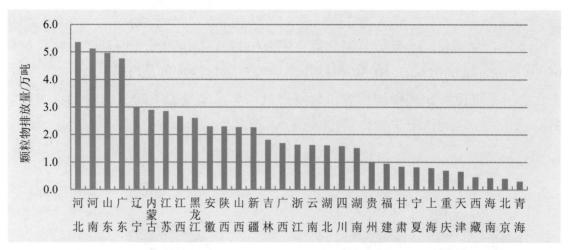

图 3-28　2012 年全国各省份机动车颗粒物排放量

3.5.2　氮氧化物排放情况

2012 年全国机动车氮氧化物排放量为 640.0 万吨。其中，汽车排放 582.9 万吨，占 91.1%；低速汽车排放 46.6 万吨，占 7.3%；摩托车排放 10.5 万吨，占 1.6%。

2012 年全国机动车氮氧化物排放量前 5 位的依次为河北、河南、广东、山东、江苏。

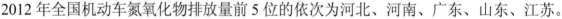
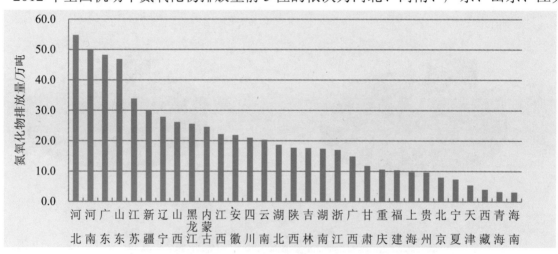

图 3-29　2012 年全国各省份机动车氮氧化物排放量

3.5.3　一氧化碳排放情况

2012 年全国机动车一氧化碳排放量为 3 471.7 万吨。其中，汽车排放 2 865.5 万吨，占 82.5%；低速汽车排放 16.0 万吨，占 0.5%；摩托车排放 590.2 万吨，占 17.0%。

2012 年全国机动车一氧化碳排放量前 5 位的省份依次为广东、河北、山东、河南、江苏。

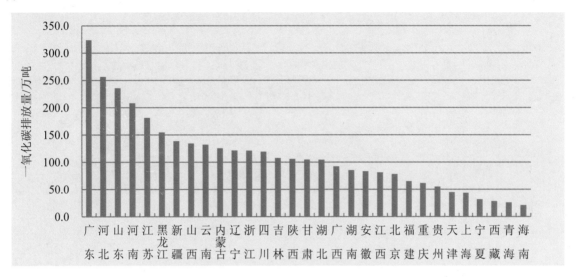

图 3-30　2012 年全国各省份机动车一氧化碳排放量

3.5.4　碳氢化合物排放情况

2012 年全国机动车碳氢化合物排放量为 438.2 万吨。其中，汽车排放 345.2 万吨，占 78.7%；低速汽车排放 17.6 万吨，占 4.1%；摩托车排放 75.4 万吨，占 17.2%。

2012 年全国机动车碳氢化合物排放量前 5 位的省份依次为广东、河北、山东、河南、江苏。

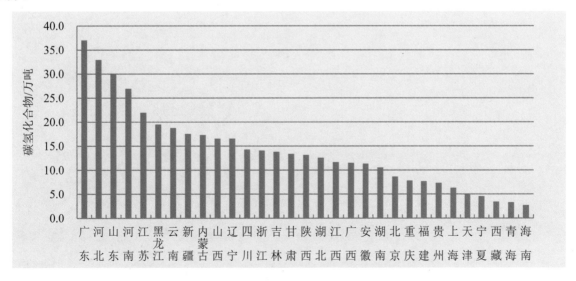

图 3-31　2012 年全国各省份机动车碳氢化合物排放量

4

工业固体废物

ANNUAL STATISTIC REPORT ON ENVIRONMENT IN CHINA
2012

4.1 一般工业固体废物产生及处理情况

2012 年，全国一般工业固体废物产生量 32.9 亿吨，比 2011 年增加 2.0%。其中，尾矿产生量为 11.0 亿吨，占全国总产量的 33.4%；粉煤灰 4.6 亿吨，占 14.0%；煤矸石 3.7 亿吨，占 11.2%；冶炼废渣 3.5 亿吨，占 10.7%。

全国一般工业固体废物综合利用量为 20.2 亿吨，比 2011 年增加 3.7%，综合利用率为 61.0%。其中，尾矿综合利用量为 3.1 亿吨，综合利用率为 28.0%；粉煤灰综合利用量为 3.8 亿吨，综合利用率为 82.1%；煤矸石综合利用量为 2.9 亿吨，综合利用率为 77.8%；冶炼废渣综合利用量为 3.3 亿吨，综合利用率为 92.2%；炉渣综合利用量为 2.4 亿吨，综合利用率为 87.9%。

2012 年，全国一般工业固体废物贮存量为 6.0 亿吨，比 2011 年减少 1.1%；处置量为 7.1 亿吨，比 2011 年增加了 0.4%；倾倒丢弃量为 144.2 万吨。

表 4-1 全国一般工业固体废物产生及处理情况　　　　　　　单位：万吨

年份	产生量	综合利用量	贮存量	处置量	倾倒丢弃量
2011	322 722.3	195 214.6	60 424.3	70 465.3	433.3
2012	329 044.3	202 461.9	59 786.3	70 744.8	144.2
变化率/%	1.96	3.71	−1.06	0.40	−66.72

注：1. "综合利用量"包括综合利用往年贮存量，"处置量"包括处置往年贮存量；
　　2. 工业固体废物综合利用率=工业固体废物综合利用量/（工业固体废物产生量＋综合利用往年贮存量）。

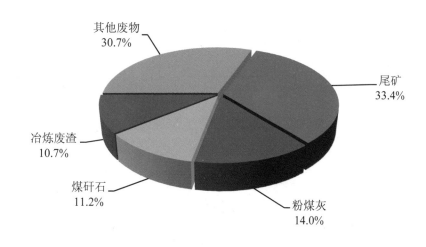

图 4-1 一般工业固体废物构成情况

尾矿产生量较大的省份依次为河北 2.7 亿吨、辽宁 1.1 亿吨、内蒙古 0.9 亿吨、江西 0.7 亿吨、四川 0.7 亿吨,其中河北、辽宁两省尾矿产生量占全国的 34.5%。

粉煤灰产生量较大的省份依次为内蒙古 4 316.2 万吨、山东 4 222.2 万吨、山西 3 922.6 万吨、河南 3 497.8 万吨和江苏 2 902.1 万吨,5 省粉煤灰产生量占全国的 40.9%。

煤矸石产生量较大的省份依次为山西 1.3 亿吨、内蒙古 0.5 亿吨、安徽 0.3 亿吨、河南 0.2 亿吨和山东 0.2 亿吨。其中,山西省煤矸石产生量占全国的 35.1%。

冶炼废渣产生量较大的省份依次为河北 7 125.5 万吨、山东 3 015.4 万吨、辽宁 2 011.0 万吨、江苏 2 593.4 万吨和山西 2 174.4 万吨,其中河北省冶炼废渣产生量占全国的 20.2%。

4.1.1 各地区一般工业固体废物产生及处理情况

2012 年一般工业固体废物产生量较大的省份为河北 4.6 亿吨,占全国工业企业产生量的 14.0%;山西 2.9 亿吨,占全国工业企业产生量的 8.8%;辽宁 2.7 亿吨,占全国工业企业产生量的 8.2%;内蒙古 2.4 亿吨,占全国工业企业产生量的 7.3%;山东 1.8 亿吨,占全国工业企业产生量的 5.5%。

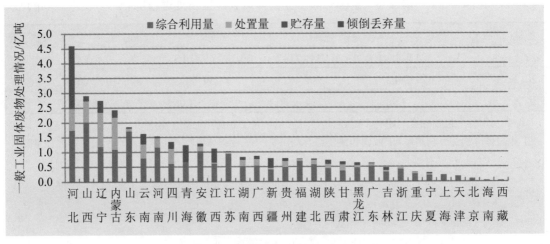

图 4-2 各地区一般工业固体废物综合利用、处置、贮存、倾倒丢弃情况

综合利用量较大的省份为山西 2.0 亿吨,主要为煤矸石,占全省工业企业综合利用量的 50.6%;河北 1.7 亿吨,主要为冶炼废渣,占全省工业企业的 40.7%;山东 1.7 亿吨,主要为粉煤灰、冶炼废渣和尾矿,占全省工业企业的 56.9%;辽宁 1.2 亿吨,主要为尾矿、冶炼废渣和粉煤灰,占全省工业企业的 50.9%;河南 1.2 亿吨,主要为粉煤灰、煤矸石和炉渣,占全省工业企业的 65.2%。这 5 个省份的一般工业固体废物综合利用量占全国工业企业的 38.6%。一般工业固体废物综合利用率较大的省份为天津、上海、山东、浙江和江苏,这 5 个省份的一般工业固体废物综合利用率均高于 90%。

处置量较大的省份为辽宁 1.2 亿吨，主要为尾矿，占全省工业企业处置量的 49.5%；内蒙古 1.1 亿吨，主要为尾矿，占全省工业企业的 65.5%；河北 0.7 亿吨，主要为尾矿，占全省工业企业的 94.1%；山西 0.7 亿吨，主要为煤矸石和尾矿，占全省工业企业的 63.2%；四川 0.5 亿吨，主要为尾矿，占全省工业企业的 87.2%。这 5 个省份的一般工业固体废物处置量占全国工业企业的 59.7%。

贮存量较大的省份为河北 2.1 亿吨，主要为尾矿，占全省工业企业贮存量的 85.0%；青海 0.5 亿吨，主要为尾矿和其他废物，占全省工业企业的 91.7%；江西 0.5 亿吨，主要为赤泥，占全省工业企业的 98.1%；辽宁 0.4 亿吨，主要为尾矿，占全省工业企业的 84.3%；云南 0.4 亿吨，主要为尾矿，占全省工业企业的 57.9%。其中，河北省的一般工业固体废物贮存量占全国工业企业的 35.5%。

倾倒丢弃量较大的省份为云南 43.1 万吨，主要为尾矿，占全省一般工业固体废物倾倒丢弃量的 93.5%；新疆 34.9 万吨，主要为尾矿，占全省工业企业的 59.0%；山西 16.1 万吨，主要为炉渣，占全省工业企业的 75.6%；贵州 14.0 万吨，主要为煤矸石，占全省工业企业的 74.1%；辽宁 10.4 万吨，主要为其他废物，占全省工业企业的 98.8%。5 个省份的工业固体废物倾倒丢弃量占全国工业企业的 82.2%。

4.1.2 工业行业固体废物产生及利用情况

2012 年，一般工业固体废物产生量较大的行业依次为黑色金属矿采选业 7.1 亿吨，占重点调查工业企业的 22.5%；电力、热力生产和供应业 6.1 亿吨，占重点调查工业企业的 19.6%；黑色金属冶炼和压延加工业 4.2 亿吨，占重点调查工业企业的 13.4%；有色金属矿采选业 4.0 亿吨，占重点调查工业企业的 12.8%；煤炭开采和洗选业 3.9 亿吨，占重点调查工业企业的 12.3%；化学原料和化学制品制造业 2.7 亿吨，占重点调查工业企业的 8.5%。

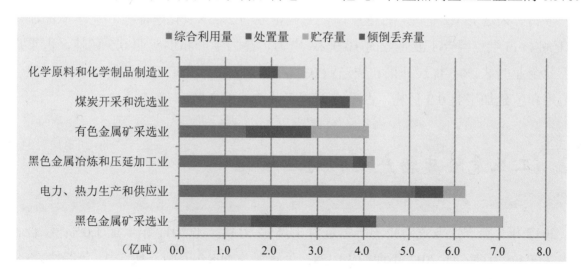

图 4-3　一般工业固体废物处理情况

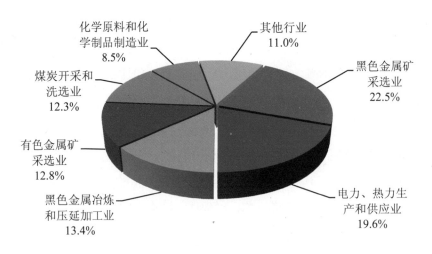

图 4-4　一般工业固体废物产生量行业构成

综合利用量较大的行业依次为电力、热力生产和供应业 5.1 亿吨，占重点调查工业企业的 26.7%；黑色金属冶炼和压延加工业 3.7 亿吨，占重点调查工业企业的 19.6%；煤炭开采和洗选业 3.0 亿吨，占重点调查工业企业的 15.9%；化学原料和化学制品制造业 1.7 亿吨，占重点调查工业企业的 9.1%；黑色金属矿采选业 1.6 亿吨，占全国的 8.1%；有色金属矿采选业 1.4 亿吨，占重点调查工业企业的 7.6%。

处置量较大的行业依次为黑色金属矿采选业 2.7 亿吨，占重点调查工业企业的 39.8%；有色金属矿采选业 1.4 亿吨，占重点调查工业企业的 20.7%；煤炭开采和洗选业 0.6 亿吨，占重点调查工业企业的 9.3%；电力、热力生产和供应业 0.6 亿吨，占重点调查工业企业的 9.3%；化学原料和化学制品制造业 0.4 亿吨，占重点调查工业企业的 5.9%；有色金属冶炼和压延加工业 0.3 亿吨，占重点调查工业企业的 4.8%。

贮存量较大的行业依次为黑色金属矿采选业 2.8 亿吨，占重点调查工业企业的 47.9%；有色金属矿采选业 1.2 亿吨，占重点调查工业企业的 21.3%；化学原料和化学制品制造业 0.6 亿吨，占重点调查工业企业的 10.0%；电力、热力生产和供应业 0.5 亿吨，占重点调查工业企业的 8.1%；煤炭开采和洗选业 0.3 亿吨，占重点调查工业企业的 4.6%；黑色金属冶炼和压延加工业 0.2 亿吨，占重点调查工业企业的 2.9%。

4.2　工业危险废物产生和处理情况

2012 年，全国工业危险废物产生量为 3 465.2 万吨，比 2011 年增加 1.0%；综合利用量为 2 004.6 万吨，比 2011 年增加 13.1%；处置量为 698.2 万吨，比 2011 年减少 23.8%；贮存量为 846.9 万吨，比 2011 年增加 2.8%；倾倒丢弃量为 16.1 吨。工业危险废物处置利用率为 76.1%，比 2011 年下降 0.4 个百分点。

表 4-2　全国工业危险废物产生及处理情况　　　　　　　　　单位：万吨

年份	产生量	综合利用量	处置量	贮存量	倾倒丢弃量
2011	3 431.2	1 773.1	916.5	823.7	0.0
2012	3 465.2	2 004.6	698.2	846.9	0.0
变化率/%	0.99	13.06	−23.82	2.82	—

注：危险废物处置利用率＝（危险废物综合利用量＋处置量）/（危险废物产生量＋综合利用往年贮存量＋处置往年贮存量）。

产生量较大危险废物种类为废碱 752.5 万吨，占重点调查工业企业工业危险废物产生量的 21.7%；石棉废物 709.0 万吨，占重点调查工业企业的 20.5%；废酸 392.4 万吨，占重点调查工业企业的 11.3%；有色金属冶炼废物 275.5 万吨，占重点调查工业企业的 8.0%；无机氰化物废物 111.5 万吨，占重点调查工业企业的 3.2%；废矿物油 102.4 万吨，占重点调查工业企业的 3.0%。

废碱产生量较大的省份为山东 490.5 万吨和新疆 107.7 万吨，两省废碱产生量占重点调查工业企业的 79.5%。

石棉废物产生量较大的省份为青海 389.1 万吨和新疆 319.4 万吨，两省石棉废物产生量占重点调查工业企业的 99.9%。

废酸产生量较大的省份为四川 83.5 万吨和江苏 50.3 万吨，两省废酸产生量占重点调查工业企业的 34.1%。

有色金属冶炼废物产生量较大的省份为云南 105.4 万吨，占重点调查工业企业的 38.2%。

无机氰化物废物产生量较大的省份为山东 81.3 万吨，占重点调查工业企业的 73.0%。

废矿物油产生量较大的省份为山东 27.3 万吨和辽宁 16.5 万吨，两省废矿物油产生量占重点调查工业企业的 42.8%。

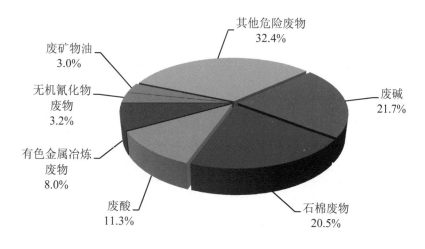

图 4-5　工业危险废物产生量构成情况

4.2.1 各地区工业危险废物产生和处理情况

2012 年各地区工业危险废物产生量较大的省份为山东 820.3 万吨，占工业企业危险废物产生量的 23.7%；新疆 444.2 万吨，占全国工业企业的 12.8%；青海 404.3 万吨，占全国工业企业的 11.7%；湖南 267.5 万吨，占全国工业企业的 7.7%；江苏 208.6 万吨，占全国工业企业的 6.0%；云南 208.0 万吨，占全国工业企业的 6.0%。

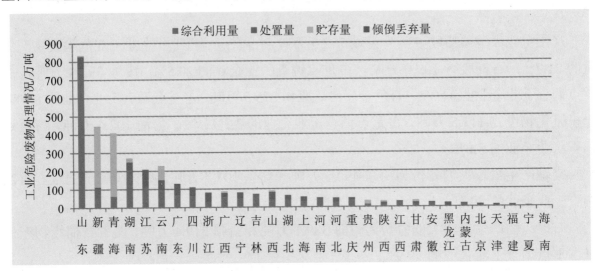

图 4-6　各地区工业危险废物综合利用、处置、贮存、倾倒丢弃情况

2012 年工业危险废物综合利用量较大的省份为山东 761.1 万吨，占全国工业企业危险废物综合利用量的 38.0%；湖南 207.7 万吨，占全国工业企业的 10.4%；江苏 110.0 万吨，占全国工业企业的 5.5%；新疆 102.4 万吨，占全国工业企业的 5.1%；云南 97.4 万吨，占全国工业企业的 4.9%。全国工业危险废物综合利用率为 57.2%。共有 12 个省份工业危险废物综合利用率超过全国平均水平，其中吉林和山东均超过 90%，河南和江西均超过 80%。

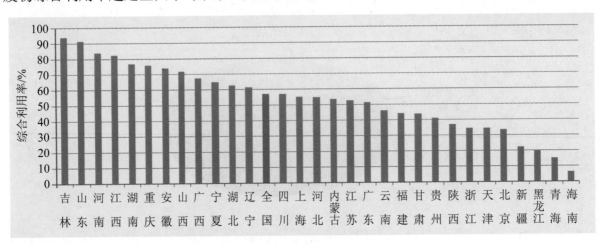

图 4-7　各地区工业危险废物综合利用率

2012 年工业危险废物处置量较大的省份为江苏 97.6 万吨，占全国工业企业工业危险废物处置量的 14.0%；山东 66.1 万吨，占全国工业企业的 9.5%；广东 62.8 万吨，占全国工业企业的 9.0%；云南 53.1 万吨，占全国工业企业的 7.6%；浙江 51.6 万吨，占全国工业企业的 7.4%。

2012 年工业危险废物贮存量较大的省份为青海 347.5 万吨，占全国工业企业工业危险废物贮存量的 41.0%；新疆 332.5 万吨，占全国工业企业的 39.3%；云南 76.5 万吨，占全国工业企业的 9.0%。

4.2.2 工业行业危险废物产生和处理情况

2012 年工业危险废物产生量较大的行业为造纸和纸制品业 715.1 万吨，占重点调查工业企业危险废物产生量的 20.6%；非金属矿采选业 712.3 万吨，占重点调查工业企业的 20.6%；化学原料和化学制品制造业 661.6 万吨，占重点调查工业企业的 19.1%；有色金属冶炼和压延加工业 453.9 万吨，占重点调查工业企业的 13.1%。

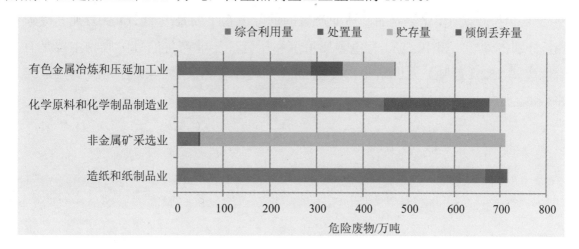

图 4-8 主要工业行业危险废物处理情况

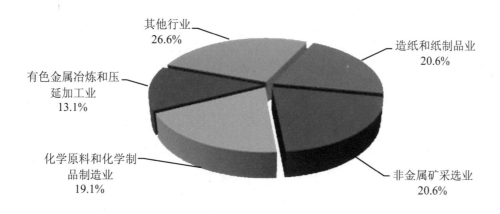

图 4-9 工业危险废物产生量行业分布情况

103

工业危险废物综合利用量较大的行业为造纸和纸制品业 667.4 万吨，占重点调查工业企业危险废物综合利用量的 33.3%，综合利用率为 93.3%；化学原料和化学制品制造业 447.1 万吨，占重点调查工业企业的 22.3%，综合利用率为 66.4%；有色金属冶炼和压延加工业 286.4 万吨，占重点调查工业企业的 14.3%，综合利用率为 61.7%。

工业危险废物处置量较大的行业为化学原料和化学制品制造业 229.2 万吨，占重点调查工业企业工业危险废物处置量的 32.8%，处置率为 32.7%；有色金属冶炼和压延加工业 68.3 万吨，占重点调查工业企业的 9.8%，处置率为 14.9%；计算机、通信和其他电子设备制造业 67.4 万吨，占重点调查工业企业的 9.7%，处置率为 45.6%。

工业危险废物贮存量较大的行业为非金属矿采选业 663.1 万吨，占重点调查工业企业工业危险废物贮存量的 78.3%；有色金属冶炼和压延加工业 113.4 万吨，占重点调查工业企业的 13.4%；化学原料和化学制品制造业 35.6 万吨，占重点调查工业企业的 4.2%。

造纸和纸制品业产生的危险废物主要是废碱 607.0 万吨，占该行业重点调查工业企业危险废物产生量的 84.9%；非金属矿采选业产生的危险废物主要是石棉废物 708.6 万吨，占 99.5%；化学原料和化学制品制造业产生的危险废物主要是废酸 269.9 万吨和废碱 82.8 万吨，分别占 40.9% 和 12.5%。有色金属冶炼和压延加工业产生的危险废物主要是有色金属冶炼废物 252.5 万吨、无机氰化物废物 58.3 万吨和含铅废物 54.7 万吨，分别占 55.6%、12.8% 和 12.1%。

5

环境污染治理投资

ANNUAL STATISTIC REPORT ON ENVIRONMENT IN CHINA
2012

5.1 总体情况

5.1.1 环境污染治理投资总额

环境污染治理投资包括老工业污染源治理、建设项目"三同时"、城市环境基础设施建设三个部分。2012 年,我国环境污染治理投资总额为 8 253.6 亿元,占国内生产总值(GDP)的 1.59%,占社会固定资产投资的 2.20%,比 2011 年增加 37.0%。其中,城市环境基础设施建设投资 5 062.7 亿元,老工业污染源治理投资 500.5 亿元,建设项目"三同时"投资 2 690.4 亿元,分别占环境污染治理投资总额的 61.3%、6.1%、32.6%。

表 5-1 全国环境污染治理投资情况   单位:亿元

年份	城市环境基础设施建设投资	老工业污染源治理投资	建设项目"三同时"环保投资	投资总额
2005	1 289.7	458.2	640.1	2 388.0
2010	4 224.2	397.0	2 033.0	6 654.2
2011	3 469.4	444.4	2 112.4	6 026.2
2012	5 062.7	500.5	2 690.4	8 253.6
变化率/%	45.9	12.6	27.4	37.0

注:从 2012 年起,城市环境基础设施建设投资中不仅仅包括城市的环境基础设施建设投资,还包括县城的相关投资,下同。

5.1.2 污染治理设施直接投资

污染治理设施直接投资是指直接用于污染治理设施、具有直接环保效益的投资,具体包括老工业污染源、建设项目"三同时",以及城市环境基础设施投资中用于污水处理及再生利用、污泥处置和垃圾处理设施的投资。因此污染治理设施直接投资的统计口径小于污染治理投资。

2012 年,我国污染治理设施直接投资总额为 3 702.1 亿元,占污染治理投资总额的 44.9%,其中城市环境基础设施投资、老工业污染源治理投资和建设项目"三同时"环保投资分别占污染治理设施直接投资的 13.8%、13.5% 和 72.7%。建设项目"三同时"环保投资是污染治理设施直接投资的主要来源。

2012 年,环境治理设施直接投资比 2011 年增加 20.3%。其中,城市环境基础设施投资比 2011 年减少了 1.6%,老工业污染源治理投资和建设项目"三同时"环保投资分别比 2011 年增加 12.6% 和 27.4%。

表 5-2　我国污染治理设施直接投资情况　　　　　　　　　　　　　　单位：亿元

年份	污染治理设施直接投资	城市环境基础设施建设投资	老工业污染源治理投资	建设项目"三同时"环保投资	占当年环境污染治理投资总额比例/%	占当年 GDP 比例/%
2005	1 346.4	248.1	458.2	640.1	56.4	0.74
2010	3 078.8	648.8	397.0	2 033.0	46.3	0.77
2011	3 076.5	519.7	444.4	2 112.4	51.1	0.65
2012	3 702.1	511.2	500.5	2 690.4	44.9	0.71
变化率/%	20.3	-1.6	12.6	27.4	—	—

5.1.3　各地区环境污染治理投资

2012 年，我国环境污染治理投资总额为 8 253.6 亿元，除海南、贵州、西藏、青海、宁夏外，其余 26 个地区环境污染治理投资总额超过 100 亿元。与 2011 年相比，除天津、上海、广东、重庆、西藏 5 个地区外，其余 26 个地区环境污染治理投资总额均有所增长。

2012 年，15 个地区环境污染治理投资占 GDP 比重超过 1.5%，上海、吉林、河南、湖南、广东、四川、西藏等 7 个地区比重较低，污染治理投资占 GDP 的比重均低于 1%。

2012 年，全国 GDP 比 2011 年增加 9.7%，环境污染治理投资增速与 GDP 相比呈现正增长，全国环境污染治理投资弹性系数（环境污染治理投资增速/GDP 增速）为 3.79。26 个地区环保投资弹性系数为正值，其中超过 2（环境污染治理投资增速高于 GDP 增速）的地区有 21 个，5 个地区弹性系数为负值（环境污染治理投资为负增长）。

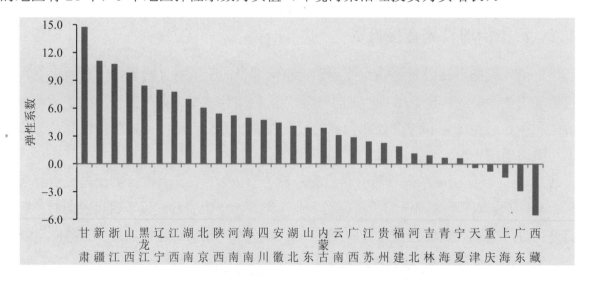

图 5-1　2012 年各地区环境污染治理弹性系数

5.1.4 污染治理设施运行费用情况

污染治理设施运行费用是指工业、城镇生活污染（废水、废气及固体废物）治污设施运行费用，不包括农村污染治理设施。2012 年，污染治理设施运行费用 2 620.6 亿元，比 2011 年减少 3.7%。

2012 年，工业废气治理设施 225 913 套（台），运行费用 1 452.3 亿元，占污染治理设施总运行费用的 55.4%。其中，脱硫设施和脱硝设施运行费用分别为 540.4 亿元和 66.0 亿元，占废气治理设施运行费用的 37.2% 和 4.5%。与 2005 年相比，废气治理设施运行费用增加 443.7%。

废水治理设施运行费用包括工业企业废水治理设施和污水处理厂两个部分。2012 年，废水治理设施运行费用 1 015.9 亿元。其中，工业废水治理设施费用 667.7 亿元，占废水治理设施运行费用的 65.7%，比 2011 年减少 8.8%；污水处理厂运行费用 348.2 亿元，占废水治理设施运行费用的 34.3%，比 2011 年增加 13.3%。与 2005 年相比，工业废水治理设施和污水处理厂运行费用分别增加 141.3% 和 372.5%。

2012 年，生活垃圾处理场和危险（医疗）废物集中处理（置）场运行费用分别为 98.5 亿元和 53.9 亿元，分别比 2011 年增加 66.4% 和 11.9%。

表 5-3 污染治理设施运行费用　　　　　　　　　　　　　　单位：亿元

年份	工业废水治理设施	工业废气治理设施	脱硫设施	脱硝设施	污水处理厂	生活垃圾处理场	危险（医疗）废物集中处理（置）场	合计	占污染治理设施直接投资比例/%
2005	276.7	267.1	—	—	73.7	—	7.1	624.6	46.39
2010	545.3	1 054.5	435.0	—	259.0	—	38.9	1 897.7	61.64
2011	732.1	1 579.5	587.3	44.5	307.2	59.2	48.2	2 726.1	88.61
2012	667.7	1 452.3	540.4	66.0	348.2	98.5	53.9	2 620.6	69.60
变化率/%	−8.8	−8.1	−8.0	48.3	13.3	66.4	11.8	−3.9	—

5.2 城市环境基础设施建设

2012 年，城市环境基础设施建设投资中，燃气工程建设投资 551.8 亿元，比 2011 年增加 66.5%；集中供热工程建设投资 798.1 亿元，比 2011 年增加 82.4%；排水工程建设投资 934.1 亿元，比 2011 年增加 21.3%；园林绿化工程建设投资 2 380.0 亿元，比 2011 年增加 53.9%；市容环境卫生工程建设投资 398.6 亿元，比 2011 年增加 3.8%。

燃气、集中供热、排水、园林绿化和市容环境卫生投资分别占城市环境基础设施建设总投资的 10.9%、15.8%、18.5%、47.0%和 7.9%，园林绿化和排水设施投资为城市环境基础设施建设投资的重点。

表 5-4　全国近年城市环境基础设施建设投资构成　　　　单位：亿元

年份	投资总额	燃气	集中供热	排水	园林绿化	市容环境卫生
2005	1 289.7	142.4	220.2	368.0	411.3	147.8
2010	4 224.2	290.8	433.2	901.6	2 297.0	301.6
2011	3 469.4	331.4	437.6	770.1	1 546.2	384.1
2012	5 062.7	551.8	798.1	934.1	2 380.0	398.6
变化率/%	45.9	66.5	82.4	21.3	53.9	3.8

5.3　老工业污染源治理投资

2012 年，老工业污染源污染治理本年施工项目 5 390 个，其中，废水、废气、固体废物、噪声及其他治理项目分别 1 801、2 144、256、105 和 1 084 个，占本年施工项目数的 33.4%、39.8%、4.8%、1.9%和 20.1%。

2012 年，老工业污染源污染治理投资中，废水、废气、固体废物、噪声及其他治理项目投资分别为 140.3 亿、257.7 亿、24.7 亿、1.2 亿和 76.5 亿元，分别占老工业污染源治理投资额的 28.0%、51.5%、5.0%、0.2%和 15.3%。与 2011 年相比，废水、固体废物、噪声治理项目投资分别减少 11.0%、21.3%和 45.5%，废气及其他污染治理项目分别增加 21.7%和 84.8%。

表 5-5　老工业污染源治理投资构成　　　　单位：万元

年份	投资总额	废水	废气	固体废物	噪声	其他
2005	458.2	133.7	213.0	27.4	3.1	81.0
2010	397.0	130.1	188.8	14.3	1.5	62.2
2011	444.4	157.7	211.7	31.4	2.2	41.4
2012	500.5	140.3	257.7	24.7	1.2	76.5
变化率/%	12.6	−11.0	21.7	−21.3	−45.5	84.8

5.4 建设项目"三同时"环保投资

2012 年，建设项目"三同时"环保投资 2 690.4 亿元，比 2011 年增加 27.4%，占建设项目投资总额的 2.7%。

表 5-6　建设项目"三同时"投资情况

年份	环保投资额/亿元	占建设项目投资总额比重/%	占全社会固定资产投资总额比重/%	占环境治理投资总额比重/%
2001	336.4	3.6	0.9	30.4
2005	640.1	4.0	0.7	26.8
2010	2 033.0	4.1	0.7	30.6
2011	2 112.4	3.1	0.7	35.1
2012	2 690.4	2.7	0.7	31.9
变化率/%	27.4	—	—	—

6

环境管理

ANNUAL STATISTIC REPORT ON ENVIRONMENT IN CHINA
2012

6.1 环保机构建设

2012 年，全国环保系统机构总数 13 225 个。其中，国家级机构 45 个，省级机构 394 个，地市级环保机构 2 102 个，县级环保机构 8 807 个，乡镇环保机构 1 877 个。各级环保行政机构 3 177 个，各级环境监察机构 2 892 个，各级环境监测机构 2 725 个。

全国环保系统共有 20.5 万人。其中，环保机关人员 5.3 万人，占环保系统总人数的 26.0%；环境监察人员 6.1 万人，占环保系统总人数的 29.7%；环境监测人员 5.7 万人，占环保系统总人数的 27.5%。

表 6-1 环保行政机构、监察机构、监测站年末实有人员情况

年份	年末实有人数/人	环保行政机构		环境监察机构		环境监测站	
		实有人数/人	占本级人员总数比例/%	实有人数/人	占本级人员总数比例/%	实有人数/人	占本级人员总数比例/%
2005	166 774	44 024	26.4	50 040	30.0	46 984	28.2
2006	170 290	44 141	25.9	52 845	31.2	47 689	28.2
2007	176 988	43 626	24.6	57 427	32.4	49 335	27.9
2008	183 555	44 847	24.4	59 477	32.1	51 753	28.3
2009	188 991	45 626	24.1	60 896	32.2	52 944	28.0
2010	193 911	45 938	23.7	62 468	32.2	54 698	28.2
2011	201 161	46 128	22.9	64 426	32.0	56 226	28.0
2012	205 334	53 286	26.0	61 081	29.7	56 554	27.5
国家级	2 839	357	12.6	481	16.9	184	6.5
省 级	14 154	2 745	19.4	1 261	8.9	3 163	22.3
地市级	45 203	9 920	21.9	9 390	20.8	16 015	35.4
县 级	135 628	40 264	29.7	49 949	36.8	37 192	27.4

6.2 环境法制及环境信访情况

2012 年，全国各级环保系统承办的人大建议数 6 093 件，政协提案数 13 374 件。全国办理环境行政处罚案件 11.7 万件，环境行政复议案件 427 件。

全国各级环保系统当年颁布地方性法规 23 件，累计有效的地方性法规 348 件；当年颁布地方政府规章 27 件，累计有效的政府规章 347 件；当年备案的地方环境标准数 19 件，累计备案的地方环境标准数 95 件。

全国各级环保系统共收到群众来信 10.7 万封，群众来访 4.3 万批次，9.6 万人次，其中，已办结来信和来访 15.9 万件。电话及网络投诉 89.2 万件，其中，已办结数为 88.8 万件。

全国发生突发环境事件 542 次，其中重大环境事件 5 次，较大环境事件 5 次，一般环境事件 532 次。

表 6-2 环境信访工作情况

年份	来信总数/封	来访批次/批	来访人次/次	来信、来访已办结数量/件	电话/网络投诉数/件	电话/网络投诉办结数/件
2005	608 245	88 237	142 360	—	—	—
2006	616 122	71 287	110 592	—	—	—
2007	123 357	43 909	77 399	—	—	—
2008	705 127	43 862	84 971	—	—	—
2009	696 134	42 170	73 798	—	—	—
2010	701 073	34 683	65 948	—	—	—
2011	201 631	53 505	107 597	251 607	852 700	834 588
2012	107 120	43 260	96 145	159 283	892 348	888 836

6.3 环境监测

2012 年，全国监测用房总面积为 263.5 万米^2，监测业务经费为 45.9 亿元。环境监测仪器 23.4 万台（套），仪器设备原值为 124.8 亿元。

全国环境空气监测点位 3 189 个，酸雨监测点位 1 672 个，沙尘天气影响环境质量监测点位数 220 个，地表水水质监测断面 8 173 个，饮用水水源地监测点位数 2 995 个，近岸海域监测点位数 645 个，开展环境噪声监测的监测点位数 24.6 万，开展生态监测的监测点位数 87 个，开展污染源监督性监测的重点企业数 57 136 个。

6.4 自然生态保护

2012 年，全国各类自然保护区共计 2 669 个。自然保护区面积 14 978.7 万公顷，约占国土面积的 14.9%。国家级、省级、地市级、县级自然保护区个数分别占全国自然保护区总数的 13.6%、32.8%、15.2%、38.4%，其面积分别占自然保护区总面积的 62.9%、27.3%、2.9%、6.9%。

全国共建设国家级生态市 4 个，国家级生态县 51 个。省级生态市 45 个，省级生态县 187 个。共建成农村生态示范建设 1 797 个，国家有机食品生产基地 138 个。

表 6-3　全国自然保护区数量　　　　　　　　　　　　　　　　单位：个

年份	自然保护区数	国家级	省级	地市级	县级
2005	2 349	243	773	421	912
2006	2 395	265	793	422	915
2007	2 531	303	780	462	986
2008	2 538	303	806	432	997
2009	2 541	319	827	416	979
2010	2 588	319	859	418	992
2011	2 640	335	870	421	1 014
2012	2 669	363	876	406	1 024

表 6-4　全国自然保护区面积　　　　　　　　　　　　　　　　单位：万公顷

年份	自然保护区面积	国家级	省级	地市级	县级
2005	14 994.9	8 898.9	4 487.0	501.5	1 107.5
2006	15 153.5	9 169.7	4 441.8	522.4	1 019.6
2007	15 188.2	9 365.6	4 260.1	537.6	1 024.9
2008	14 894.3	9 120.3	4 240.2	497.1	1 036.8
2009	14 774.7	9 267.1	4 004.5	471.2	1 031.9
2010	14 944.1	9 267.6	4 174.8	468.2	1 033.4
2011	14 971.1	9 315.3	4 152.6	472.4	1 030.9
2012	14 978.7	9 414.6	4 090.9	432.5	1 040.8

6.5　环境影响评价及环保验收

2012 年，共审批建设项目环境影响评价文件 42.8 万个，其中编制报告书的项目 2.9 万个，填报报告表的项目 17.9 万个，编制登记表的项目 21.9 万个。

2012 年，全国完成环保验收项目 13.2 万个，其中环保验收一次合格的项目 12.9 万个。执行"三同时"的建设项目用于环保工程的实际投资为 2 690.4 亿元，占项目总投资的 2.7%。

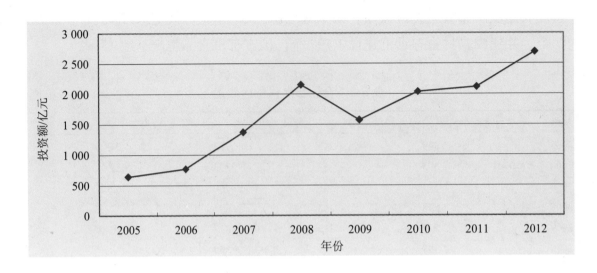

图 6-1　全国实际执行"三同时"建设项目环保投资情况

6.6　污染源控制及管理

　　除废水、废气中主要污染物的治理投入和工程设施建设运行外（详见第 2、3 章主要污染物排放与治理情况及第 5 章环境污染治理投资），2012 年，全国完成清洁生产审核企业 11 971 家，其中进行强制性审核的企业 8 490 家。应开展监测的重金属污染防控重点企业 4 525 家，其中重金属排放达标的重点企业 3 786 家。已发放危险废物经营许可证 1 528 个，其中具有医疗废物经营范围的许可证 245 个。

6.7　污染源自动监控及排污收费

　　2012 年，全国已实施自动监控的国家重点监控企业 9 215 个，其中已实施自动监控的水排放口 7 293 个，气排放口 6 765 个。实施自动监控国家重点监控企业中，COD 监控设备与环境保护部门稳定联网的企业 4 503 个，NH_3-N 监控设备与环境保护部门稳定联网的企业 3 194 个，SO_2 监控设备与环境保护部门稳定联网的企业 4 314 个，NO_x 监控设备与环境保护部门稳定联网的企业 4 106 个。

　　全国排污费解缴入库单位共 35.1 万户，入库金额 188.9 亿元。

6.8 环境宣教

2012 年，各级环保系统共建成环境教育基地 1 845 个，组织开展社会环境宣传教育活动 10 209 次，参与社会环境宣传教育活动的人数达到 4 664.4 万人。

7

全国辐射环境水平

ANNUAL STATISTIC REPORT ON ENVIRONMENT IN CHINA
2012

2012 年，全国辐射环境质量总体良好。环境电离辐射水平保持在天然本底的涨落范围内，核设施、核技术利用项目周围环境电离辐射水平总体未见明显变化；环境电磁辐射水平总体情况较好，电磁辐射发射设施周围环境电磁辐射水平总体未见明显变化。

7.1 环境电离辐射

全国环境 γ 辐射空气吸收剂量率，气溶胶、沉降物总 α 和总 β 活度浓度，空气中氡活度浓度均为正常环境水平。长江、黄河、珠江、松花江、淮河、海河、辽河、浙闽片河流、西南诸河、西北诸河、重点湖（水库）人工放射性核素活度浓度与历年相比未见明显变化，天然放射性核素活度浓度与 1983—1990 年全国环境天然放射性水平调查结果处于同一水平。12 个集中式饮用水源地总 α 和总 β 活度浓度均低于《生活饮用水卫生标准》（GB 5749—2006）规定的限值。近岸海域人工放射性核素锶-90 和铯-137 活度浓度均低于《海水水质标准》（GB 3097—1997）规定的限值。土壤中人工放射性核素活度浓度与历年相比未见明显变化，天然放射性核素活度浓度与 1983—1990 年全国环境天然放射性水平调查结果处于同一水平。

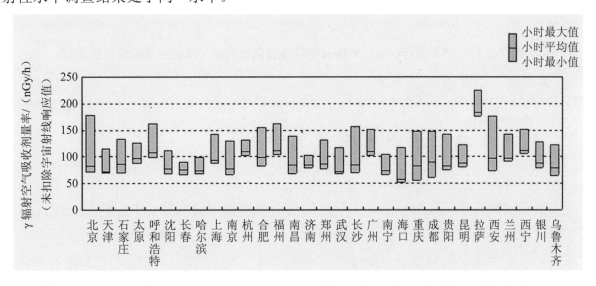

图 7-1 2012 年直辖市及省会城市辐射环境自动监测站 γ 辐射空气吸收剂量率

7.2 运行核电厂周围环境电离辐射

秦山核电基地各核电厂、大亚湾/岭澳核电厂、田湾核电厂外围各辐射环境自动监测站实时连续γ辐射空气吸收剂量率（未扣除宇宙射线响应值）年均值分别为101.1 nGy/h、124.8 nGy/h和100.1 nGy/h，均在当地的天然本底水平涨落范围内。秦山核电基地周围关键居民点空气、降水、地表水及部分生物样品中氚活度浓度，大亚湾/岭澳核电厂和田湾核电厂排放口附近海域海水氚活度浓度与核电厂运行前本底值相比有所升高，但对公众造成的辐射剂量远低于国家规定的剂量限值。核电厂外围各种环境介质中除氚外其余放射性核素活度浓度与历年相比未见明显变化。

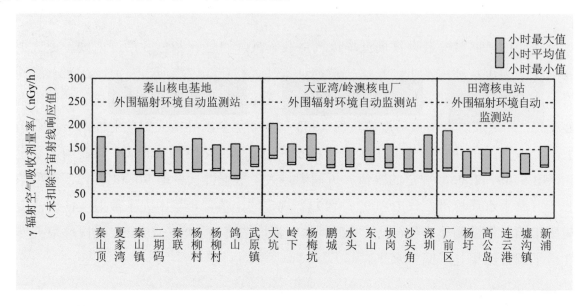

图 7-2　2012 年运行核电厂外围辐射环境自动监测站γ辐射空气吸收剂量率

7.3　其他反应堆周围环境电离辐射

中国原子能科学研究院、清华大学核能与新能源技术研究院、中国核动力研究设计院等研究设施外围环境γ辐射空气吸收剂量率，气溶胶、沉降物、地表水、地下水、土壤和生物样品中放射性核素活度浓度与历年相比未见明显变化；饮用地下水总α和总β活度浓度低于《生活饮用水卫生标准》规定的限值。

7.4　核燃料循环设施和废物处置设施周围环境电离辐射

兰州铀浓缩有限公司、陕西铀浓缩有限公司、包头核燃料元件厂、中核建中核燃料元件公司、中核四〇四有限公司等核燃料循环设施及西北低中放废物处置场、北龙低中放废物处置场外围环境 γ 辐射空气吸收剂量率为正常环境水平，环境介质中也未监测到由上述企业生产、加工、贮存、处理、运输等活动引起的放射性核素活度浓度升高。

7.5　铀矿冶周围环境电离辐射

铀矿冶设施周围环境空气中氡活度浓度，气溶胶、沉降物总 α 和总 β 活度浓度，地下水和生物样品中放射性核素铀和镭-226 活度浓度未见异常。

7.6　电磁辐射设施周围环境电磁辐射

环境电磁辐射水平总体情况较好。开展监测的移动通信基站天线周围环境敏感点的电磁辐射水平低于《电磁辐射防护规定》规定的公众照射导出限值；开展监测的输变电设施周围环境敏感点工频电场强度和磁感应强度均低于《500 kV 超高压送变电工程电磁辐射环境影响评价技术规范》中的居民区工频电场评价标准和公众全天辐射时的工频磁场限值。

简 要 说 明

一、本年报资料根据全国 31 个省、自治区、直辖市及新疆生产建设兵团环境统计资料汇总整理而成，未包括香港特别行政区、澳门特别行政区以及台湾省数据。

二、本年报主要反映中国环境污染排放、治理及环境管理情况。主要内容包括废水及污染物的排放与治理情况，废气及污染物排放与治理情况，一般工业固体废物、危险废物（医疗废物）、生活垃圾的产生、综合利用及处理处置情况，环境污染治理投资以及环境管理等情况。

三、调查范围

本年报数据根据"十二五"环境统计报表制度调查收集，调查范围较"十一五"有所扩大。"十二五"环境统计报表制度调查范围包括工业污染源、农业污染源、城镇生活污染源、机动车、集中式污染治理设施和环境管理 6 方面内容。

1. 工业污染源调查范围：辖区内有污染物排放的所有工业企业；

2. 城镇生活污染源调查范围：城市和集镇内居民在日常生活及各种活动中产生、排放的污染物情况；

3. 机动车调查范围：辖区内所有机动车；

4. 农业污染源调查范围：畜禽养殖业、水产养殖业和种植业；

5. 集中式污染治理设施调查范围：辖区内所有集中式污染治理设施，包括污水处理厂、垃圾处理厂（场）、危险废物（医疗废物）处理（处置）厂；

6. 环境管理反映环保系统自身能力建设、业务工作进展及成果等情况，是在以往环境统计报表制度中专业年报基础上简化而来，调查范围主要包括环保机构数/人数、环境信访与环境法制、环境保护能力建设投资、环境污染源控制与管理、环境监测、污染源自动监控、排污费征收、自然生态保护与建设、环境影响评价、建设项目竣工环境保护验收情况、突发环境事件、环境宣教等 12 个方面的内容。

四、本年报特点

与"十一五"环境统计报表制度相比，本年报特点是：

1. 新增了农业污染源调查内容，农业污染源调查内容包括畜禽养殖业、水产养殖业和种植业；

2．细化了机动车污染调查统计，调查载客汽车、载货汽车、三轮汽车及低速载货汽车、摩托车的总颗粒物、氮氧化物、一氧化碳、碳氢化合物等污染物排放量；

3．新增了生活垃圾处理厂（场）调查内容，调查范围为垃圾填埋厂（场）、垃圾堆肥厂（场）、垃圾焚烧厂（场）和其他方式处理垃圾的处理厂（场）；

4．删除了医院污染排放情况调查表，因城镇生活污染源报表中的人均污染物排放量指标均已包含医院污染物排放。

五、调查及核算方法

1．主要污染物排放总量核算

废水污染物排放总量＝工业污染源排放量＋农业污染源排放量＋城镇生活污染源排放量＋集中式污染治理设施排放量（集中式污水处理厂除外）；

废气污染物排放总量＝工业污染源排放量＋城镇生活污染源排放量＋机动车排放量（仅限 NO_x 指标）＋集中式污染治理设施排放量。

2．工业污染源：工业污染源采取对重点调查工业企业逐个发表调查汇总，非重点调查工业企业采用比率估算法的方式核算；工业污染排放总量为重点调查工业企业与非重点调查工业企业排放量之和；工业污染治理投资的调查对象为调查年度正式施工的，且没有纳入"三同时"项目管理的老工业源污染治理投资项目，以及调查年度内完成"三同时"环境保护竣工验收的工业类建设项目。

3．农业污染源：农业污染源包括畜禽养殖业、水产养殖业和种植业，以县（区）为基本单位进行调查；畜禽养殖业中的规模化养殖场（小区）采用逐场（小区）发表调查，根据饲养量、产污系数及去除率等测算污染物排放量；养殖专业户的污染物排放量根据饲养量和排污强度核算；水产养殖业污染物排放量根据第一次全国污染源普查及总量减排核定减少水产围网养殖面积核算；种植业污染物排放量与第一次全国污染源普查数据保持一致。

4．城镇生活污染源：以地市为基本调查单位，污染物产生量依据相关部门的统计数据和产污系数核算，排放量为产生量与集中式污水处理厂生活污染物的去除量之差。

5．机动车：以地市为基本调查单位，根据"遵循基数、核清增量、核实减量"原则核算污染物排放量。

6．集中式污染治理设施：逐家发表调查汇总。

六、其他需要说明的问题

1．2011年环境统计年报中采用水利部流域代码对流域进行汇总，流域由松花江、辽河、海河、黄河、淮河、长江、珠江、东南诸河、西南诸河和西北诸河十大水系组成。

自2012年起，根据《重点流域水污染防治"十二五"规划》中的流域分区，增加了对松花江、辽河、海河、黄河中上游、淮河、长江中下游等重点规划流域的汇总，各流域汇总范围与水利部相应流域并不完全重合，因此数据有所不同。

2．本年报中所指集中式污染治理设施的排放量，仅指生活垃圾处理厂（场）和危险（医疗）废物集中处置厂（场）的渗滤液和焚烧废气中的污染物。污水处理厂仅作为污染治理设施，不产生和排放污染物。

主要环境统计指标解释附后。

8

各地区环境统计

ANNUAL STATISTIC REPORT ON ENVIRONMENT IN CHINA
2012

各地区主要污染物排放情况（一）
Discharge of Key Pollutants by Region（1）

（2012）

单位：亿吨 (100 million tons)

年 份 地 区	Year Region	废水排放总量 Total Volume of Waste Water Discharged	工业 Industrial	生活 Household	集中式 Centralized Treatment
2011		659.19	230.87	427.92	0.40
2012		684.76	221.59	462.69	0.49
北 京	Beijing	14.03	0.92	13.10	0.01
天 津	Tianjin	8.28	1.91	6.37	0.00
河 北	Hebei	30.58	12.26	18.30	0.01
山 西	Shanxi	13.43	4.81	8.62	0.00
内蒙古	Inner Mongolia	10.24	3.36	6.88	0.00
辽 宁	Liaoning	23.88	8.72	15.15	0.01
吉 林	Jilin	11.95	4.48	7.46	0.01
黑龙江	Heilongjiang	16.26	5.84	10.42	0.00
上 海	Shanghai	21.92	4.64	17.25	0.04
江 苏	Jiangsu	59.82	23.61	36.18	0.03
浙 江	Zhejiang	42.10	17.54	24.50	0.05
安 徽	Anhui	25.43	6.72	18.70	0.02
福 建	Fujian	25.63	10.63	14.97	0.03
江 西	Jiangxi	20.12	6.79	13.31	0.03
山 东	Shandong	47.91	18.36	29.51	0.03
河 南	Henan	40.37	13.74	26.62	0.01
湖 北	Hubei	29.02	9.16	19.83	0.03
湖 南	Hunan	30.42	9.71	20.67	0.04
广 东	Guangdong	83.86	18.61	65.19	0.05
广 西	Guangxi	24.56	11.07	13.48	0.01
海 南	Hainan	3.71	0.75	2.96	0.01
重 庆	Chongqing	13.24	3.06	10.17	0.01
四 川	Sichuan	28.37	7.00	21.34	0.02
贵 州	Guizhou	9.15	2.34	6.80	0.01
云 南	Yunnan	15.40	4.28	11.11	0.01
西 藏	Tibet	0.47	0.04	0.43	0.00
陕 西	Shaanxi	12.87	3.80	9.06	0.01
甘 肃	Gansu	6.28	1.92	4.36	0.00
青 海	Qinghai	2.20	0.89	1.31	0.00
宁 夏	Ningxia	3.89	1.65	2.24	0.00
新 疆	Xinjiang	9.38	2.97	6.40	0.00

各地区主要污染物排放情况（二）
Discharge of Key Pollutants by Region（2）
（2012）

单位：万吨 　　　（10 000 tons）

年　份 地　区	Year Region	化学需氧量 排放总量 Total Volume of COD Discharged	工业 Industrial	农业 Agricultural	生活 Household	集中式 Centralized Treatment
	2011	2 499.86	354.80	1 186.11	938.84	20.11
	2012	2 423.73	338.45	1 153.80	912.75	18.73
北　京	Beijing	18.65	0.63	7.83	9.57	0.63
天　津	Tianjin	22.94	2.65	11.39	8.85	0.05
河　北	Hebei	134.91	19.18	91.83	23.11	0.79
山　西	Shanxi	47.68	8.67	18.06	20.69	0.26
内蒙古	Inner Mongolia	88.39	9.08	61.92	17.16	0.23
辽　宁	Liaoning	130.59	10.22	86.33	33.41	0.64
吉　林	Jilin	78.75	7.63	50.39	19.48	1.24
黑龙江	Heilongjiang	149.88	9.82	105.18	34.69	0.19
上　海	Shanghai	24.26	2.62	3.31	17.75	0.58
江　苏	Jiangsu	119.70	23.14	38.77	57.25	0.54
浙　江	Zhejiang	78.62	18.57	20.40	38.92	0.74
安　徽	Anhui	92.43	8.94	38.49	44.02	0.98
福　建	Fujian	66.00	9.06	21.41	35.13	0.41
江　西	Jiangxi	74.83	10.07	23.90	39.82	1.04
山　东	Shandong	192.12	13.94	134.41	43.30	0.47
河　南	Henan	139.36	17.90	80.26	40.30	0.90
湖　北	Hubei	108.66	13.53	47.55	46.12	1.45
湖　南	Hunan	126.34	14.96	56.73	53.04	1.61
广　东	Guangdong	180.29	23.85	60.01	95.04	1.39
广　西	Guangxi	78.03	19.96	21.57	36.11	0.39
海　南	Hainan	19.74	1.25	10.31	8.02	0.15
重　庆	Chongqing	40.28	4.92	12.36	22.94	0.06
四　川	Sichuan	126.87	11.84	53.98	60.48	0.57
贵　州	Guizhou	33.30	6.47	6.32	19.97	0.53
云　南	Yunnan	54.86	16.96	7.42	29.17	1.30
西　藏	Tibet	2.58	0.09	0.39	2.08	0.02
陕　西	Shaanxi	53.62	9.85	19.66	23.45	0.66
甘　肃	Gansu	38.93	9.30	14.48	14.99	0.16
青　海	Qinghai	10.38	4.18	2.24	3.75	0.20
宁　夏	Ningxia	22.80	10.54	10.17	2.01	0.08
新　疆	Xinjiang	67.92	18.63	36.72	12.12	0.45

各地区主要污染物排放情况（三）
Discharge of Key Pollutants by Region（3）
（2012）

单位：万吨 （10 000 tons）

年 份 地 区	Year Region	氨氮排放总量 Total Volume of Ammonia Nitrogen Discharged	工业 Industrial	农业 Agricultural	生活 Household	集中式 Centralized Treatment
	2011	260.44	28.12	82.65	147.66	2.00
	2012	253.59	26.41	80.62	144.63	1.93
北 京	Beijing	2.05	0.04	0.47	1.48	0.06
天 津	Tianjin	2.54	0.33	0.59	1.61	0.00
河 北	Hebei	11.07	1.58	4.49	4.94	0.06
山 西	Shanxi	5.69	0.83	1.26	3.58	0.02
内蒙古	Inner Mongolia	5.27	1.11	1.22	2.91	0.02
辽 宁	Liaoning	10.75	0.84	3.42	6.40	0.09
吉 林	Jilin	5.63	0.44	1.77	3.29	0.13
黑龙江	Heilongjiang	9.28	0.58	3.45	5.22	0.02
上 海	Shanghai	4.74	0.23	0.35	4.14	0.03
江 苏	Jiangsu	15.31	1.63	3.91	9.70	0.07
浙 江	Zhejiang	11.23	1.21	2.70	7.25	0.06
安 徽	Anhui	10.61	0.84	3.78	5.90	0.09
福 建	Fujian	9.32	0.68	3.33	5.28	0.04
江 西	Jiangxi	9.11	1.02	2.97	5.03	0.08
山 东	Shandong	16.86	1.10	7.41	8.30	0.05
河 南	Henan	14.98	1.30	6.40	7.16	0.11
湖 北	Hubei	12.89	1.48	4.66	6.59	0.16
湖 南	Hunan	16.13	2.58	6.26	7.14	0.15
广 东	Guangdong	22.41	1.49	5.83	14.95	0.14
广 西	Guangxi	8.26	0.86	2.67	4.69	0.03
海 南	Hainan	2.25	0.09	0.92	1.23	0.01
重 庆	Chongqing	5.34	0.31	1.30	3.72	0.02
四 川	Sichuan	14.07	0.56	5.74	7.70	0.07
贵 州	Guizhou	3.87	0.36	0.79	2.66	0.06
云 南	Yunnan	5.86	0.43	1.18	4.09	0.17
西 藏	Tibet	0.32	0.01	0.05	0.26	0.00
陕 西	Shaanxi	6.19	0.87	1.54	3.69	0.09
甘 肃	Gansu	4.10	1.39	0.57	2.13	0.01
青 海	Qinghai	0.98	0.20	0.09	0.67	0.01
宁 夏	Ningxia	1.74	0.86	0.22	0.66	0.01
新 疆	Xinjiang	4.72	1.18	1.28	2.24	0.03

各地区主要污染物排放情况（四）
Discharge of Key Pollutants by Region（4）

（2012）

单位：万吨 (10 000 tons)

年　份 地　区	Year Region	二氧化硫 排放总量 Total Volume of Sulphur Dioxide Discharged	工业 Industrial	生活 Household	集中式 Centralized Treatment
	2011	2 217.91	2 017.23	200.39	0.29
	2012	2 117.63	1 911.71	205.66	0.26
北　京	Beijing	9.38	5.93	3.45	0.00
天　津	Tianjin	22.45	21.55	0.90	0.01
河　北	Hebei	134.12	123.87	10.24	0.01
山　西	Shanxi	130.18	119.46	10.71	0.00
内蒙古	Inner Mongolia	138.49	124.15	14.34	0.00
辽　宁	Liaoning	105.87	97.90	7.96	0.00
吉　林	Jilin	40.35	35.23	5.11	0.00
黑龙江	Heilongjiang	51.43	39.73	11.70	0.00
上　海	Shanghai	22.82	19.34	3.48	0.01
江　苏	Jiangsu	99.20	95.92	3.25	0.03
浙　江	Zhejiang	62.58	61.09	1.47	0.02
安　徽	Anhui	51.96	46.98	4.94	0.04
福　建	Fujian	37.13	35.24	1.89	0.00
江　西	Jiangxi	56.77	55.15	1.62	0.00
山　东	Shandong	174.88	154.38	20.48	0.02
河　南	Henan	127.59	112.99	14.60	0.00
湖　北	Hubei	62.24	54.86	7.37	0.00
湖　南	Hunan	64.50	59.33	5.16	0.00
广　东	Guangdong	79.92	77.15	2.72	0.06
广　西	Guangxi	50.41	47.16	3.25	0.00
海　南	Hainan	3.41	3.30	0.11	0.00
重　庆	Chongqing	56.48	50.98	5.50	0.00
四　川	Sichuan	86.44	79.40	7.02	0.02
贵　州	Guizhou	104.11	83.71	20.40	0.00
云　南	Yunnan	67.22	62.26	4.96	0.00
西　藏	Tibet	0.42	0.13	0.29	0.00
陕　西	Shaanxi	84.38	74.70	9.67	0.00
甘　肃	Gansu	57.25	47.99	9.24	0.02
青　海	Qinghai	15.39	12.91	2.48	0.00
宁　夏	Ningxia	40.66	38.44	2.23	0.00
新　疆	Xinjiang	79.61	70.47	9.14	0.00

各地区主要污染物排放情况（五）
Discharge of Key Pollutants by Region（5）

（2012）

单位：万吨 （10 000 tons）

年 份 地 区	Year Region	氮氧化物 排放总量 Total Volume of Nitrogen Oxide Discharged	工业 Industrial	生活 Household	机动车 Vehicle	集中式 Centralized Treatment
	2011	2 404.27	1 729.71	36.62	637.59	0.35
	2012	2 337.76	1 658.05	39.31	640.03	0.37
北 京	Beijing	17.75	8.53	1.20	7.99	0.03
天 津	Tianjin	33.42	27.56	0.44	5.41	0.02
河 北	Hebei	176.11	119.48	1.84	54.80	0.00
山 西	Shanxi	124.40	95.07	3.15	26.16	0.01
内蒙古	Inner Mongolia	141.89	114.70	2.62	24.57	0.00
辽 宁	Liaoning	103.63	74.15	1.61	27.86	0.00
吉 林	Jilin	57.59	38.74	1.18	17.66	0.00
黑龙江	Heilongjiang	78.06	48.13	4.37	25.57	0.00
上 海	Shanghai	40.16	28.48	1.91	9.75	0.02
江 苏	Jiangsu	147.96	113.36	0.65	33.90	0.05
浙 江	Zhejiang	80.88	63.50	0.33	17.04	0.02
安 徽	Anhui	92.13	69.28	0.93	21.89	0.03
福 建	Fujian	46.72	36.09	0.24	10.39	0.00
江 西	Jiangxi	57.71	35.25	0.30	22.16	0.00
山 东	Shandong	173.90	123.82	3.20	46.87	0.02
河 南	Henan	162.59	110.00	2.43	50.15	0.01
湖 北	Hubei	64.00	44.09	1.21	18.68	0.01
湖 南	Hunan	60.72	42.48	0.79	17.45	0.00
广 东	Guangdong	130.34	80.91	1.13	48.23	0.07
广 西	Guangxi	49.83	34.56	0.36	14.90	0.01
海 南	Hainan	10.34	7.19	0.02	3.12	0.01
重 庆	Chongqing	38.27	27.21	0.45	10.61	0.00
四 川	Sichuan	65.90	43.87	1.00	21.00	0.03
贵 州	Guizhou	56.35	45.89	0.81	9.65	0.00
云 南	Yunnan	54.43	33.53	0.60	20.31	0.00
西 藏	Tibet	4.43	0.40	0.03	4.00	0.00
陕 西	Shaanxi	80.81	60.45	2.65	17.71	0.00
甘 肃	Gansu	47.34	34.10	1.44	11.79	0.00
青 海	Qinghai	12.61	8.80	0.56	3.25	0.00
宁 夏	Ningxia	45.54	37.88	0.27	7.38	0.00
新 疆	Xinjiang	81.95	50.57	1.58	29.80	0.00

各地区主要污染物排放情况（六）
Discharge of Key Pollutants by Region（6）
（2012）

单位：万吨 （10 000 tons）

年　份 地　区	Year Region	烟（粉）尘排放 总量 Total Volume of Soot and dust Discharged	工业 Industrial	生活 Household	机动车 Vehicle	集中式 Centralized Treatment
2011		1 278.82	1 100.88	114.76	62.93	0.25
2012		1 234.31	1 029.31	142.67	62.14	0.19
北　京	Beijing	6.68	3.08	3.19	0.41	0.00
天　津	Tianjin	8.41	5.90	1.84	0.66	0.00
河　北	Hebei	123.59	105.57	12.64	5.36	0.01
山　西	Shanxi	107.09	94.81	10.01	2.27	0.00
内蒙古	Inner Mongolia	83.30	66.77	13.64	2.89	0.00
辽　宁	Liaoning	72.63	62.63	6.99	3.00	0.00
吉　林	Jilin	26.48	19.55	5.11	1.81	0.00
黑龙江	Heilongjiang	69.93	45.01	22.31	2.61	0.00
上　海	Shanghai	8.71	6.37	1.55	0.79	0.01
江　苏	Jiangsu	44.32	39.60	1.85	2.85	0.02
浙　江	Zhejiang	25.40	23.32	0.43	1.63	0.02
安　徽	Anhui	46.21	35.16	8.72	2.30	0.03
福　建	Fujian	25.26	23.40	0.92	0.94	0.00
江　西	Jiangxi	35.74	32.18	0.88	2.67	0.00
山　东	Shandong	69.53	52.59	11.97	4.96	0.00
河　南	Henan	58.52	49.28	4.11	5.13	0.00
湖　北	Hubei	34.97	28.16	5.19	1.61	0.01
湖　南	Hunan	34.07	29.88	2.67	1.51	0.01
广　东	Guangdong	32.83	26.75	1.27	4.76	0.05
广　西	Guangxi	29.97	26.85	1.42	1.69	0.00
海　南	Hainan	1.66	1.07	0.17	0.43	0.00
重　庆	Chongqing	18.23	16.61	0.91	0.70	0.00
四　川	Sichuan	29.58	26.78	1.21	1.58	0.01
贵　州	Guizhou	29.45	25.65	2.82	0.98	0.00
云　南	Yunnan	39.06	35.95	1.49	1.62	0.00
西　藏	Tibet	0.66	0.10	0.10	0.46	0.00
陕　西	Shaanxi	46.21	38.55	5.36	2.29	0.00
甘　肃	Gansu	20.76	15.66	4.25	0.84	0.00
青　海	Qinghai	15.64	13.37	1.96	0.31	0.00
宁　夏	Ningxia	19.83	18.07	0.94	0.82	0.00
新　疆	Xinjiang	69.61	60.60	6.75	2.26	0.00

各地区工业废水排放及处理情况（一）
Discharge and Treatment of Industrial Waste Water by Region（1）

（2012）

单位：万吨 (10 000 tons)

年　份 地　区	Year Region	工业废水排放量 Total Volume of Industrial Waste Water Discharged	直接排入环境的 Direct discharged to environment	排入污水处理厂的 Discharged to centralized Waste water treatment	工业废水处理量 Industrial Waste Water treated
2011		2 308 743	1 839 635	469 108	5 805 511
2012		2 215 857	1 732 817	483 039	5 274 705
北　京	Beijing	9 190	2 232	6 958	11 081
天　津	Tianjin	19 117	8 027	11 090	40 464
河　北	Hebei	122 645	86 974	35 671	776 119
山　西	Shanxi	48 108	44 732	3 376	204 376
内蒙古	Inner Mongolia	33 618	18 070	15 548	104 767
辽　宁	Liaoning	87 168	71 969	15 198	235 649
吉　林	Jilin	44 842	38 464	6 379	99 305
黑龙江	Heilongjiang	58 355	52 926	5 429	92 346
上　海	Shanghai	46 359	19 728	26 631	78 777
江　苏	Jiangsu	236 094	157 759	78 336	405 492
浙　江	Zhejiang	175 416	77 542	97 874	256 554
安　徽	Anhui	67 175	59 167	8 009	206 430
福　建	Fujian	106 319	94 063	12 257	173 788
江　西	Jiangxi	67 871	66 432	1 439	175 329
山　东	Shandong	183 634	118 744	64 890	365 429
河　南	Henan	137 356	120 979	16 377	176 636
湖　北	Hubei	91 609	80 109	11 499	281 273
湖　南	Hunan	97 133	94 524	2 609	269 496
广　东	Guangdong	186 126	149 391	36 735	287 598
广　西	Guangxi	110 671	107 263	3 408	345 307
海　南	Hainan	7 465	6 530	935	9 144
重　庆	Chongqing	30 611	29 466	1 145	44 824
四　川	Sichuan	69 984	63 669	6 315	207 455
贵　州	Guizhou	23 399	23 399	0	92 722
云　南	Yunnan	42 811	41 496	1 315	143 693
西　藏	Tibet	353	353	0	556
陕　西	Shaanxi	38 037	31 854	6 183	63 704
甘　肃	Gansu	19 188	17 688	1 500	33 001
青　海	Qinghai	8 917	8 917	0	19 785
宁　夏	Ningxia	16 548	15 202	1 346	25 385
新　疆	Xinjiang	29 738	25 150	4 589	48 221

各地区工业废水排放及处理情况（二）

Discharge and Treatment of Industrial Waste Water by Region（2）

（2012）

年　份 地　区	Year Region	废水治理设施数/套 Number of Facilities for Treatment of Waste Water （set）	废水治理设施 治理能力/ （万吨/日） Capacity of Facilities for Treatment of Waste Water （10 000 tons/day）	废水治理设施 运行费用/ 万元 Annual Expenditure for Operation （10 000 yuan）
	2011	91 506	31 406	7 321 459.5
	2012	85 673	26 620	6 677 024.6
北　京	Beijing	521	65	51 338.7
天　津	Tianjin	967	171	127 294.3
河　北	Hebei	4 780	3 725	550 249.2
山　西	Shanxi	3 567	1 013	257 998.7
内蒙古	Inner Mongolia	1 102	659	100 782.2
辽　宁	Liaoning	2 387	1 212	259 558.7
吉　林	Jilin	677	485	92 430.7
黑龙江	Heilongjiang	1 199	1 214	233 221.7
上　海	Shanghai	1 802	321	328 406.1
江　苏	Jiangsu	7 495	1 953	966 723.4
浙　江	Zhejiang	8 572	1 362	504 155.7
安　徽	Anhui	2 412	1 021	208 954.2
福　建	Fujian	3 461	736	161 724.5
江　西	Jiangxi	2 488	811	158 461.8
山　东	Shandong	5 317	2 007	532 290.2
河　南	Henan	3 610	970	230 013.2
湖　北	Hubei	2 110	1 023	176 058.4
湖　南	Hunan	3 131	1 122	178 616.7
广　东	Guangdong	10 608	1 655	518 526.4
广　西	Guangxi	2 392	1 275	182 340.2
海　南	Hainan	364	40	33 048.1
重　庆	Chongqing	1 578	395	79 899.4
四　川	Sichuan	4 255	999	284 376.8
贵　州	Guizhou	1 663	443	58 325.8
云　南	Yunnan	4 878	934	121 067.0
西　藏	Tibet	30	6	662.4
陕　西	Shaanxi	2 206	333	92 997.5
甘　肃	Gansu	631	207	36 721.7
青　海	Qinghai	185	66	14 512.3
宁　夏	Ningxia	377	159	40 555.0
新　疆	Xinjiang	908	238	95 713.6

各地区工业废水排放及处理情况（三）

Discharge and Treatment of Industrial Waste Water by Region（3）

（2012）

单位：吨   （ton）

年 份 地 区	Year Region	工业废水中污染物产生量 Amount of Pollutants Produced in the Industrial Waste Water				
		化学需氧量 COD	氨氮 Ammonia Nitrogen	石油类 Petroleum	挥发酚 Volatile Phenol	氰化物 Cyanide
2011		34 475 339. 7	1 734 151. 3	375 986. 2	70 560. 5	6 220. 9
2012		24 305 917. 4	1 460 908. 1	296 813. 7	65 801. 3	5 180. 2
北 京	Beijing	61 769. 1	2 725. 3	3 298. 4	876. 9	5. 9
天 津	Tianjin	130 525. 3	7 368. 7	1 135. 9	2. 6	0. 5
河 北	Hebei	1 798 954. 4	98 923. 6	21 833. 8	11 702. 5	558. 8
山 西	Shanxi	442 388. 7	97 828. 4	8 180. 5	7 785. 5	363. 4
内蒙古	Inner Mongolia	538 003. 2	54 915. 2	3 693. 1	1 661. 7	37. 7
辽 宁	Liaoning	508 531. 7	54 472. 1	16 175. 9	1 479. 3	152. 8
吉 林	Jilin	595 902. 4	17 539. 5	2 017. 0	874. 2	19. 7
黑龙江	Heilongjiang	1 161 474. 6	34 380. 8	39 539. 5	5 271. 8	84. 6
上 海	Shanghai	312 601. 6	11 889. 5	7 690. 6	1 066. 4	217. 7
江 苏	Jiangsu	2 071 536. 1	76 072. 4	13 124. 4	3 894. 7	378. 6
浙 江	Zhejiang	2 437 630. 0	85 121. 7	21 593. 4	286. 9	762. 1
安 徽	Anhui	707 430. 1	33 766. 7	7 784. 1	1 795. 3	288. 6
福 建	Fujian	949 105. 5	41 558. 9	11 335. 1	288. 9	130. 9
江 西	Jiangxi	481 916. 4	41 670. 6	5 917. 0	2 514. 3	66. 8
山 东	Shandong	2 801 201. 0	118 603. 6	39 199. 2	5 487. 7	269. 9
河 南	Henan	1 764 115. 6	52 294. 7	41 903. 6	2 539. 6	119. 4
湖 北	Hubei	589 765. 1	35 872. 9	4 316. 0	794. 5	58. 8
湖 南	Hunan	626 236. 4	66 562. 6	4 402. 2	1 180. 6	358. 6
广 东	Guangdong	1 472 798. 8	50 465. 1	4 968. 2	144. 8	628. 3
广 西	Guangxi	1 175 146. 1	65 962. 3	3 761. 1	2 080. 9	19. 9
海 南	Hainan	92 870. 1	2 202. 2	162. 6	0. 1	0. 0
重 庆	Chongqing	288 043. 8	28 232. 9	2 727. 8	849. 6	84. 5
四 川	Sichuan	669 543. 7	38 438. 4	3 981. 4	962. 2	77. 6
贵 州	Guizhou	183 282. 2	51 891. 9	2 936. 7	1 146. 7	68. 7
云 南	Yunnan	669 714. 5	77 034. 0	3 090. 9	2 178. 5	118. 1
西 藏	Tibet	6 537. 4	269. 4	4. 2	0. 0	0. 0
陕 西	Shaanxi	507 776. 1	39 479. 7	6 753. 6	5 008. 1	164. 6
甘 肃	Gansu	207 768. 6	18 924. 6	2 372. 4	197. 7	3. 1
青 海	Qinghai	83 378. 9	2 997. 0	1 262. 3	4. 3	31. 2
宁 夏	Ningxia	497 532. 7	72 394. 9	3 189. 4	538. 6	22. 5
新 疆	Xinjiang	472 437. 1	81 048. 4	8 463. 6	3 186. 6	87. 0

各地区工业废水排放及处理情况（四）

Discharge and Treatment of Industrial Waste Water by Region（4）

（2012）

单位：吨 （ton）

年　份 地　区	Year Region	工业废水中污染物产生量 Amount of Pollutants Produced in the Industrial Waste Water					
		汞 Mercury	镉 Cadmium	六价铬 Hexavalent Chromium	总铬 Total Chrome	铅 Lead	砷 Arsenic
2011		23.916	2 580.017	3 550.042	8 456.407	3 793.586	9 977.714
2012		19.961	2 318.538	3 273.441	7 322.903	3 260.838	11 300.867
北　京	Beijing	0.000	0.001	10.270	11.038	0.269	2.344
天　津	Tianjin	0.039	0.024	8.561	13.436	15.271	0.949
河　北	Hebei	0.042	1.269	312.680	381.079	11.112	42.040
山　西	Shanxi	0.363	15.515	15.211	15.791	25.149	59.400
内蒙古	Inner Mongolia	0.954	85.503	1.153	1.272	93.535	247.933
辽　宁	Liaoning	0.134	41.802	18.415	22.901	41.636	39.446
吉　林	Jilin	0.215	0.018	11.570	11.580	0.292	2.857
黑龙江	Heilongjiang	0.160	0.051	6.597	7.280	6.592	1.515
上　海	Shanghai	0.003	0.027	101.576	130.581	1.380	0.818
江　苏	Jiangsu	0.400	2.495	287.800	462.678	21.124	6.344
浙　江	Zhejiang	0.030	4.253	1 575.204	2 905.764	7.249	31.335
安　徽	Anhui	2.253	13.573	19.557	26.983	57.277	556.702
福　建	Fujian	0.084	12.406	411.547	763.195	30.142	50.075
江　西	Jiangxi	0.312	398.974	47.569	51.618	95.817	5 826.814
山　东	Shandong	0.953	271.392	48.787	132.628	267.017	851.823
河　南	Henan	1.511	267.055	19.553	197.229	344.904	115.570
湖　北	Hubei	0.426	8.433	50.332	92.570	20.301	97.804
湖　南	Hunan	4.477	203.135	14.048	38.398	494.243	474.135
广　东	Guangdong	0.135	1.811	125.204	1 599.586	16.689	11.337
广　西	Guangxi	0.517	93.656	26.111	55.723	96.953	149.229
海　南	Hainan	0.000	0.014	0.000	16.896	0.003	0.014
重　庆	Chongqing	0.000	0.001	85.731	163.953	6.990	2.998
四　川	Sichuan	0.968	45.370	52.045	135.526	71.586	488.677
贵　州	Guizhou	0.149	5.006	0.636	3.351	9.688	21.971
云　南	Yunnan	1.040	342.599	3.989	4.774	562.597	943.367
西　藏	Tibet	0.059	0.029	0.000	6.138	0.703	14.894
陕　西	Shaanxi	0.385	74.986	9.743	12.104	112.006	63.383
甘　肃	Gansu	0.251	394.856	3.298	41.878	819.014	1 002.478
青　海	Qinghai	0.439	28.130	1.192	1.303	22.075	77.432
宁　夏	Ningxia	0.710	3.757	1.077	1.434	4.873	86.819
新　疆	Xinjiang	2.953	2.396	3.984	14.216	4.351	30.364

各地区工业废水排放及处理情况（五）

Discharge and Treatment of Industrial Waste Water by Region（5）

（2012）

单位：吨 (ton)

年　份 地　区	Year Region	工业废水中污染物排放量 Amount of Pollutants Discharged in the Industrial Waste Water				
		化学需氧量 COD	氨氮 Ammonia Nitrogen	石油类 Petroleum	挥发酚 Volatile Phenol	氰化物 Cyanide
2011		3 547 997.6	281 201.7	20 589.1	2 410.5	215.4
2012		3 384 539.3	264 142.3	17 327.2	1 481.4	171.8
北　京	Beijing	6 265.8	352.0	49.1	0.2	0.1
天　津	Tianjin	26 533.3	3 294.5	138.2	1.2	0.0
河　北	Hebei	191 779.1	15 825.6	977.7	119.0	12.8
山　西	Shanxi	86 669.5	8 335.4	1 197.8	727.0	30.3
内蒙古	Inner Mongolia	90 802.0	11 146.4	793.2	216.3	1.3
辽　宁	Liaoning	102 213.2	8 366.8	698.4	29.2	4.0
吉　林	Jilin	76 319.4	4 388.0	290.7	3.5	2.2
黑龙江	Heilongjiang	98 187.3	5 846.8	292.7	6.3	1.6
上　海	Shanghai	26 159.5	2 281.0	646.2	2.7	2.2
江　苏	Jiangsu	231 447.1	16 259.4	1 199.9	51.8	15.5
浙　江	Zhejiang	185 671.3	12 089.5	677.2	23.2	7.6
安　徽	Anhui	89 423.8	8 354.0	730.6	5.3	6.3
福　建	Fujian	90 552.5	6 757.9	421.2	9.1	6.0
江　西	Jiangxi	100 699.5	10 207.9	569.1	8.9	7.8
山　东	Shandong	139 404.1	10 981.6	1 080.9	38.0	6.0
河　南	Henan	178 987.1	13 026.8	1 140.6	135.0	16.1
湖　北	Hubei	135 278.1	14 769.9	965.4	15.4	7.8
湖　南	Hunan	149 622.2	25 802.0	771.8	18.2	11.2
广　东	Guangdong	238 503.3	14 937.3	684.3	11.3	10.9
广　西	Guangxi	199 618.3	8 574.2	283.5	14.8	4.9
海　南	Hainan	12 542.6	883.2	3.4	0.0	0.0
重　庆	Chongqing	49 157.2	3 077.2	353.1	9.5	1.5
四　川	Sichuan	118 377.9	5 591.2	418.8	1.3	1.5
贵　州	Guizhou	64 718.6	3 605.9	460.1	1.0	1.5
云　南	Yunnan	169 632.9	4 299.2	402.0	1.9	1.3
西　藏	Tibet	923.9	82.6	0.5	0.0	0.0
陕　西	Shaanxi	98 456.1	8 653.0	736.1	2.5	5.6
甘　肃	Gansu	93 006.6	13 919.2	263.5	1.9	0.2
青　海	Qinghai	41 846.1	2 032.2	314.5	1.3	0.0
宁　夏	Ningxia	105 438.1	8 609.1	178.5	10.6	2.0
新　疆	Xinjiang	186 302.6	11 792.7	588.0	14.7	3.7

各地区工业废水排放及处理情况（六）

Discharge and Treatment of Industrial Waste Water by Region（6）

（2012）

单位：吨 （ton）

| 年 份
地 区 | Year
Region | 工业废水中污染物排放量
Amount of Pollutants Discharged in the Industrial Waste Water | | | | | |
		汞 Mercury	镉 Cadmium	六价铬 Hexavalent Chromium	总铬 Total Chrome	铅 Lead	砷 Arsenic
2011		1.214	35.096	106.186	290.264	150.831	145.169
2012		1.082	26.662	70.375	188.636	97.137	127.692
北 京	Beijing	0.000	0.001	0.326	0.429	0.054	0.012
天 津	Tianjin	0.004	0.010	0.169	0.454	1.004	0.019
河 北	Hebei	0.001	0.001	2.870	5.914	0.260	0.030
山 西	Shanxi	0.005	0.794	0.466	0.491	0.432	0.568
内蒙古	Inner Mongolia	0.075	0.404	0.004	0.012	2.992	5.066
辽 宁	Liaoning	0.001	0.036	0.510	0.679	0.479	0.331
吉 林	Jilin	0.002	0.004	0.108	0.115	0.089	0.995
黑龙江	Heilongjiang	0.000	0.001	0.368	0.368	0.019	0.002
上 海	Shanghai	0.000	0.001	1.011	2.773	0.243	0.073
江 苏	Jiangsu	0.100	0.022	4.534	11.292	2.224	0.557
浙 江	Zhejiang	0.005	0.193	9.355	19.446	0.396	0.162
安 徽	Anhui	0.003	0.104	2.278	3.451	1.611	4.971
福 建	Fujian	0.020	0.260	2.765	11.722	3.032	1.211
江 西	Jiangxi	0.088	2.187	17.148	17.325	6.571	8.651
山 东	Shandong	0.012	0.988	0.528	7.065	0.653	2.170
河 南	Henan	0.014	1.293	1.003	32.554	4.555	1.340
湖 北	Hubei	0.209	0.623	11.557	12.402	3.187	9.553
湖 南	Hunan	0.226	13.467	2.077	18.051	38.476	53.466
广 东	Guangdong	0.016	0.778	8.959	28.352	4.784	0.730
广 西	Guangxi	0.045	1.397	0.719	1.747	5.390	6.630
海 南	Hainan	0.000	0.001	0.000	0.131	0.002	0.013
重 庆	Chongqing	0.000	0.000	0.204	0.507	0.073	1.358
四 川	Sichuan	0.068	0.127	0.840	3.478	1.482	2.541
贵 州	Guizhou	0.022	0.115	0.080	0.146	0.252	0.549
云 南	Yunnan	0.008	1.636	0.014	0.026	8.822	10.469
西 藏	Tibet	0.000	0.000	0.000	0.000	0.000	8.942
陕 西	Shaanxi	0.028	0.616	0.295	1.689	1.628	0.637
甘 肃	Gansu	0.089	1.299	0.381	5.008	6.774	3.745
青 海	Qinghai	0.009	0.117	0.007	0.009	0.711	1.449
宁 夏	Ningxia	0.004	0.023	0.109	0.338	0.087	0.167
新 疆	Xinjiang	0.028	0.166	1.690	2.661	0.856	1.280

各地区工业废气排放及处理情况（一）

Discharge and Treatment of Industrial Waste Gas by Region（1）

（2012）

年　份 地　区	Year Region	工业废气排放总量 （标态）/亿米³ Total Volume of Industrial Waste Gas Emission（100 millon cu.m）	工业废气中污染物排放量/吨 Volume of Pollutants Emission in the Industrial Waste Gas（ton）		
			二氧化硫 Sulphur Dioxide	氮氧化物 Nitrogen Oxide	烟（粉）尘 Soot and Dust
	2011	674 509	20 172 265	17 297 077	11 008 839
	2012	635 519	19 117 055	16 580 515	10 293 082
北　京	Beijing	3 264	59 330	85 331	30 844
天　津	Tianjin	9 032	215 481	275 553	59 036
河　北	Hebei	67 647	1 238 737	1 194 755	1 055 732
山　西	Shanxi	38 124	1 194 634	950 748	948 115
内蒙古	Inner Mongolia	28 133	1 241 475	1 146 979	667 689
辽　宁	Liaoning	31 917	979 025	741 514	626 344
吉　林	Jilin	10 316	352 337	387 416	195 510
黑龙江	Heilongjiang	10 445	397 284	481 257	450 097
上　海	Shanghai	13 361	193 405	284 764	63 732
江　苏	Jiangsu	48 623	959 210	1 133 633	395 978
浙　江	Zhejiang	23 967	610 882	634 950	233 228
安　徽	Anhui	29 645	469 776	692 801	351 639
福　建	Fujian	14 739	352 389	360 879	233 989
江　西	Jiangxi	14 814	551 502	352 508	321 830
山　东	Shandong	45 420	1 543 766	1 238 173	525 896
河　南	Henan	35 002	1 129 858	1 100 046	492 771
湖　北	Hubei	19 513	548 591	440 941	281 577
湖　南	Hunan	15 887	593 342	424 789	298 832
广　东	Guangdong	27 078	771 467	809 107	267 498
广　西	Guangxi	27 611	471 621	345 556	268 524
海　南	Hainan	1 960	33 036	71 850	10 660
重　庆	Chongqing	8 360	509 788	272 094	166 142
四　川	Sichuan	21 910	793 965	438 656	267 820
贵　州	Guizhou	14 312	837 101	458 917	256 539
云　南	Yunnan	14 955	622 599	335 271	359 517
西　藏	Tibet	114	1 316	3 986	955
陕　西	Shaanxi	14 767	747 018	604 455	385 522
甘　肃	Gansu	13 900	479 938	341 040	156 646
青　海	Qinghai	5 508	129 094	88 009	133 698
宁　夏	Ningxia	9 324	384 378	378 835	180 716
新　疆	Xinjiang	15 870	704 710	505 701	606 003

各地区工业废气排放及处理情况（二）
Discharge and Treatment of Industrial Waste Gas by Region（2）
（2012）

单位：吨 (ton)

年 份 地 区	Year Region	工业废气中污染物产生量 Volume of Pollutants Emission in the Industrial Waste Gas		
		二氧化硫 Sulphur Dioxide	氮氧化物 Nitrogen Oxide	烟（粉）尘 Soot and Dust
2011		59 847 513	18 159 854	818 909 810
2012		59 328 890	17 932 874	777 994 342
北 京	Beijing	159 493	97 827	4 219 833
天 津	Tianjin	623 909	297 400	7 758 866
河 北	Hebei	3 232 634	1 276 454	62 118 989
山 西	Shanxi	3 685 739	1 021 546	56 230 755
内蒙古	Inner Mongolia	4 019 535	1 201 549	47 922 188
辽 宁	Liaoning	2 234 164	768 162	30 295 091
吉 林	Jilin	568 163	396 156	14 977 068
黑龙江	Heilongjiang	580 919	489 171	17 378 794
上 海	Shanghai	535 339	325 286	7 254 212
江 苏	Jiangsu	3 720 115	1 338 194	39 707 107
浙 江	Zhejiang	2 167 547	721 919	26 481 356
安 徽	Anhui	2 155 513	746 567	39 184 134
福 建	Fujian	976 728	410 160	16 997 420
江 西	Jiangxi	2 464 973	365 711	16 884 725
山 东	Shandong	5 004 972	1 297 358	62 888 056
河 南	Henan	3 070 767	1 156 795	65 934 261
湖 北	Hubei	1 982 555	455 510	24 164 940
湖 南	Hunan	1 777 669	533 554	27 128 112
广 东	Guangdong	2 583 146	969 987	20 910 930
广 西	Guangxi	1 693 576	372 445	19 518 104
海 南	Hainan	114 835	85 670	1 777 183
重 庆	Chongqing	1 285 476	272 968	17 468 228
四 川	Sichuan	2 015 608	470 396	29 873 618
贵 州	Guizhou	2 626 811	472 670	19 761 043
云 南	Yunnan	2 177 964	347 873	23 431 930
西 藏	Tibet	1 333	3 988	348 455
陕 西	Shaanxi	2 035 179	678 947	21 667 470
甘 肃	Gansu	2 822 005	355 624	12 718 413
青 海	Qinghai	226 508	88 034	3 881 906
宁 夏	Ningxia	1 275 437	405 763	21 141 046
新 疆	Xinjiang	1 510 279	509 189	17 970 109

各地区工业废气排放及处理情况（三）

Discharge and Treatment of Industrial Waste Gas by Region（3）

（2012）

单位：套 （set）

年 份 地 区	Year Region	废气治理 设施数 Facilities for Treatment of Waste Gas	脱硫设施数 Desulfurization Facilities	脱硝设施数 Denitrification Facilities	除尘设施数 Dedusting Facilities	废气治理设施处理 能力（标态） （万米³/时） Capacity of Facilities for Treatment of Waste Gas （10 000 cu.m/hour）
	2011	216 457	20 539	497	169 788	1 568 592
	2012	225 913	20 968	832	177 985	1 649 353
北 京	Beijing	3 004	406	51	2 176	10 901
天 津	Tianjin	3 911	954	9	2 378	22 382
河 北	Hebei	16 924	1 329	45	14 995	182 217
山 西	Shanxi	12 706	3 392	35	8 995	62 972
内蒙古	Inner Mongolia	7 400	477	10	6 868	77 655
辽 宁	Liaoning	11 585	968	14	10 352	86 497
吉 林	Jilin	3 627	102	12	3 203	31 065
黑龙江	Heilongjiang	5 006	178	7	4 819	37 098
上 海	Shanghai	4 795	330	37	2 698	32 703
江 苏	Jiangsu	17 641	1 714	99	11 153	99 232
浙 江	Zhejiang	17 771	1 468	61	12 036	159 773
安 徽	Anhui	5 848	320	20	5 048	40 258
福 建	Fujian	7 780	136	45	6 812	85 065
江 西	Jiangxi	5 426	340	28	4 815	26 452
山 东	Shandong	15 915	2 691	39	12 316	116 317
河 南	Henan	10 184	979	30	8 865	76 908
湖 北	Hubei	6 136	434	24	5 550	93 354
湖 南	Hunan	5 734	436	21	5 120	34 007
广 东	Guangdong	18 667	1 740	126	10 300	78 241
广 西	Guangxi	6 418	474	6	5 594	38 313
海 南	Hainan	555	16	4	494	3 003
重 庆	Chongqing	4 319	248	0	3 338	19 990
四 川	Sichuan	8 384	500	34	7 254	50 885
贵 州	Guizhou	3 413	412	14	2 651	22 084
云 南	Yunnan	6 618	329	16	5 190	32 704
西 藏	Tibet	250	0	0	159	348
陕 西	Shaanxi	4 701	258	24	4 022	30 906
甘 肃	Gansu	3 237	96	11	3 105	39 977
青 海	Qinghai	1 229	19	2	1 203	7 772
宁 夏	Ningxia	1 766	68	6	1 686	17 215
新 疆	Xinjiang	4 963	154	2	4 790	33 060

各地区工业废气排放及处理情况（四）
Discharge and Treatment of Industrial Waste Gas by Region（4）

（2012）

单位：万元 （10 000 yuan）

年 份 地 区	Year Region	废气治理设施 运行费用 Annul Expenditure for Operation	脱硫设施运行费用 Annul Expenditure for Desulfurization Facilities	脱硝设施运行费用 Annul Expenditure for Denitrification Facilities	除尘设施运行费用 Annul Expenditure for Dedusting Facilities
	2011	15 794 757. 8	5 872 661. 3	444 155. 0	8 186 504. 2
	2012	14 522 519. 8	5 404 357. 7	660 273. 6	6 474 310. 3
北 京	Beijing	87 848. 1	28 094. 9	19 977. 4	29 421. 2
天 津	Tianjin	257 708. 9	90 104. 0	10 023. 0	149 592. 9
河 北	Hebei	1 304 985. 9	369 399. 4	27 706. 5	874 505. 7
山 西	Shanxi	693 788. 5	364 335. 5	37 099. 6	274 120. 4
内蒙古	Inner Mongolia	597 282. 2	258 672. 5	12 441. 3	311 479. 6
辽 宁	Liaoning	607 390. 5	172 351. 9	13 385. 3	392 731. 4
吉 林	Jilin	147 837. 9	40 782. 3	5 498. 1	87 494. 1
黑龙江	Heilongjiang	128 714. 6	37 783. 7	1 751. 0	87 074. 1
上 海	Shanghai	405 020. 1	144 314. 4	28 390. 3	149 908. 4
江 苏	Jiangsu	2 243 764. 3	473 732. 6	111 154. 6	414 103. 1
浙 江	Zhejiang	990 296. 1	421 215. 0	83 560. 6	416 739. 4
安 徽	Anhui	465 839. 0	165 109. 8	22 317. 1	251 451. 8
福 建	Fujian	305 612. 1	128 012. 8	27 501. 8	134 351. 8
江 西	Jiangxi	332 532. 1	156 566. 3	9 831. 4	160 523. 1
山 东	Shandong	1 142 868. 9	561 380. 6	26 911. 0	510 105. 3
河 南	Henan	564 513. 9	232 768. 1	29 807. 0	269 306. 3
湖 北	Hubei	354 541. 8	129 621. 1	7 999. 1	216 270. 9
湖 南	Hunan	306 442. 4	112 171. 0	16 264. 2	159 595. 8
广 东	Guangdong	906 533. 0	447 588. 0	94 642. 9	260 541. 0
广 西	Guangxi	272 190. 9	79 345. 7	4 446. 0	167 560. 6
海 南	Hainan	55 571. 5	21 531. 1	6 027. 0	16 387. 1
重 庆	Chongqing	202 828. 0	75 734. 5	0. 0	117 883. 7
四 川	Sichuan	429 004. 0	145 679. 1	5 301. 4	251 201. 3
贵 州	Guizhou	361 191. 2	205 284. 7	1 610. 6	108 892. 3
云 南	Yunnan	367 775. 1	126 600. 7	9 426. 8	182 301. 2
西 藏	Tibet	1 702. 3	0. 0	0. 0	1 167. 1
陕 西	Shaanxi	276 399. 6	123 544. 1	16 588. 4	116 488. 0
甘 肃	Gansu	206 345. 4	93 478. 2	8 836. 0	86 893. 2
青 海	Qinghai	85 717. 6	14 664. 3	307. 2	70 736. 1
宁 夏	Ningxia	206 959. 4	111 569. 3	21 348. 0	74 022. 1
新 疆	Xinjiang	213 314. 5	72 922. 1	120. 0	131 461. 3

各地区一般工业固体废物产生及处置利用情况
Generation and Utilization of Industrial Solid Wastes by Region
（2012）

单位：万吨 （10 000 tons）

年 份 地 区	Year Region	一般工业固体废物产生量 Industrial Solid Wastes Generated	一般工业固体废物综合利用量 Industrial Solid Wastes Utilized	一般工业固体废物处置量 Industrial Solid Wastes Disposed	一般工业固体废物贮存量 Stock of Industrial Solid Wastes	一般工业固体废物倾倒丢弃量 Industrial Solid Wastes Discharged	一般工业固体废物综合利用率/% Industrial Solid Wastes Utilization rate （%）
	2011	322 772	195 215	70 465	60 424	433	59.9
	2012	329 044	202 462	70 745	59 786	144	61.0
北 京	Beijing	1 104	872	219	13	0	79.0
天 津	Tianjin	1 820	1 816	7	0	0	99.6
河 北	Hebei	45 576	17 361	7 439	21 210	0	37.8
山 西	Shanxi	29 031	20 235	7 132	1 759	16	69.4
内蒙古	Inner Mongolia	24 226	10 925	10 943	2 426	5	45.0
辽 宁	Liaoning	27 280	11 862	11 655	3 948	10	43.4
吉 林	Jilin	4 731	3 198	541	993	0	67.6
黑龙江	Heilongjiang	6 313	4 646	808	918	0	73.3
上 海	Shanghai	2 199	2 140	56	10	0	97.0
江 苏	Jiangsu	10 224	9 342	630	303	0	91.1
浙 江	Zhejiang	4 461	4 083	314	67	0	91.5
安 徽	Anhui	12 022	10 266	1 705	895	0	81.5
福 建	Fujian	7 720	6 887	764	84	0	89.1
江 西	Jiangxi	11 134	6 071	388	4 692	2	54.5
山 东	Shandong	18 343	17 073	1 061	487	0	91.8
河 南	Henan	15 250	11 597	3 204	560	2	75.5
湖 北	Hubei	7 611	5 737	1 561	377	1	74.7
湖 南	Hunan	8 116	5 188	2 144	955	1	62.6
广 东	Guangdong	5 965	5 198	814	253	3	83.1
广 西	Guangxi	7 964	5 369	2 218	1 063	0	63.3
海 南	Hainan	386	238	59	116	0	57.6
重 庆	Chongqing	3 115	2 569	475	97	5	81.6
四 川	Sichuan	13 187	6 052	5 099	2 279	2	45.7
贵 州	Guizhou	7 835	4 839	2 067	938	14	61.6
云 南	Yunnan	16 038	7 938	4 767	3 513	43	48.9
西 藏	Tibet	366	6	27	349	0	1.6
陕 西	Shaanxi	7 215	4 422	1 457	1 360	2	61.1
甘 肃	Gansu	6 671	3 593	2 110	992	0	53.7
青 海	Qinghai	12 301	6 831	6	5 483	0	55.4
宁 夏	Ningxia	2 961	2 044	540	398	0	68.6
新 疆	Xinjiang	7 880	4 063	533	3 251	35	51.6

各地区工业危险废物产生及处置利用情况
Generation and Utilization of Hazardous Wastes by Region

（2012）

单位：吨 (ton)

年 份 Year 地 区 Region	危险废物 产生量 Hazardous Wastes Generated	危险废物 综合利用量 Hazardous Wastes Utilized	危险废物 处置量 Hazardous Wastes Disposed	危险废物 贮存量 Stock of Hazardous Wastes	危险废物倾 倒丢弃量 Hazardous Wastes Discharged	危险废物处置 利用率/% Hazardous Wastes Disposition and Utilization rate （%）
2011	34 312 203	17 730 522	9 164 762	8 237 304	96	76.5
2012	34 652 427	20 046 450	6 982 060	8 469 139	16	76.1
北 京 Beijing	134 130	44 990	89 130	19	0	100.0
天 津 Tianjin	114 710	39 589	75 121	0	0	100.0
河 北 Hebei	491 781	270 283	219 074	3 174	0	99.4
山 西 Shanxi	185 254	133 259	50 855	1 574	0	99.2
内蒙古 Inner Mongolia	699 936	416 517	398 401	78 105	0	91.3
辽 宁 Liaoning	732 061	494 613	305 432	3 513	2	99.6
吉 林 Jilin	712 138	670 078	42 006	220	0	100.0
黑龙江 Heilongjiang	213 281	42 049	169 931	1 392	0	99.3
上 海 Shanghai	549 568	303 444	245 994	1 420	0	99.7
江 苏 Jiangsu	2 085 866	1 099 832	976 449	27 941	0	98.7
浙 江 Zhejiang	805 810	279 356	515 555	19 671	0	97.6
安 徽 Anhui	247 402	183 242	62 117	2 915	0	98.8
福 建 Fujian	103 617	46 176	57 311	1 699	4	98.4
江 西 Jiangxi	305 645	252 234	50 957	3 256	0	98.9
山 东 Shandong	8 203 093	7 611 337	661 400	62 915	0	99.2
河 南 Henan	500 988	422 904	78 135	4 176	0	99.2
湖 北 Hubei	634 997	398 614	236 993	4 054	0	99.4
湖 南 Hunan	2 674 846	2 076 545	416 563	213 700	0	92.1
广 东 Guangdong	1 301 703	670 089	628 154	3 972	0	99.7
广 西 Guangxi	787 884	554 501	228 935	76 121	0	91.1
海 南 Hainan	15 317	951	15 579	311	0	98.2
重 庆 Chongqing	490 253	371 766	118 048	452	0	99.9
四 川 Sichuan	1 104 370	631 382	466 340	9 007	0	99.2
贵 州 Guizhou	338 683	139 063	3 559	196 142	0	42.1
云 南 Yunnan	2 080 373	974 428	531 269	764 721	2	66.3
西 藏 Tibet	0	0	0	0	0	—
陕 西 Shaanxi	306 531	115 381	121 103	76 299	0	75.6
甘 肃 Gansu	290 904	130 393	114 377	100 685	0	70.9
青 海 Qinghai	4 043 366	612 982	1 241	3 475 126	0	15.0
宁 夏 Ningxia	56 227	36 569	7 989	11 681	0	79.2
新 疆 Xinjiang	4 441 695	1 023 884	94 042	3 324 878	8	25.2

各地区汇总工业企业概况（一）

Summarization of Industrial Enterprises Investigeted by Region（1）

（2012）

年 份 地 区	Year Region	汇总工业 企业数/个 Number of Industrial Enterprises Investigated （unit）	工业总产值 （现价）/ 亿元 Gross Industrial Output Value （current rate） （100 million yuan）	工业炉窑 数/台 Number of Industrial Furnaces （unit）	工业锅炉 数/台 Number of Industrial Boilers （unit）	35 蒸吨及 以上的 Beyond 35 tons of steam	20（含）～ 35 蒸吨之 间的 Between 20 and 35 tons of steam	10（含）～ 20 蒸吨之 间的 Between 20 and 35 tons of steam	10 蒸吨 以下的 Below 10 tons of steam
	2011	153 027	450 660.2	101 705	104 009	13 013	7 619	13 984	69 393
	2012	147 996	368 365.3	99 205	103 469	13 683	7 400	13 726	68 660
北 京	Beijing	896	6 174.9	317	1 760	308	268	342	842
天 津	Tianjin	2 021	11 723.4	702	2 202	419	307	345	1 131
河 北	Hebei	10 074	21 007.1	6 377	6 393	753	336	858	4 446
山 西	Shanxi	6 367	11 939.5	5 927	6 101	736	449	708	4 208
内蒙古	Inner Mongolia	3 025	7 767.0	3 716	4 088	771	447	784	2 086
辽 宁	Liaoning	6 316	17 513.9	6 575	7 097	1 185	976	1 288	3 648
吉 林	Jilin	1 466	5 175.1	643	2 881	594	265	386	1 636
黑龙江	Heilongjiang	1 927	6 667.9	3 081	4 929	734	510	812	2 873
上 海	Shanghai	2 158	18 166.1	1 285	2 122	160	70	160	1 732
江 苏	Jiangsu	11 107	41 678.1	4 508	6 940	943	282	668	5 047
浙 江	Zhejiang	13 541	27 148.4	4 789	7 997	658	243	891	6 205
安 徽	Anhui	8 403	11 309.9	4 564	2 733	266	83	302	2 082
福 建	Fujian	5 740	9 578.3	2 687	3 439	153	122	417	2 747
江 西	Jiangxi	5 118	5 945.3	3 721	2 102	123	64	229	1 686
山 东	Shandong	7 791	38 287.6	4 916	6 527	1 792	482	837	3 416
河 南	Henan	6 550	12 653.7	5 763	4 161	553	269	481	2 858
湖 北	Hubei	3 699	12 001.9	2 475	2 451	217	113	331	1 790
湖 南	Hunan	4 800	7 863.3	4 578	2 200	215	66	225	1 694
广 东	Guangdong	14 890	35 660.5	4 922	7 759	522	278	1 079	5 880
广 西	Guangxi	3 517	7 924.8	2 490	1 843	364	191	205	1 083
海 南	Hainan	460	1 362.5	193	304	44	37	24	199
重 庆	Chongqing	3 107	6 814.8	1 947	1 456	112	53	125	1 166
四 川	Sichuan	7 548	11 006.4	7 229	3 682	223	176	338	2 945
贵 州	Guizhou	3 801	4 102.2	4 269	1 542	136	69	92	1 245
云 南	Yunnan	4 105	7 077.3	2 323	1 668	222	162	202	1 082
西 藏	Tibet	86	54.2	27	27	1	0	1	25
陕 西	Shaanxi	3 760	8 299.6	2 806	2 397	305	197	406	1 489
甘 肃	Gansu	2 290	3 750.1	2 612	1 830	204	266	409	951
青 海	Qinghai	611	1 483.4	804	478	40	57	88	293
宁 夏	Ningxia	884	2 170.4	1 433	1 190	241	144	198	607
新 疆	Xinjiang	1 938	6 057.7	1 526	3 170	689	418	495	1 568

各地区汇总工业企业概况（二）

Summarization of Industrial Enterprises Investigeted by Region（2）

（2012）

年　份 地　区	Year Region	工业用水 总量/万吨 Quantity of Water Used by Industry （10 000 tons）	取水量 Fresh Water	重　复 用水量 Recycled Water	工业煤炭消耗量/ 万吨 Total Amount of Industrial Coal Consumed （10 000 tons）	燃料煤 Coal Used as Fuel
	2011	39 622 524	6 678 557	32 943 967	385 802	288 960
	2012	38 493 488	5 020 781	33 472 707	382 786	290 056
北　京	Beijing	325 063	18 462	306 601	1 612	1 513
天　津	Tianjin	1 004 487	40 473	964 013	4 583	4 319
河　北	Hebei	4 459 948	267 648	4 192 300	28 989	18 232
山　西	Shanxi	2 291 757	138 783	2 152 974	32 336	19 531
内蒙古	Inner Mongolia	1 418 432	91 228	1 327 204	32 958	23 994
辽　宁	Liaoning	1 802 847	310 855	1 491 992	17 370	13 745
吉　林	Jilin	860 731	96 214	764 517	10 848	7 735
黑龙江	Heilongjiang	806 428	150 298	656 130	13 012	9 230
上　海	Shanghai	860 514	81 673	778 841	5 419	4 087
江　苏	Jiangsu	3 337 257	520 370	2 816 887	25 898	23 356
浙　江	Zhejiang	2 076 202	691 502	1 384 700	14 424	13 602
安　徽	Anhui	1 447 606	122 742	1 324 864	13 495	10 809
福　建	Fujian	603 619	101 677	501 941	7 412	6 177
江　西	Jiangxi	653 969	141 475	512 494	6 265	4 976
山　东	Shandong	3 363 742	321 185	3 042 557	31 122	23 617
河　南	Henan	2 526 444	229 084	2 297 360	27 436	17 754
湖　北	Hubei	1 479 297	248 432	1 230 865	9 207	6 188
湖　南	Hunan	955 794	161 003	794 791	7 718	5 873
广　东	Guangdong	1 516 063	303 882	1 212 181	17 608	16 334
广　西	Guangxi	1 417 405	343 978	1 073 426	6 483	5 186
海　南	Hainan	220 177	13 147	207 030	926	923
重　庆	Chongqing	556 557	136 872	419 685	4 140	3 390
四　川	Sichuan	878 194	115 556	762 638	8 233	6 322
贵　州	Guizhou	703 033	41 357	661 676	8 189	7 052
云　南	Yunnan	896 611	84 284	812 327	8 575	6 594
西　藏	Tibet	1 772	666	1 106	40	38
陕　西	Shaanxi	502 841	76 327	426 515	12 746	9 441
甘　肃	Gansu	262 438	53 260	209 178	5 690	4 928
青　海	Qinghai	148 263	15 354	132 909	1 551	1 252
宁　夏	Ningxia	460 894	36 680	424 214	9 004	6 642
新　疆	Xinjiang	655 102	66 312	588 791	9 495	7 216

各地区汇总工业企业概况（三）

Summarization of Industrial Enterprises Investigeted by Region（3）

（2012）

年　份 地　区	Year Region	燃料油消耗量/ 万吨 Total Amount of Fuel Oil Consumed （10 000 tons）	焦炭消耗量/ 万吨 Total Amount of Coke Consumed （10 000 tons）	天然气消耗量/ 亿米^3 Total Amount of Natural Gas Consumed（100 million cu.m）	其他燃料消耗量/ 万吨标煤 Total Amount of other Fuel Consumed （10 000 tons of standard coal）
	2011	1 613	28 698	1 542	16 466
	2012	1 912	28 442	2 126	12 479
北　京	Beijing	40	1	37	141
天　津	Tianjin	9	520	19	175
河　北	Hebei	14	7 083	75	1 005
山　西	Shanxi	3	1 708	17	611
内蒙古	Inner Mongolia	6	1 130	21	350
辽　宁	Liaoning	156	2 285	55	877
吉　林	Jilin	22	482	17	119
黑龙江	Heilongjiang	24	306	29	265
上　海	Shanghai	43	119	37	642
江　苏	Jiangsu	688	2 087	576	1 224
浙　江	Zhejiang	134	406	132	219
安　徽	Anhui	13	803	61	670
福　建	Fujian	55	486	38	173
江　西	Jiangxi	18	948	3	324
山　东	Shandong	89	2 789	175	1 032
河　南	Henan	36	926	42	419
湖　北	Hubei	36	361	191	120
湖　南	Hunan	6	432	123	167
广　东	Guangdong	382	455	81	631
广　西	Guangxi	51	744	12	848
海　南	Hainan	2	0	61	58
重　庆	Chongqing	2	245	25	197
四　川	Sichuan	4	941	58	1 190
贵　州	Guizhou	7	293	5	40
云　南	Yunnan	9	1 056	34	180
西　藏	Tibet	12	0	0	0
陕　西	Shaanxi	15	240	8	137
甘　肃	Gansu	10	462	72	176
青　海	Qinghai	1	148	36	30
宁　夏	Ningxia	2	167	6	27
新　疆	Xinjiang	23	818	80	433

各地区工业污染防治投资情况（一）

Treatment Investment for Industrial Pollution by Region（1）

（2012）

单位：个 (unit)

年 份　Year 地 区　Region	汇总工业企业数 Number of Industrial Enterprises Collected	本年施工项目总数 Numer of Projects under Construction	工业废水治理项目 Treatment of Waste Water	工业废气治理项目 Treatment of Waste Gas	脱硫治理项目 Treatment of desulfurization	脱硝治理项目 Treatment of Denitration	工业固体废物治理项目 Treatment of Solid Wastes	噪声治理项目 Treatment of Noise Pollution	其他治理项目 Treatment of Other Pollution
2011	6 500	9 257	3 738	3 478	956	98	577	175	1 289
2012	6 203	5 390	1 801	2 144	501	293	256	105	1 084
北　京　Beijing	41	51	10	27	11	1	2	1	11
天　津　Tianjin	110	135	50	50	9	3	2	8	25
河　北　Hebei	191	159	35	102	46	16	1	0	21
山　西　Shanxi	294	318	72	131	47	17	38	7	70
内蒙古　Inner Mongolia	96	126	44	63	18	5	4	0	15
辽　宁　Liaoning	110	82	16	34	7	5	6	1	25
吉　林　Jilin	66	36	7	20	3	4	1	1	7
黑龙江　Heilongjiang	59	57	11	41	9	5	4	0	1
上　海　Shanghai	122	142	39	47	10	3	6	8	42
江　苏　Jiangsu	431	357	130	166	27	31	8	8	45
浙　江　Zhejiang	687	559	196	205	22	26	11	4	143
安　徽　Anhui	167	105	40	54	8	17	2	3	6
福　建　Fujian	338	302	109	86	18	19	43	4	60
江　西　Jiangxi	104	80	39	29	8	1	4	3	5
山　东　Shandong	554	528	177	235	87	24	16	9	91
河　南　Henan	186	158	67	50	19	2	9	6	26
湖　北　Hubei	166	128	39	53	9	5	4	8	24
湖　南　Hunan	220	143	50	52	9	16	7	0	34
广　东　Guangdong	637	615	169	243	47	23	6	14	183
广　西　Guangxi	194	207	95	40	8	2	16	1	55
海　南　Hainan	38	31	20	6	1	2	0	0	5
重　庆　Chongqing	168	111	49	36	5	8	3	4	19
四　川　Sichuan	281	212	94	61	10	6	14	5	38
贵　州　Guizhou	203	86	35	36	7	13	3	0	12
云　南　Yunnan	308	256	65	119	7	5	26	5	41
西　藏　Tibet	11	10	2	3	0	0	2	0	3
陕　西　Shaanxi	181	182	73	49	20	14	13	4	43
甘　肃　Gansu	71	57	29	14	2	2	0	1	13
青　海　Qinghai	29	25	7	8	1	5	1	0	9
宁　夏　Ningxia	61	60	9	40	8	4	4	0	7
新　疆　Xinjiang	79	72	23	44	18	9	0	0	5

各地区工业污染防治投资情况（二）

Treatment Investment for Industrial Pollution by Region（2）

（2012）

单位：个 （unit）

年　份 地　区	Year Region	本年竣工项目总数 Number of Projects Completed	工业废水治理项目 Treatment of Waste Water	工业废气治理项目 Treatment of Waste Gas	脱硫治理项目 Treatment of desulfurization	脱硝治理项目 Treatment of Denitration	工业固体废物治理项目 Treatment of Solid Wastes	噪声治理项目 Treatment of Noise Pollution	其他治理项目 Treatment of Other Pollution
2011		7 005	2 633	2 787	753	62	420	146	1 019
2012		5 565	2 017	2 153	513	209	229	113	1 053
北　京	Beijing	52	13	25	8	1	2	1	11
天　津	Tianjin	73	20	26	5	3	2	6	19
河　北	Hebei	172	45	104	39	12	2	0	21
山　西	Shanxi	326	70	140	45	8	32	6	78
内蒙古	Inner Mongolia	109	50	43	13	2	2	1	13
辽　宁	Liaoning	110	33	34	7	2	6	2	35
吉　林	Jilin	36	5	21	3	4	1	1	8
黑龙江	Heilongjiang	56	9	40	8	5	4	0	3
上　海	Shanghai	158	41	54	10	2	6	8	49
江　苏	Jiangsu	385	158	164	24	24	7	9	47
浙　江	Zhejiang	564	207	192	22	17	7	6	152
安　徽	Anhui	122	46	59	7	14	2	3	12
福　建	Fujian	273	102	69	11	15	42	7	53
江　西	Jiangxi	90	45	24	4	0	7	3	11
山　东	Shandong	550	220	217	79	15	15	10	88
河　南	Henan	151	53	53	23	2	6	7	32
湖　北	Hubei	109	38	45	9	4	5	6	15
湖　南	Hunan	154	55	65	16	14	8	1	25
广　东	Guangdong	629	185	281	65	20	9	15	139
广　西	Guangxi	203	92	42	10	0	13	1	55
海　南	Hainan	49	21	17	1	1	1	0	10
重　庆	Chongqing	150	78	43	8	6	4	4	21
四　川	Sichuan	235	115	69	10	6	10	6	35
贵　州	Guizhou	191	122	45	15	9	5	2	17
云　南	Yunnan	257	72	131	9	2	14	3	37
西　藏	Tibet	12	3	3	0	0	3	0	3
陕　西	Shaanxi	160	64	46	22	12	9	4	37
甘　肃	Gansu	57	30	11	1	2	1	1	14
青　海	Qinghai	15	4	5	2	2	1	0	5
宁　夏	Ningxia	63	13	42	9	4	3	0	5
新　疆	Xinjiang	54	8	43	28	1	0	0	3

各地区工业污染防治投资情况（三）

Treatment Investment for Industrial Pollution by Region（3）

（2012）

单位：万元 （10 000 yuan）

年 份 Year 地 区 Region	施工项目本年完成投资 Investment Completed in the Treatment of Industrial Pollution This Year	工业废水治理项目 Treatment of Waste Water	工业废气治理项目 Treatment of Waste Gas	脱硫治理项目 Treatment of desulfurization	脱硝治理项目 Treatment of Denitration	工业固体废物治理项目 Treatment of Solid Wastes	噪声治理项目 Treatment of Noise Pollution	其他治理项目 Treatment of Other Pollution
2011	4 443 610. 1	1 577 471. 1	2 116 810. 6	1 127 347. 8	126 755. 7	313 875. 3	21 622. 5	413 830. 7
2012	5 004 572. 7	1 403 447. 5	2 577 138. 7	801 975. 2	909 416. 3	247 499. 3	11 626. 8	764 860. 4
北 京 Beijing	32 840. 4	3 011. 9	24 652. 3	3 026. 3	210. 0	1 011. 0	40. 0	4 125. 3
天 津 Tianjin	125 558. 6	11 305. 9	42 459. 5	21 441. 2	11 965. 0	1 301. 0	1 207. 9	69 284. 4
河 北 Hebei	236 289. 7	52 178. 3	181 167. 3	105 101. 9	38 906. 9	86. 2	0. 0	2 858. 0
山 西 Shanxi	323 269. 5	30 523. 1	170 664. 5	50 297. 5	89 108. 6	40 200. 7	238. 8	81 642. 4
内蒙古 Inner Mongolia	189 715. 4	38 699. 0	123 390. 7	55 400. 0	42 130. 3	18 197. 0	5. 0	9 423. 7
辽 宁 Liaoning	119 447. 4	27 776. 9	56 167. 6	20 516. 0	15 064. 0	2 354. 4	111. 0	33 037. 5
吉 林 Jilin	57 268. 9	14 701. 7	30 422. 4	11 778. 5	13 095. 0	1 095. 0	350. 0	10 699. 7
黑龙江 Heilongjiang	39 287. 3	7 349. 7	27 660. 5	9 642. 2	3 586. 5	2 501. 0	0. 0	1 776. 0
上 海 Shanghai	115 915. 1	5 336. 1	55 455. 4	1 090. 9	9 450. 0	432. 0	560. 2	54 131. 5
江 苏 Jiangsu	390 143. 5	73 571. 8	265 738. 7	52 219. 4	122 086. 4	12 988. 9	322. 8	37 521. 4
浙 江 Zhejiang	283 022. 8	101 840. 0	138 422. 0	22 630. 5	82 169. 0	1 457. 0	1 405. 0	39 898. 8
安 徽 Anhui	127 349. 6	21 475. 9	100 710. 8	17 030. 0	64 002. 3	2 335. 0	183. 0	2 644. 9
福 建 Fujian	237 634. 6	102 882. 9	107 724. 7	11 385. 0	40 852. 5	7 576. 4	381. 8	19 068. 8
江 西 Jiangxi	39 478. 1	16 574. 6	17 575. 8	7 610. 6	2 500. 0	1 624. 0	15. 9	3 687. 7
山 东 Shandong	670 633. 4	263 796. 5	303 864. 7	136 325. 1	41 344. 7	35 350. 4	1 305. 7	66 316. 2
河 南 Henan	148 347. 3	30 452. 0	77 995. 3	35 777. 9	26 500. 0	3 952. 4	425. 2	35 522. 5
湖 北 Hubei	148 964. 0	35 375. 8	75 490. 2	12 070. 9	48 248. 3	1 476. 0	1 114. 2	35 507. 8
湖 南 Hunan	179 560. 6	48 312. 1	78 513. 4	22 288. 8	37 167. 0	4 156. 8	280. 0	48 298. 4
广 东 Guangdong	280 995. 9	57 889. 4	190 206. 7	45 937. 0	63 154. 5	17 347. 9	755. 3	14 796. 7
广 西 Guangxi	85 643. 8	43 969. 2	25 546. 0	12 486. 0	1 871. 0	8 782. 3	2. 0	7 344. 2
海 南 Hainan	48 279. 4	25 042. 9	21 206. 5	10 360. 0	5 600. 0	180. 0	0. 0	1 850. 0
重 庆 Chongqing	38 226. 5	17 510. 5	16 530. 9	2 941. 6	11 713. 7	1 471. 2	309. 0	2 404. 9
四 川 Sichuan	110 607. 6	53 646. 4	48 615. 4	15 096. 7	16 148. 1	2 892. 2	473. 7	4 979. 9
贵 州 Guizhou	124 662. 7	21 040. 7	67 346. 3	43 737. 5	19 265. 0	2 362. 9	1 034. 0	32 878. 7
云 南 Yunnan	197 259. 3	101 066. 0	56 597. 7	17 233. 9	3 337. 6	5 733. 3	575. 3	33 287. 0
西 藏 Tibet	1 775. 0	922. 0	174. 0	0. 0	0. 0	615. 0	0. 0	64. 0
陕 西 Shaanxi	271 265. 9	112 221. 1	112 463. 6	12 118. 5	27 888. 0	11 406. 4	480. 1	34 694. 7
甘 肃 Gansu	210 984. 5	29 509. 6	62 219. 7	3 070. 7	45 790. 0	53 498. 6	51. 0	65 705. 6
青 海 Qinghai	21 880. 1	3 262. 6	14 129. 6	6 378. 6	5 413. 0	1 513. 5	0. 0	2 974. 4
宁 夏 Ningxia	69 159. 7	14 298. 3	43 543. 0	11 632. 4	11 624. 0	3 601. 0	0. 0	7 717. 5
新 疆 Xinjiang	79 106. 2	37 904. 8	40 483. 4	25 349. 4	9 225. 0	0. 0	0. 0	718. 0

各地区工业污染防治投资情况（四）

Treatment Investment for Industrial Pollution by Region（4）

（2012）

年　份 地　区	Year Region	本年竣工项目新增设计处理能力 Capacity for Treatment of Industrial Pollution New-added		
		治理废水/ （万吨/日） Treatment of Waste Water （10 000 tons/day）	治理废气（标态）/ （万米³/时） Treatment of Waste Gas （10 000 cu.m/hour）	治理固体废物/ （万吨/日） Treatment of Solid Wastes （10 000 tons/day）
2011		1 493.5	52 513.0	495.7
2012		485.8	50 060.7	71.8
北　京	Beijing	0.9	265.6	0.1
天　津	Tianjin	4.0	721.4	0.0
河　北	Hebei	12.5	3 498.6	0.3
山　西	Shanxi	18.9	3 163.4	2.4
内蒙古	Inner Mongolia	8.0	2 249.5	0.2
辽　宁	Liaoning	3.9	292.8	2.1
吉　林	Jilin	6.3	589.0	0.2
黑龙江	Heilongjiang	2.8	279.3	0.1
上　海	Shanghai	1.1	654.7	0.0
江　苏	Jiangsu	28.5	4 277.9	0.3
浙　江	Zhejiang	22.5	3 604.2	0.2
安　徽	Anhui	4.2	2 047.0	0.2
福　建	Fujian	34.4	2 416.7	0.4
江　西	Jiangxi	4.1	553.1	0.1
山　东	Shandong	55.8	8 674.5	1.2
河　南	Henan	11.8	2 911.7	14.3
湖　北	Hubei	6.8	1 388.8	0.4
湖　南	Hunan	18.1	1 407.5	13.0
广　东	Guangdong	24.6	3 217.5	25.5
广　西	Guangxi	28.7	1 482.1	0.7
海　南	Hainan	7.1	133.2	0.0
重　庆	Chongqing	3.8	282.1	0.0
四　川	Sichuan	10.5	934.5	0.3
贵　州	Guizhou	8.7	1 656.2	0.3
云　南	Yunnan	37.9	726.8	3.3
西　藏	Tibet	0.3	0.0	0.0
陕　西	Shaanxi	15.2	537.8	2.5
甘　肃	Gansu	4.4	509.3	3.6
青　海	Qinghai	1.2	320.2	0.1
宁　夏	Ningxia	2.4	783.4	0.2
新　疆	Xinjiang	96.4	481.9	0.0

各地区农业污染排放情况（一）
Discharge of Agricultural Pollution by Region
（2012）

单位：万吨 （10 000 tons）

年 份 地 区	Year Region	农业污染物排放（流失）总量 Total Amount of Discharge of Agricultural Pollution			
		化学需氧量 COD	氨氮 Ammonial Nitrogen	总氮 Total Nitrogen	总磷 Total Phosphorus
2011		1 186.1	82.7	454.7	56.3
2012		1 153.8	80.6	451.4	48.9
北 京	Beijing	7.8	0.5	3.3	0.4
天 津	Tianjin	11.4	0.6	3.3	0.4
河 北	Hebei	91.8	4.5	36.0	3.9
山 西	Shanxi	18.1	1.3	8.4	0.8
内蒙古	Inner Mongolia	61.9	1.2	26.7	2.1
辽 宁	Liaoning	86.3	3.4	20.5	2.8
吉 林	Jilin	50.4	1.8	12.9	1.6
黑龙江	Heilongjiang	105.2	3.4	24.8	2.4
上 海	Shanghai	3.3	0.3	1.5	0.2
江 苏	Jiangsu	38.8	3.9	17.5	1.9
浙 江	Zhejiang	20.4	2.7	9.2	1.1
安 徽	Anhui	38.5	3.8	18.1	2.0
福 建	Fujian	21.4	3.3	9.6	1.2
江 西	Jiangxi	23.9	3.0	11.4	1.4
山 东	Shandong	134.4	7.4	56.3	6.1
河 南	Henan	80.3	6.4	41.9	4.8
湖 北	Hubei	47.6	4.7	19.5	2.3
湖 南	Hunan	56.7	6.3	22.1	2.4
广 东	Guangdong	60.0	5.8	19.5	2.5
广 西	Guangxi	21.6	2.7	11.6	1.3
海 南	Hainan	10.3	0.9	4.1	0.5
重 庆	Chongqing	12.4	1.3	5.4	0.6
四 川	Sichuan	54.0	5.7	22.1	2.5
贵 州	Guizhou	6.3	0.8	4.7	0.4
云 南	Yunnan	7.4	1.2	7.6	0.7
西 藏	Tibet	0.4	0.0	0.6	0.0
陕 西	Shaanxi	19.7	1.5	8.6	0.8
甘 肃	Gansu	14.5	0.6	4.7	0.4
青 海	Qinghai	2.2	0.1	0.7	0.1
宁 夏	Ningxia	10.2	0.2	2.7	0.2
新 疆	Xinjiang	36.7	1.3	16.2	1.2

各地区农业污染排放情况（二）
Discharge of Agricultural Pollution by Region（2）

（2012）

单位：吨 　　　（ton）

年　份 地　区	Year Region	化学需氧量排放总量 Total Amount of COD Discharged	水产养殖业 Aquiculture	畜禽养殖业 Livestock and Poultry	规模化畜禽养殖场/小区 Large-scale Farms	养殖专业户 Small-scale Farms
2011		11 861 057	556 485	11 304 573	3 884 218	7 420 355
2012		11 537 999	548 418	10 989 581	3 794 478	7 195 103
北　京	Beijing	78 292	3 833	74 459	42 799	31 660
天　津	Tianjin	113 889	15 405	98 484	33 471	65 013
河　北	Hebei	918 344	8 814	909 530	416 218	493 312
山　西	Shanxi	180 623	246	180 377	78 378	102 000
内蒙古	Inner Mongolia	619 236	775	618 462	145 460	473 001
辽　宁	Liaoning	863 316	13 114	850 201	167 539	682 663
吉　林	Jilin	503 950	1 022	502 928	107 228	395 700
黑龙江	Heilongjiang	1 051 771	4 743	1 047 028	318 280	728 748
上　海	Shanghai	33 090	5 561	27 529	14 597	12 932
江　苏	Jiangsu	387 692	55 607	332 085	112 365	219 720
浙　江	Zhejiang	203 995	23 961	180 033	79 116	100 917
安　徽	Anhui	384 905	13 711	371 195	187 477	183 717
福　建	Fujian	214 105	32 828	181 277	83 459	97 817
江　西	Jiangxi	239 025	12 978	226 047	114 704	111 343
山　东	Shandong	1 344 093	9 232	1 334 861	385 724	949 137
河　南	Henan	802 598	8 523	794 076	442 601	351 475
湖　北	Hubei	475 521	141 762	333 759	152 033	181 726
湖　南	Hunan	567 301	35 285	532 016	183 829	348 186
广　东	Guangdong	600 064	114 508	485 556	168 162	317 394
广　西	Guangxi	215 683	12 408	203 275	56 419	146 856
海　南	Hainan	103 069	13 269	89 800	22 191	67 609
重　庆	Chongqing	123 595	2 821	120 774	39 985	80 790
四　川	Sichuan	539 793	8 065	531 728	193 366	338 362
贵　州	Guizhou	63 205	1 089	62 115	17 404	44 712
云　南	Yunnan	74 243	3 168	71 075	19 156	51 919
西　藏	Tibet	3 855	0	3 855	436	3 419
陕　西	Shaanxi	196 624	911	195 713	80 285	115 428
甘　肃	Gansu	144 776	380	144 396	17 433	126 963
青　海	Qinghai	22 423	1	22 422	6 862	15 560
宁　夏	Ningxia	101 704	2 501	99 203	24 096	75 107
新　疆	Xinjiang	367 218	1 895	365 323	83 407	281 917

各地区农业污染排放情况（三）
Discharge of Agricultural Pollution by Region（3）
（2012）

单位：吨 （ton）

年 份 地 区	Year Region	氨氮排放总量 Total Amount of Ammonia Nitrogen Discharged	种植业 Crops	水产养殖业 Aquiculture	畜禽养殖业 Livestock and Poultry	规模化畜禽养 殖场/小区 Large-scale Farms	养殖专业户 Small-scale Farms
	2011	826 529	151 424	23 143	651 962	362 832	289 130
	2012	806 216	151 566	23 400	631 250	347 878	283 372
北 京	Beijing	4 732	372	188	4 172	3 000	1 172
天 津	Tianjin	5 946	488	713	4 744	2 047	2 697
河 北	Hebei	44 864	5 779	767	38 318	21 748	16 570
山 西	Shanxi	12 553	2 818	24	9 711	5 897	3 815
内蒙古	Inner Mongolia	12 213	3 146	28	9 038	4 259	4 778
辽 宁	Liaoning	34 226	2 487	908	30 830	12 431	18 399
吉 林	Jilin	17 686	2 717	65	14 905	5 931	8 973
黑龙江	Heilongjiang	34 465	7 495	696	26 275	9 993	16 282
上 海	Shanghai	3 469	908	54	2 507	1 632	875
江 苏	Jiangsu	39 144	13 723	1 807	23 615	9 772	13 843
浙 江	Zhejiang	27 041	5 720	1 493	19 828	12 684	7 144
安 徽	Anhui	37 823	9 106	472	28 245	18 857	9 388
福 建	Fujian	33 266	3 876	6 983	22 407	13 742	8 665
江 西	Jiangxi	29 694	5 590	164	23 940	17 836	6 104
山 东	Shandong	74 135	8 940	764	64 431	31 063	33 368
河 南	Henan	64 011	4 900	418	58 693	44 025	14 668
湖 北	Hubei	46 623	13 737	1 253	31 633	20 038	11 594
湖 南	Hunan	62 631	8 856	362	53 413	30 577	22 836
广 东	Guangdong	58 273	9 165	3 631	45 477	26 787	18 690
广 西	Guangxi	26 746	9 333	762	16 651	8 008	8 643
海 南	Hainan	9 174	2 291	754	6 129	2 480	3 649
重 庆	Chongqing	12 962	2 425	60	10 477	5 874	4 603
四 川	Sichuan	57 449	9 758	281	47 410	22 382	25 027
贵 州	Guizhou	7 883	3 485	373	4 025	1 787	2 238
云 南	Yunnan	11 797	7 533	109	4 154	2 282	1 872
西 藏	Tibet	457	407	0	50	22	28
陕 西	Shaanxi	15 443	3 075	44	12 324	5 692	6 632
甘 肃	Gansu	5 704	1 586	20	4 098	1 109	2 988
青 海	Qinghai	889	140	0	749	452	297
宁 夏	Ningxia	2 165	412	120	1 634	598	1 036
新 疆	Xinjiang	12 752	1 297	87	11 368	4 872	6 496

各地区农业污染排放情况（四）
Discharge of Agricultural Pollution by Region（4）

（2012）

单位：吨 　　　（ton）

年　份 地　区	Year Region	总氮排放总量 Total Amount of Total Nitrogen Discharged	种植业 Crops	水产养殖业 Aquiculture	畜禽养殖业 Livestock and Poultry	规模化畜禽养 殖场/小区 Large-scale Farms	养殖专业户 Small-scale Farms
	2011	4 546 677	1 495 332	85 589	2 965 756	1 503 427	1 462 329
	2012	4 513 733	1 578 163	81 668	2 853 902	1 424 957	1 428 945
北　京	Beijing	32 580	8 262	629	23 688	18 452	5 236
天　津	Tianjin	32 924	9 098	2 298	21 528	10 305	11 224
河　北	Hebei	360 360	132 309	2 273	225 778	154 014	71 763
山　西	Shanxi	84 464	36 616	96	47 752	30 119	17 633
内蒙古	Inner Mongolia	266 955	25 292	109	241 555	41 729	199 826
辽　宁	Liaoning	205 329	18 140	2 870	184 319	68 045	116 274
吉　林	Jilin	128 962	20 126	177	108 659	50 422	58 238
黑龙江	Heilongjiang	247 530	27 426	378	219 727	82 047	137 679
上　海	Shanghai	15 460	5 885	383	9 192	6 728	2 464
江　苏	Jiangsu	174 719	88 580	7 144	78 995	38 489	40 506
浙　江	Zhejiang	92 298	39 750	3 833	48 716	28 165	20 550
安　徽	Anhui	181 210	89 156	2 630	89 424	57 758	31 666
福　建	Fujian	95 684	28 143	11 245	56 295	33 998	22 297
江　西	Jiangxi	113 817	41 510	1 650	70 657	50 114	20 543
山　东	Shandong	562 704	261 322	2 228	299 154	157 883	141 272
河　南	Henan	418 649	149 893	1 512	267 244	210 777	56 467
湖　北	Hubei	194 583	76 669	15 226	102 688	67 452	35 236
湖　南	Hunan	220 546	68 584	4 463	147 498	76 851	70 647
广　东	Guangdong	194 586	66 403	13 512	114 671	56 827	57 845
广　西	Guangxi	115 852	66 649	2 304	46 900	19 789	27 111
海　南	Hainan	41 170	20 387	2 036	18 747	6 478	12 269
重　庆	Chongqing	54 139	22 089	355	31 695	16 518	15 177
四　川	Sichuan	221 369	72 790	2 171	146 407	71 279	75 128
贵　州	Guizhou	46 785	31 050	844	14 891	5 821	9 071
云　南	Yunnan	76 056	58 777	361	16 918	8 154	8 764
西　藏	Tibet	5 710	4 897	0	813	133	680
陕　西	Shaanxi	85 777	49 491	214	36 073	19 314	16 758
甘　肃	Gansu	47 458	20 433	63	26 963	6 053	20 911
青　海	Qinghai	6 688	1 453	0	5 235	2 238	2 997
宁　夏	Ningxia	27 311	5 431	384	21 496	8 743	12 752
新　疆	Xinjiang	162 059	31 554	280	130 225	20 265	109 960

各地区农业污染排放情况（五）
Discharge of Agricultural Pollution by Region（5）
（2012）

单位：吨 （ton）

年 份 地 区	Year Region	总磷排放总量 Total Amount of Total phosphorus Discharged	种植业 Crops	水产养殖业 Aquiculture	畜禽养殖业 Livestock and Poultry	规模化畜禽养 殖场/小区 Large-scale Farms	养殖专业户 Small-scale Farms
	2011	563 340	115 987	16 388	430 965	249 224	181 740
	2012	488 530	108 716	16 513	363 301	202 877	160 424
北 京	Beijing	4 362	388	124	3 850	3 114	736
天 津	Tianjin	3 616	498	452	2 666	1 413	1 254
河 北	Hebei	38 605	7 100	463	31 042	22 016	9 026
山 西	Shanxi	8 052	2 085	21	5 946	3 988	1 958
内蒙古	Inner Mongolia	20 709	1 419	76	19 214	4 256	14 958
辽 宁	Liaoning	27 628	1 176	326	26 126	10 859	15 267
吉 林	Jilin	15 793	1 054	27	14 712	7 530	7 182
黑龙江	Heilongjiang	23 691	1 979	214	21 498	9 176	12 322
上 海	Shanghai	1 804	427	76	1 301	928	373
江 苏	Jiangsu	18 564	4 353	1 580	12 631	6 875	5 756
浙 江	Zhejiang	10 621	3 387	632	6 602	3 961	2 641
安 徽	Anhui	20 058	5 009	787	14 261	9 599	4 662
福 建	Fujian	12 092	2 975	1 921	7 197	4 913	2 284
江 西	Jiangxi	13 620	4 157	308	9 155	6 489	2 666
山 东	Shandong	60 994	13 730	480	46 785	25 929	20 856
河 南	Henan	48 135	7 529	284	40 322	32 671	7 650
湖 北	Hubei	23 343	5 473	2 968	14 903	10 628	4 275
湖 南	Hunan	23 702	6 287	836	16 580	9 255	7 325
广 东	Guangdong	24 957	7 383	2 851	14 723	5 623	9 100
广 西	Guangxi	13 395	6 491	430	6 473	2 505	3 968
海 南	Hainan	5 037	2 224	314	2 499	951	1 549
重 庆	Chongqing	6 400	2 178	68	4 155	2 344	1 811
四 川	Sichuan	25 169	7 066	836	17 267	9 388	7 880
贵 州	Guizhou	4 459	2 751	178	1 531	694	837
云 南	Yunnan	7 390	5 148	74	2 168	1 093	1 075
西 藏	Tibet	397	329	0	67	15	52
陕 西	Shaanxi	7 536	3 201	42	4 293	2 733	1 560
甘 肃	Gansu	3 847	1 376	14	2 457	699	1 758
青 海	Qinghai	545	79	0	466	240	226
宁 夏	Ningxia	2 111	300	76	1 735	757	978
新 疆	Xinjiang	11 893	1 162	56	10 675	2 233	8 442

各地区城镇生活污染排放及处理情况（一）

Discharge and treatment of Household Pollution by Region（1）

（2012）

年 份 地 区	Year Region	城镇人口/ 万人 Urban Population （10 000 people）	生活煤炭消费量/ 万吨 Total Amount of Household Coal Consumed （10 000 tons）	生活用水总量/ 万吨 Quantity of Water Consumed by Household （10 000 tons）	居民家庭用水 总量 Water Consumed in Household	公共服务用水 总量 Water Consumed by Public Services
	2011	69 258.2	15 909.6	4 983 749.9	4 115 515.5	868 234.4
	2012	71 707.8	16 164.5	5 455 716.8	4 552 151.6	903 565.2
北　京	Beijing	1 783.7	461.5	156 113.0	89 254.0	66 859.0
天　津	Tianjin	1 152.5	230.5	68 329.3	54 329.3	14 000.0
河　北	Hebei	3 384.2	902.8	208 250.2	171 971.8	36 278.4
山　西	Shanxi	1 851.0	1 297.4	110 980.0	86 579.5	24 400.5
内蒙古	Inner Mongolia	1 437.6	1 167.7	82 853.8	67 625.1	15 228.7
辽　宁	Liaoning	2 881.8	811.7	187 509.7	147 958.4	39 551.4
吉　林	Jilin	1 477.0	584.2	87 281.9	62 198.6	25 083.3
黑龙江	Heilongjiang	2 157.1	1 897.6	130 219.0	115 339.1	14 879.9
上　海	Shanghai	2 125.0	171.7	191 700.0	176 340.0	15 360.0
江　苏	Jiangsu	4 992.6	216.1	426 678.9	338 459.7	88 219.2
浙　江	Zhejiang	3 461.1	123.1	298 086.0	245 231.7	52 854.3
安　徽	Anhui	2 783.2	394.7	216 019.5	186 019.1	30 000.4
福　建	Fujian	2 234.2	115.0	173 983.9	158 269.7	15 714.2
江　西	Jiangxi	2 140.0	88.7	157 035.2	136 955.9	20 079.3
山　东	Shandong	5 078.8	1 452.2	335 619.2	294 140.8	41 478.4
河　南	Henan	3 996.1	1 168.8	307 689.1	264 604.4	43 084.7
湖　北	Hubei	3 091.8	772.6	227 072.6	200 914.2	26 158.4
湖　南	Hunan	3 098.4	382.4	243 092.1	213 546.3	29 545.7
广　东	Guangdong	7 140.0	181.4	772 947.0	678 589.9	94 357.2
广　西	Guangxi	2 038.0	183.3	158 143.2	136 565.2	21 577.9
海　南	Hainan	458.2	8.7	36 398.1	28 151.0	8 247.1
重　庆	Chongqing	1 678.1	147.4	120 134.2	106 853.1	13 281.2
四　川	Sichuan	3 643.9	449.4	249 939.9	184 967.1	64 972.9
贵　州	Guizhou	1 269.5	490.3	86 920.2	68 330.8	18 589.4
云　南	Yunnan	1 793.7	244.7	135 040.7	110 512.2	24 528.6
西　藏	Tibet	94.0	10.2	5 100.2	4 438.6	661.6
陕　西	Shaanxi	1 875.4	659.0	110 886.6	89 034.6	21 851.9
甘　肃	Gansu	999.0	529.5	53 053.5	45 272.8	7 780.7
青　海	Qinghai	271.9	235.8	16 060.1	11 289.2	4 770.9
宁　夏	Ningxia	328.0	104.8	25 305.6	20 152.0	5 153.6
新　疆	Xinjiang	992.0	681.6	77 274.1	58 257.5	19 016.6

各地区城镇生活污染排放及处理情况（二）
Discharge and treatment of Household Pollution by Region（2）

（2012）

年　份 地　区	Year Region	城镇生活污水 排放量/ 万吨 Amount of Household Waste Water Discharged （10 000 tons）	城镇生活化学需 氧量产生量/ 吨 Amount of Household COD Generated（ton）	城镇生活氨氮 产生量/吨 Amount of Household Ammonia Nitrogen Generated（ton）	城镇生活化学需 氧量排放量/ 吨 Amount of Household COD Discharged（ton）	城镇生活氨氮 排放量/吨 Amount of Household Ammonia Nitrogen Discharged（ton）
	2011	4 279 159.1	16 984 202.1	2 166 719.0	9 388 423.1	1 476 626.1
	2012	4 626 897.1	17 325 170.1	2 220 738.5	9 127 509.5	1 446 275.8
北　京	Beijing	130 984.7	514 451.0	63 166.7	95 661.0	14 762.0
天　津	Tianjin	63 650.6	289 313.9	39 471.9	88 544.0	16 138.0
河　北	Hebei	183 043.2	735 385.0	98 715.0	231 111.4	49 438.2
山　西	Shanxi	86 161.9	421 001.7	57 995.3	206 902.5	35 768.9
内蒙古	Inner Mongolia	68 786.0	325 597.1	41 109.1	171 558.8	29 115.9
辽　宁	Liaoning	151 495.1	672 166.7	92 414.0	334 062.0	64 017.5
吉　林	Jilin	74 606.7	337 803.8	43 568.6	194 824.6	32 917.6
黑龙江	Heilongjiang	104 212.6	476 277.2	63 145.5	346 893.9	52 222.4
上　海	Shanghai	172 530.0	589 475.0	75 235.6	177 504.0	41 369.8
江　苏	Jiangsu	361 835.4	1 260 419.2	164 072.3	572 502.1	97 047.6
浙　江	Zhejiang	245 049.1	912 822.7	118 036.0	389 175.1	72 502.4
安　徽	Anhui	186 980.0	655 673.1	77 974.7	440 215.1	59 018.2
福　建	Fujian	149 680.1	546 585.0	70 974.3	351 262.9	52 757.6
江　西	Jiangxi	133 058.8	530 526.6	60 890.8	398 211.8	50 306.1
山　东	Shandong	295 135.1	1 197 084.9	160 427.0	433 029.0	82 968.1
河　南	Henan	266 169.9	890 545.2	115 859.4	402 981.1	71 645.7
湖　北	Hubei	198 337.0	701 500.9	87 937.6	461 247.4	65 918.8
湖　南	Hunan	206 679.9	737 262.8	94 625.2	530 397.3	71 431.3
广　东	Guangdong	651 882.6	1 824 866.5	234 655.6	950 426.3	149 475.2
广　西	Guangxi	134 786.0	491 710.0	59 928.2	361 103.0	46 937.0
海　南	Hainan	29 586.9	125 061.2	14 890.8	80 233.5	12 275.8
重　庆	Chongqing	101 676.7	416 506.9	53 288.4	229 447.5	37 215.3
四　川	Sichuan	213 442.8	909 423.0	110 747.8	604 793.0	76 986.2
贵　州	Guizhou	68 003.9	289 562.6	37 141.3	199 737.4	26 635.3
云　南	Yunnan	111 088.2	464 436.8	54 377.3	291 729.2	40 892.0
西　藏	Tibet	4 328.5	22 139.1	2 712.6	20 803.2	2 640.6
陕　西	Shaanxi	90 625.8	421 224.3	54 449.8	234 538.8	36 901.1
甘　肃	Gansu	43 588.7	208 115.4	27 706.3	149 912.7	21 271.6
青　海	Qinghai	13 070.6	58 988.4	7 984.9	37 477.0	6 730.3
宁　夏	Ningxia	22 388.6	74 216.5	9 337.3	20 061.9	6 617.3
新　疆	Xinjiang	64 031.7	225 027.4	27 898.7	121 162.0	22 352.1

各地区城镇生活污染排放及处理情况（三）

Discharge and treatment of Household Pollution by Region （3）

（2012）

单位：吨 (ton)

年 份 地 区	Year Region	城镇生活二氧化硫排放量 Amount of Household Sulphur Dioxide Emission	城镇生活氮氧化物排放量 Amount of Household Nitrogen Oxide Emission	城镇生活烟尘排放量 Amount of Household Soot Emission
	2011	2 003 949. 5	366 236. 5	1 147 647. 1
	2012	2 056 646. 1	393 122. 9	1 426 733. 5
北 京	Beijing	34 485. 1	11 991. 2	31 867. 5
天 津	Tianjin	8 959. 0	4 447. 0	18 400. 0
河 北	Hebei	102 365. 6	18 352. 4	126 443. 0
山 西	Shanxi	107 098. 1	31 524. 0	100 071. 6
内蒙古	Inner Mongolia	143 449. 5	26 210. 6	136 376. 9
辽 宁	Liaoning	79 638. 1	16 126. 0	69 908. 4
吉 林	Jilin	51 143. 5	11 821. 9	51 144. 7
黑龙江	Heilongjiang	117 014. 2	43 683. 3	223 104. 7
上 海	Shanghai	34 754. 3	19 138. 3	15 453. 9
江 苏	Jiangsu	32 468. 3	6 541. 0	18 530. 0
浙 江	Zhejiang	14 654. 1	3 255. 7	4 280. 1
安 徽	Anhui	49 405. 3	9 322. 7	87 153. 0
福 建	Fujian	18 850. 6	2 382. 3	9 198. 1
江 西	Jiangxi	16 176. 1	2 984. 2	8 811. 2
山 东	Shandong	204 847. 3	31 953. 6	119 725. 8
河 南	Henan	146 009. 5	24 323. 6	41 104. 8
湖 北	Hubei	73 747. 6	12 132. 1	51 931. 1
湖 南	Hunan	51 615. 1	7 888. 5	26 716. 7
广 东	Guangdong	27 205. 1	11 265. 5	12 670. 3
广 西	Guangxi	32 466. 7	3 633. 3	14 247. 8
海 南	Hainan	1 078. 3	212. 3	1 674. 6
重 庆	Chongqing	54 983. 9	4 487. 3	9 139. 0
四 川	Sichuan	70 238. 9	10 029. 9	12 067. 7
贵 州	Guizhou	203 982. 7	8 139. 6	28 197. 0
云 南	Yunnan	49 599. 9	5 974. 2	14 922. 2
西 藏	Tibet	2 868. 3	291. 9	1 019. 5
陕 西	Shaanxi	96 732. 6	26 511. 4	53 586. 6
甘 肃	Gansu	92 380. 2	14 421. 8	42 492. 0
青 海	Qinghai	24 758. 2	5 577. 3	19 614. 7
宁 夏	Ningxia	22 252. 6	2 748. 0	9 430. 1
新 疆	Xinjiang	91 417. 5	15 752. 2	67 450. 7

各地区机动车污染排放情况
Discharge of Motor Vehicle Pollution by Region
（2012）

单位：万吨 （10 000 tons）

年 份　Year 地 区　Region	机动车污染物排放总量 Total Amount of Discharge of Motor Vehicle Pollution			
	总颗粒物 Total Particulate	氮氧化物 Nitrogen Oxide	一氧化碳 Carbon Monoxide	碳氢化合物 Hydrocarbon
2011	62.924	637.585	3 466.570	440.475
2012	62.139	640.029	3 471.669	438.158
北　京　Beijing	0.407	7.985	78.114	8.641
天　津　Tianjin	0.659	5.405	44.715	5.092
河　北　Hebei	5.362	54.797	256.329	32.928
山　西　Shanxi	2.266	26.162	134.215	16.531
内蒙古　Inner Mongolia	2.894	24.570	125.466	17.285
辽　宁　Liaoning	2.996	27.865	121.419	16.528
吉　林　Jilin	1.808	17.660	107.651	13.792
黑龙江　Heilongjiang	2.607	25.566	154.487	19.482
上　海　Shanghai	0.791	9.748	43.534	6.332
江　苏　Jiangsu	2.850	33.896	181.389	21.972
浙　江　Zhejiang	1.632	17.040	121.125	14.072
安　徽　Anhui	2.296	21.886	83.301	11.329
福　建　Fujian	0.944	10.391	65.128	7.699
江　西　Jiangxi	2.673	22.158	81.294	11.655
山　东　Shandong	4.960	46.865	235.751	30.091
河　南　Henan	5.128	50.147	208.198	26.941
湖　北　Hubei	1.609	18.685	104.421	12.565
湖　南　Hunan	1.512	17.453	85.365	10.537
广　东　Guangdong	4.760	48.233	323.715	37.015
广　西　Guangxi	1.692	14.898	92.222	11.493
海　南　Hainan	0.427	3.122	21.074	2.685
重　庆　Chongqing	0.698	10.607	61.603	7.835
四　川　Sichuan	1.584	20.997	119.267	14.278
贵　州　Guizhou	0.976	9.648	54.757	7.351
云　南　Yunnan	1.620	20.309	132.063	18.755
西　藏　Tibet	0.462	4.003	28.299	3.420
陕　西　Shaanxi	2.294	17.715	105.883	13.149
甘　肃　Gansu	0.842	11.787	104.665	13.369
青　海　Qinghai	0.308	3.247	25.987	3.274
宁　夏　Ningxia	0.819	7.382	31.693	4.548
新　疆　Xinjiang	2.263	29.802	138.540	17.514

各地区城镇污水处理情况（一）
Urban Waste Water treatment by Region（1）

（2012）

年　份 地　区	Year Region	污水处理厂数/ 座 Number of Urban Waste Water Treatment Plants （unit）	污水处理厂 设计处理能力/ （万吨/日） Treatment Capacity （10 000 tons/day）	本年运行费用/ 万元 Annul Expenditure for Operation （10 000 yuan）	污水处理厂累计 完成投资/万元 Total Investment of Urban Waste Water Treatment （10 000 yuan）	新增固定资产/ 万元 Newly-added Fixed Assets （10 000 yuan）
	2011	3 974	13 991	3 071 639.0	29 021 719.6	2 309 446.1
	2012	4 628	15 314	3 482 302.1	34 077 389.5	1 758 523.4
北　京	Beijing	130	431	109 625.0	942 199.9	7 641.5
天　津	Tianjin	57	269	70 103.3	652 905.2	72 965.1
河　北	Hebei	247	826	174 302.2	1 885 584.3	103 950.6
山　西	Shanxi	170	303	68 765.2	853 957.0	36 433.9
内蒙古	Inner Mongolia	127	302	69 038.3	1 043 388.4	70 088.3
辽　宁	Liaoning	127	605	127 024.1	1 034 924.4	49 242.2
吉　林	Jilin	51	269	48 836.6	620 695.7	47 071.7
黑龙江	Heilongjiang	64	295	59 292.0	692 842.9	14 678.2
上　海	Shanghai	55	682	145 061.5	1 350 103.0	10 278.6
江　苏	Jiangsu	611	1 434	384 600.0	3 807 064.9	245 984.4
浙　江	Zhejiang	324	1 125	349 980.3	2 659 385.9	151 086.4
安　徽	Anhui	129	483	94 244.3	1 063 581.4	95 521.9
福　建	Fujian	147	413	74 527.6	685 793.5	28 211.4
江　西	Jiangxi	121	302	70 942.1	547 616.5	15 449.4
山　东	Shandong	333	1 231	311 593.1	2 254 805.2	162 794.0
河　南	Henan	197	761	124 900.1	1 398 321.6	90 440.2
湖　北	Hubei	157	601	116 184.8	1 249 419.2	39 351.0
湖　南	Hunan	134	515	101 939.1	1 402 246.2	48 280.2
广　东	Guangdong	410	1 949	415 609.4	3 723 963.4	170 419.1
广　西	Guangxi	111	383	86 878.0	961 421.1	12 548.1
海　南	Hainan	51	133	15 634.2	306 584.3	2 401.4
重　庆	Chongqing	131	274	98 203.4	693 787.5	28 418.5
四　川	Sichuan	285	507	130 958.7	1 134 200.6	64 030.4
贵　州	Guizhou	111	167	27 588.7	405 983.2	11 680.2
云　南	Yunnan	91	266	52 994.2	752 967.8	22 232.8
西　藏	Tibet	2	6	2 181.0	18 210.0	0.0
陕　西	Shaanxi	114	309	61 489.8	834 148.4	66 797.1
甘　肃	Gansu	44	147	29 474.0	317 582.6	3 979.5
青　海	Qinghai	20	40	9 598.8	120 603.3	1 213.0
宁　夏	Ningxia	19	70	14 460.5	142 992.1	6 178.6
新　疆	Xinjiang	58	219	36 271.7	520 109.8	79 155.7

各地区城镇污水处理情况（二）
Urban Waste Water treatment by Region（2）
（2012）

单位：万吨 (10 000 tons)

年 份 地 区	Year Region	污水实际 处理量 Quantity of Waste Water Treated	污水再生 利用量 Waste Water Recycled	工业用水量 Waste Water Recycled for Industry	市政用水量 Waste Water Recycled for Municipal Services	景观用水量 Waste Water Recycled for Landscape
	2011	4 028 972	96 262	60 402	12 461	23 399
	2012	4 161 551	120 597	74 015	12 544	34 037
北 京	Beijing	125 645	9 624	4 677	1 754	3 192
天 津	Tianjin	68 581	2 707	2 018	338	351
河 北	Hebei	219 066	20 969	13 377	2 819	4 773
山 西	Shanxi	71 400	6 862	6 194	156	511
内蒙古	Inner Mongolia	65 209	12 373	9 663	1 463	1 246
辽 宁	Liaoning	164 289	8 517	7 262	44	1 211
吉 林	Jilin	65 499	630	630	0	0
黑龙江	Heilongjiang	61 630	14	1	12	1
上 海	Shanghai	202 336	0	0	0	0
江 苏	Jiangsu	377 716	7 324	3 418	241	3 665
浙 江	Zhejiang	300 329	10 104	7 405	1 785	914
安 徽	Anhui	148 130	30	28	0	2
福 建	Fujian	115 342	306	188	109	8
江 西	Jiangxi	92 485	526	314	11	202
山 东	Shandong	343 108	10 070	6 727	1 039	2 304
河 南	Henan	226 399	7 189	4 822	392	1 975
湖 北	Hubei	169 828	115	53	46	16
湖 南	Hunan	149 354	13	2	3	8
广 东	Guangdong	545 949	13 866	1 656	278	11 933
广 西	Guangxi	96 337	71	21	32	17
海 南	Hainan	28 328	711	7	452	252
重 庆	Chongqing	83 105	0	0	0	0
四 川	Sichuan	150 324	497	33	61	404
贵 州	Guizhou	51 571	168	0	168	0
云 南	Yunnan	72 794	857	274	551	31
西 藏	Tibet	2 038	0	0	0	0
陕 西	Shaanxi	69 998	2 755	1 612	148	995
甘 肃	Gansu	27 032	2 824	2 210	587	27
青 海	Qinghai	8 913	676	676	0	0
宁 夏	Ningxia	17 875	800	746	54	0
新 疆	Xinjiang	40 943	0	0	0	0

各地区城镇污水处理情况（三）
Urban Waste Water treatment by Region（3）
（2012）

单位：万吨 　　　　　　　　　　　　　　　　　　　　　　　　　　　　　　　　　　（10 000 tons）

年　份 地　区	Year Region	污泥产生量 Quantity of Sludge Generated	污泥处置量 Quantity of Sludge Disposed	土地利用量 Landuse	填埋处置量 Landfill	建筑材料利 用量 As Building Material	焚烧处置量 Incineration	污泥倾倒丢 弃量 Quantity of Sludge Discharged
	2011	2 267.54	2 267.24	354.60	1 270.38	228.59	413.67	0.12
	2012	2 418.55	2 418.54	453.57	1 266.51	235.38	463.08	0.01
北　京	Beijing	117.03	117.03	82.16	25.59	2.34	6.93	0.00
天　津	Tianjin	34.27	34.27	1.74	31.14	1.15	0.25	0.00
河　北	Hebei	144.84	144.84	17.23	123.50	0.50	3.61	0.00
山　西	Shanxi	37.42	37.42	19.18	15.69	0.06	2.49	0.00
内蒙古	Inner Mongolia	18.79	18.79	1.60	17.03	0.07	0.09	0.00
辽　宁	Liaoning	88.30	88.30	0.58	75.33	6.84	5.55	0.00
吉　林	Jilin	28.23	28.23	0.03	26.52	0.00	1.68	0.00
黑龙江	Heilongjiang	24.54	24.54	0.18	23.21	0.02	1.13	0.00
上　海	Shanghai	124.44	124.44	4.53	107.69	6.68	5.54	0.00
江　苏	Jiangsu	254.54	254.54	13.41	38.23	44.02	158.88	0.00
浙　江	Zhejiang	319.18	319.18	28.28	58.67	66.79	165.44	0.00
安　徽	Anhui	71.58	71.58	17.13	31.25	5.07	18.12	0.00
福　建	Fujian	55.92	55.92	12.35	28.91	6.69	7.97	0.00
江　西	Jiangxi	26.00	26.00	2.51	15.69	5.11	2.69	0.00
山　东	Shandong	211.05	211.05	62.91	80.11	14.15	53.89	0.00
河　南	Henan	157.78	157.78	33.21	113.70	6.70	4.17	0.00
湖　北	Hubei	47.13	47.13	1.82	41.35	0.30	3.66	0.00
湖　南	Hunan	52.50	52.50	0.19	44.24	7.58	0.49	0.00
广　东	Guangdong	304.14	304.14	105.22	146.17	35.40	17.35	0.00
广　西	Guangxi	22.97	22.97	11.55	6.64	4.76	0.02	0.00
海　南	Hainan	6.03	6.02	5.78	0.24	0.00	0.00	0.01
重　庆	Chongqing	38.31	38.31	2.99	19.34	15.98	0.00	0.00
四　川	Sichuan	70.24	70.24	6.42	57.86	5.10	0.86	0.00
贵　州	Guizhou	12.74	12.74	0.17	12.52	0.06	0.00	0.00
云　南	Yunnan	34.02	34.02	8.71	25.25	0.00	0.06	0.00
西　藏	Tibet	36.14	36.14	0.03	36.11	0.00	0.00	0.00
陕　西	Shaanxi	38.96	38.96	7.14	31.27	0.00	0.56	0.00
甘　肃	Gansu	20.07	20.07	3.09	16.98	0.00	0.00	0.00
青　海	Qinghai	5.63	5.63	0.00	5.63	0.00	0.00	0.00
宁　夏	Ningxia	9.99	9.99	2.73	5.61	0.00	1.66	0.00
新　疆	Xinjiang	5.77	5.77	0.72	5.04	0.01	0.00	0.00

各地区城镇污水处理情况（四）
Urban Waste Water treatment by Region（4）
（2012）

单位：吨 (ton)

年　份　Year 地　区　Region	污染物去除量 Quantity of Pollutants Removed by Urban Waste Water Treatment						
	化学需氧量 COD	氨氮 Ammonia Nitrogen	油类 Oil	总氮 Total Nitrogen	总磷 Total Phosphorus	挥发酚 Volatile Phenols	氰化物 Cyanide
2011	10 148 983.5	857 544.0	54 812.8	674 199.6	150 369.9	966.5	1 222.6
2012	10 130 710.1	886 256.1	45 667.9	731 117.6	124 650.3	998.6	702.4
北　京　Beijing	430 303.5	52 609.6	7 298.9	56 603.4	5 393.6	16.3	2.4
天　津　Tianjin	228 078.1	22 648.4	753.2	25 903.2	1 984.9	0.6	0.1
河　北　Hebei	623 333.7	59 336.7	266.1	26 521.8	18 701.5	0.0	2.2
山　西　Shanxi	239 205.8	22 517.5	1 554.7	17 679.2	2 087.4	25.5	1.4
内蒙古　Inner Mongolia	181 582.4	14 923.6	1 012.6	15 720.4	1 090.5	2.1	11.5
辽　宁　Liaoning	352 446.8	28 766.9	1 716.8	22 159.2	2 305.0	13.3	0.0
吉　林　Jilin	166 749.8	8 664.8	941.8	8 347.9	1 774.6	32.5	3.2
黑龙江　Heilongjiang	154 578.2	15 594.6	322.1	11 513.9	2 039.9	0.8	1.2
上　海　Shanghai	569 078.9	43 450.9	1 332.0	43 062.4	5 990.7	163.3	14.8
江　苏　Jiangsu	933 035.9	79 401.3	3 535.5	48 217.3	9 026.1	234.5	23.0
浙　江　Zhejiang	1 190 800.0	70 199.9	4 957.7	43 076.6	7 471.0	0.0	323.8
安　徽　Anhui	224 230.2	20 245.0	642.0	21 267.0	3 160.4	39.0	0.0
福　建　Fujian	272 211.1	21 581.5	909.4	20 423.0	2 386.2	0.0	158.0
江　西　Jiangxi	127 944.8	11 421.0	173.8	9 713.0	1 247.8	0.0	0.0
山　东　Shandong	1 020 316.8	92 901.7	517.6	66 180.3	9 515.0	62.1	23.6
河　南　Henan	564 624.8	52 458.1	3 470.4	64 581.3	6 726.4	58.0	3.7
湖　北　Hubei	250 384.7	22 532.2	524.1	9 472.1	13 773.4	151.6	3.0
湖　南　Hunan	230 126.1	21 027.2	313.4	15 801.4	2 028.4	16.3	0.6
广　东　Guangdong	940 121.5	88 607.5	2 550.4	71 064.8	13 119.5	6.8	128.2
广　西　Guangxi	135 652.1	15 498.0	2 226.4	11 834.7	1 893.6	0.0	0.0
海　南　Hainan	48 298.5	2 530.0	28.6	3 006.7	666.8	0.6	0.0
重　庆　Chongqing	201 319.2	16 180.9	1 078.7	24 113.9	2 701.6	0.2	0.0
四　川　Sichuan	304 630.0	33 761.1	2 267.7	34 830.9	2 846.9	71.5	1.4
贵　州　Guizhou	90 210.0	10 558.3	246.6	5 953.0	386.8	5.0	0.0
云　南　Yunnan	161 241.6	14 141.2	604.8	14 836.1	1 543.7	0.4	0.0
西　藏　Tibet	1 336.0	72.0	198.0	81.0	12.6	0.0	0.0
陕　西　Shaanxi	195 885.8	18 953.3	1 811.1	17 604.7	1 882.2	0.0	0.0
甘　肃　Gansu	77 923.1	7 728.9	405.6	11 765.6	1 241.7	83.8	0.4
青　海　Qinghai	23 535.5	1 156.7	0.2	66.2	8.0	0.0	0.0
宁　夏　Ningxia	43 980.8	3 568.3	911.6	3 997.4	837.0	9.1	0.0
新　疆　Xinjiang	147 544.3	13 218.7	3 095.8	5 719.4	807.2	5.1	0.0

各地区生活垃圾处理情况（一）
Centralized treatment of Garbage by Region（1）

（2012）

年　份 地　区	Year Region	生活垃圾处理厂数/ 座 Number of Garbage Treatment Plants（unit）	本年运行费用/ 万元 Annul Expenditure for Operation （10 000 yuan）	生活垃圾处理厂累计 完成投资/万元 Total Investment of Garbage Treatment Plants （10 000 yuan）	新增固定资产/ 万元 Newly-added Fixed Assets （10 000 yuan）
	2011	2 039	591 506.9	6 676 883.6	658 537.0
	2012	2 125	985 337.9	9 472 672.2	465 385.1
北　京	Beijing	19	75 005.9	320 297.2	21 690.3
天　津	Tianjin	11	21 647.3	250 456.4	2 138.0
河　北	Hebei	132	32 524.1	416 127.1	12 712.5
山　西	Shanxi	74	21 437.7	172 832.6	5 161.7
内蒙古	Inner Mongolia	59	25 741.0	239 927.5	19 901.3
辽　宁	Liaoning	35	31 431.8	151 044.1	13 192.0
吉　林	Jilin	47	27 875.0	179 069.9	12 657.1
黑龙江	Heilongjiang	26	8 183.8	111 124.3	1 340.0
上　海	Shanghai	24	82 735.4	378 839.6	8 900.4
江　苏	Jiangsu	65	59 707.2	638 735.6	12 485.7
浙　江	Zhejiang	100	94 219.7	861 431.8	91 411.0
安　徽	Anhui	128	33 587.8	268 065.5	63 239.1
福　建	Fujian	90	14 185.6	335 426.4	10 045.0
江　西	Jiangxi	84	17 431.6	141 035.8	3 706.0
山　东	Shandong	85	49 735.6	678 449.0	15 317.2
河　南	Henan	121	36 984.8	415 514.3	13 996.2
湖　北	Hubei	99	31 539.4	745 017.4	5 956.0
湖　南	Hunan	95	70 457.7	478 822.0	27 051.6
广　东	Guangdong	137	70 438.9	579 759.8	29 795.2
广　西	Guangxi	53	14 371.2	332 816.7	14 548.8
海　南	Hainan	22	5 903.5	105 682.9	2 797.2
重　庆	Chongqing	44	23 813.8	282 435.2	3 772.1
四　川	Sichuan	105	37 000.0	420 835.6	16 597.5
贵　州	Guizhou	33	8 767.5	157 072.5	12 162.0
云　南	Yunnan	93	24 297.7	221 326.8	18 806.7
西　藏	Tibet	10	1 050.3	22 392.4	350.0
陕　西	Shaanxi	72	13 066.7	199 696.0	4 222.1
甘　肃	Gansu	65	34 457.4	72 773.1	1 023.4
青　海	Qinghai	52	3 283.6	58 895.6	2 954.0
宁　夏	Ningxia	20	3 517.8	49 326.4	1 083.6
新　疆	Xinjiang	125	10 938.4	187 442.8	16 371.4

各地区生活垃圾处理情况（二）
Centralized treatment of Garbage by Region（2）

（2012）

单位：吨 （ton）

年 份 地 区	Year Region	渗滤液中污染物排放量 Amount of Pollutants Discharged in Landfill Leachate					
		化学需氧量 COD	氨氮 Ammonia Nitrogen	油类 Oil	总磷 Total Phosphorus	挥发酚 Volatile Phenol	氰化物 Cyanide
	2011	199 699.907	20 008.380	410.006	288.898	20.024	2.509
	2012	185 815.765	19 222.846	153.986	254.067	19.932	0.859
北 京	Beijing	6 279.541	634.833	2.362	11.562	0.341	0.016
天 津	Tianjin	460.320	37.610	0.030	0.090	0.000	0.000
河 北	Hebei	7 906.141	602.049	8.373	12.478	1.257	0.032
山 西	Shanxi	2 577.369	247.030	4.185	4.557	0.218	0.007
内蒙古	Inner Mongolia	2 338.712	199.753	5.071	4.398	0.535	0.006
辽 宁	Liaoning	5 411.829	902.286	6.198	9.516	0.907	0.023
吉 林	Jilin	12 432.870	1 332.992	10.733	12.295	1.448	0.044
黑龙江	Heilongjiang	1 897.423	219.845	0.654	1.007	0.179	0.005
上 海	Shanghai	5 805.310	294.070	3.439	6.268	1.252	0.024
江 苏	Jiangsu	5 377.496	681.069	4.918	7.155	0.345	0.021
浙 江	Zhejiang	7 231.541	637.829	6.940	12.409	0.695	0.048
安 徽	Anhui	9 792.531	945.010	8.680	12.149	0.257	0.063
福 建	Fujian	4 117.321	398.998	13.247	6.014	0.244	0.037
江 西	Jiangxi	10 407.536	842.313	9.846	16.206	1.958	0.087
山 东	Shandong	4 702.015	497.465	5.546	7.208	0.640	0.031
河 南	Henan	8 985.520	1 086.990	6.525	12.818	1.445	0.028
湖 北	Hubei	14 514.271	1 629.819	6.957	16.567	1.565	0.059
湖 南	Hunan	16 101.760	1 469.101	12.236	24.126	1.749	0.108
广 东	Guangdong	13 659.666	1 425.987	5.826	13.275	0.051	0.042
广 西	Guangxi	3 897.650	306.777	2.381	2.667	0.120	0.018
海 南	Hainan	1 510.070	149.975	0.751	2.211	0.271	0.010
重 庆	Chongqing	600.260	155.120	1.330	0.931	0.000	0.005
四 川	Sichuan	5 720.803	700.850	4.971	8.908	0.411	0.034
贵 州	Guizhou	5 329.480	625.180	1.701	3.912	0.036	0.015
云 南	Yunnan	12 976.500	1 650.764	7.932	21.273	2.002	0.030
西 藏	Tibet	180.348	12.830	0.128	0.242	0.024	0.001
陕 西	Shaanxi	6 629.980	921.210	5.159	8.753	0.160	0.030
甘 肃	Gansu	1 631.965	113.600	0.795	1.190	0.665	0.003
青 海	Qinghai	2 009.802	126.040	1.457	0.946	0.079	0.002
宁 夏	Ningxia	826.733	57.530	0.487	0.523	0.025	0.001
新 疆	Xinjiang	4 503.002	317.921	5.128	12.413	1.053	0.028

各地区生活垃圾处理情况（三）
Centralized treatment of Garbage by Region（3）
（2012）

单位：千克 (kg)

年 份	Year	渗滤液中污染物排放量					
地 区	Region	汞 Mercury	镉 Cadmium	六价铬 Hexavalent Chromium	总铬 Total Chromium	铅 Plumbum	砷 Arsenic
2011		164.803	792.959	209.529	2 687.263	4 299.244	1 404.626
2012		135.359	581.102	151.097	1 405.849	2 188.515	787.178
北 京	Beijing	0.494	16.788	0.245	31.066	162.206	9.197
天 津	Tianjin	0.010	0.030	0.000	0.090	0.350	0.170
河 北	Hebei	3.773	25.850	0.928	49.671	117.788	36.134
山 西	Shanxi	1.143	5.061	0.643	10.619	21.297	5.534
内蒙古	Inner Mongolia	2.168	12.076	0.520	22.977	58.360	16.153
辽 宁	Liaoning	9.450	20.249	2.860	61.864	78.607	18.832
吉 林	Jilin	3.508	23.051	1.802	46.980	110.047	19.870
黑龙江	Heilongjiang	1.545	3.082	1.941	8.518	11.941	2.495
上 海	Shanghai	2.820	14.571	0.140	42.801	77.929	26.391
江 苏	Jiangsu	13.746	16.704	7.925	45.342	89.610	26.900
浙 江	Zhejiang	5.464	20.130	9.626	72.913	97.584	35.101
安 徽	Anhui	5.785	28.166	18.772	96.083	125.758	91.135
福 建	Fujian	2.605	86.779	9.217	44.621	58.437	17.316
江 西	Jiangxi	7.301	28.658	6.856	109.584	171.242	54.019
山 东	Shandong	3.476	24.687	4.882	38.715	82.219	22.007
河 南	Henan	6.246	24.773	4.514	50.863	115.636	33.606
湖 北	Hubei	10.296	33.348	10.037	82.381	105.236	42.194
湖 南	Hunan	11.149	49.758	21.596	117.565	131.450	58.674
广 东	Guangdong	17.246	16.495	8.698	87.634	70.847	28.726
广 西	Guangxi	1.510	8.048	8.675	22.740	28.838	9.020
海 南	Hainan	0.954	3.821	0.160	4.651	14.146	5.160
重 庆	Chongqing	0.614	2.313	0.744	5.207	15.052	4.170
四 川	Sichuan	5.040	20.106	4.135	126.725	163.554	100.765
贵 州	Guizhou	4.614	8.626	0.616	26.166	36.308	10.869
云 南	Yunnan	5.070	19.544	12.950	68.531	94.068	26.261
西 藏	Tibet	0.080	0.644	0.000	1.288	3.222	0.805
陕 西	Shaanxi	2.980	15.163	11.916	38.708	64.863	18.643
甘 肃	Gansu	1.052	4.515	0.000	7.548	12.870	3.813
青 海	Qinghai	0.754	2.705	0.019	5.358	13.422	3.338
宁 夏	Ningxia	0.147	1.000	0.680	1.976	4.678	1.485
新 疆	Xinjiang	4.319	44.361	0.000	76.664	50.950	58.395

各地区生活垃圾处理情况（四）

Centralized treatment of Garbage by Region（4）

（2012）

单位：吨 (ton)

年　份 地　区	Year Region	焚烧废气中污染物排放量 Volume of Pollutants Emission in the Incineration Gas		
		二氧化硫 Sulfur Dioxide	氮氧化物 Nitrogen Oxide	烟尘 Soot
	2011	1 375.74	2 089.58	1 460.98
	2012	1 585.30	2 402.22	1 046.61
北　京	Beijing	34.03	300.72	40.98
天　津	Tianjin	18.50	150.00	8.90
河　北	Hebei	78.58	0.00	23.43
山　西	Shanxi	20.08	61.36	13.64
内蒙古	Inner Mongolia	0.00	0.00	0.00
辽　宁	Liaoning	37.96	7.72	32.84
吉　林	Jilin	0.00	0.00	0.00
黑龙江	Heilongjiang	0.00	0.00	0.00
上　海	Shanghai	0.00	0.01	0.01
江　苏	Jiangsu	162.20	193.32	35.30
浙　江	Zhejiang	85.50	132.57	142.16
安　徽	Anhui	394.95	267.06	186.54
福　建	Fujian	4.09	7.72	2.00
江　西	Jiangxi	0.00	0.00	0.00
山　东	Shandong	61.79	127.99	24.78
河　南	Henan	21.50	38.00	11.10
湖　北	Hubei	0.00	0.00	0.00
湖　南	Hunan	0.00	0.00	0.00
广　东	Guangdong	374.21	598.17	431.98
广　西	Guangxi	34.18	77.04	9.07
海　南	Hainan	22.47	109.85	0.85
重　庆	Chongqing	0.00	0.00	0.00
四　川	Sichuan	229.31	330.66	80.40
贵　州	Guizhou	0.00	0.00	0.00
云　南	Yunnan	5.96	0.03	2.63
西　藏	Tibet	0.00	0.00	0.00
陕　西	Shaanxi	0.00	0.00	0.00
甘　肃	Gansu	0.00	0.00	0.00
青　海	Qinghai	0.00	0.00	0.00
宁　夏	Ningxia	0.00	0.00	0.00
新　疆	Xinjiang	0.00	0.00	0.00

各地区危险（医疗）废物集中处置情况（一）
Centralized Treatment of Hazardous（Medical）Wastes by Region（1）
（2012）

年 份 地 区 Year Region	危险废物集中处置厂数/个 Number of Centralized Hazardous Wastes Treatment Plants（unit）	医疗废物集中处置厂数/个 Number of Centralized Medical Wastes Treatment Plants（unit）	本年运行费用/万元 Annul Expenditure for Operation（10 000 yuan）	危险（医疗）废物集中处置厂累计完成投资/万元 Total Investment of Hazardous/Medical Wastes Treatment Plants（10 000 yuan）	新增固定资产/万元 Newly-added Fixed Assets（10 000 yuan）
2011	644	260	481 785.9	1 908 958.4	189 987.1
2012	722	236	539 438.5	2 563 959.5	191 780.3
北 京 Beijing	2	1	31 359.7	78 221.6	2 669.2
天 津 Tianjin	13	0	17 075.2	37 252.3	1 456.5
河 北 Hebei	32	12	7 906.7	38 148.9	1 574.5
山 西 Shanxi	3	9	2 311.7	12 558.0	496.7
内蒙古 Inner Mongolia	3	14	2 405.5	28 440.2	62.0
辽 宁 Liaoning	39	7	27 267.8	105 197.4	5 956.0
吉 林 Jilin	2	5	3 319.0	12 959.0	148.0
黑龙江 Heilongjiang	4	8	3 416.8	16 454.1	397.0
上 海 Shanghai	36	2	62 794.3	182 626.5	24 433.0
江 苏 Jiangsu	153	3	88 201.1	376 529.8	21 930.1
浙 江 Zhejiang	71	8	53 981.5	236 401.0	31 537.1
安 徽 Anhui	8	12	7 592.6	37 322.8	1 996.6
福 建 Fujian	7	6	6 447.1	40 438.1	3 313.0
江 西 Jiangxi	25	9	8 016.0	182 026.3	6 434.9
山 东 Shandong	54	12	19 122.0	97 990.1	13 718.6
河 南 Henan	2	16	8 372.1	23 758.6	1 005.9
湖 北 Hubei	33	12	15 997.5	130 019.3	17 585.5
湖 南 Hunan	41	11	11 828.6	151 771.0	5 656.1
广 东 Guangdong	95	16	72 401.8	354 739.9	19 871.3
广 西 Guangxi	15	8	3 107.7	20 773.5	1 116.0
海 南 Hainan	2	1	2 136.9	19 132.2	1 249.0
重 庆 Chongqing	8	3	11 234.9	56 440.2	2 740.7
四 川 Sichuan	19	16	13 564.4	45 528.2	2 443.4
贵 州 Guizhou	7	4	3 002.8	32 058.5	560.0
云 南 Yunnan	2	7	4 675.1	21 004.4	115.9
西 藏 Tibet	0	0	0.0	0.0	0.0
陕 西 Shaanxi	17	6	8 340.0	30 495.1	755.6
甘 肃 Gansu	5	12	26 526.4	67 147.2	20 981.8
青 海 Qinghai	5	1	10 129.0	51 335.0	80.0
宁 夏 Ningxia	6	2	1 796.3	37 626.6	470.0
新 疆 Xinjiang	13	13	5 108.1	39 563.8	1 026.0

各地区危险（医疗）废物集中处置情况（二）
Centralized Treatment of Hazardous（Medical）Wastes by Region（2）
（2012）

年 份　Year 地 区　Region	危险废物设计 处置能力/ （吨/日） Treatment Capacity （ton/day）	危险废物实际 处置量/吨 Volume of Hazardous Wastes Disposed （ton）	工业危险废物 处置量 Industrial Hazardous Wastes	医疗废物 处置量 Medical Hazardous Wastes	其他危险废物 处置量 Other Hazardous Wastes	危险废物综合 利用量/吨 Volume of Hazardous Wastes Utilized （ton）
2011	352 541	2 600 177	1 933 930	497 782	168 465	3 330 420
2012	59 805	3 406 914	2 231 828	502 732	673 254	3 751 721
北　京　Beijing	731	70 751	65 227	5 524	0	52 647
天　津　Tianjin	618	58 731	40 833	17 898	0	141 333
河　北　Hebei	972	51 190	38 136	12 922	132	8 015
山　西　Shanxi	147	14 529	598	13 931	0	2 475
内蒙古　Inner Mongolia	198	9 040	0	9 040	0	0
辽　宁　Liaoning	2 465	152 374	117 935	13 778	20 662	258 119
吉　林　Jilin	232	29 438	19 166	8 284	1 988	2 018
黑龙江　Heilongjiang	180	11 544	79	11 458	7	199
上　海　Shanghai	2 082	167 895	136 184	26 709	5 002	142 569
江　苏　Jiangsu	10 620	375 471	311 708	31 362	32 401	875 968
浙　江　Zhejiang	4 743	451 533	243 443	44 356	163 734	329 630
安　徽　Anhui	397	32 156	20 955	11 201	0	27 001
福　建　Fujian	644	27 494	17 428	10 066	0	2 027
江　西　Jiangxi	933	13 767	3 449	10 176	143	114 240
山　东　Shandong	3 436	121 903	85 302	36 601	0	244 672
河　南　Henan	163	31 712	0	31 712	0	3 100
湖　北　Hubei	4 346	57 545	34 393	20 905	2 247	43 879
湖　南　Hunan	1 203	177 758	163 230	14 528	0	105 439
广　东　Guangdong	9 898	537 944	474 526	62 053	1 365	704 082
广　西　Guangxi	1 318	18 873	6 721	9 552	2 600	9 890
海　南　Hainan	85	11 044	6 350	4 694	0	0
重　庆　Chongqing	2 334	42 588	25 768	9 205	7 614	317 840
四　川　Sichuan	1 099	38 875	12 728	25 423	724	113 625
贵　州　Guizhou	385	10 797	3 011	7 771	15	3 051
云　南　Yunnan	73	15 747	3 450	12 247	50	1 305
西　藏　Tibet	0	0	0	0	0	0
陕　西　Shaanxi	2 803	55 090	32 227	22 863	0	57 819
甘　肃　Gansu	1 284	324 875	318 720	6 127	27	109 504
青　海　Qinghai	1 384	427 075	0	2 517	424 558	17 222
宁　夏　Ningxia	3 798	11 938	2 080	1 790	8 068	2 528
新　疆　Xinjiang	1 175	57 237	48 182	8 039	1 016	61 522

各地区危险（医疗）废物集中处置情况（三）
Centralized Treatment of Hazardous（Medical）Wastes by Region（3）
（2012）

年 份 地 区	Year Region	渗滤液中污染物排放量 Amount of Pollutants Discharged in Landfill Leachate					
		化学需氧量/ 吨 COD（ton）	氨氮/吨 Ammonia Nitrogen（ton）	油类/吨 Oil（ton）	总磷/吨 Total Phosphorus （ton）	挥发酚/千克 Volatile Phenol （kg）	氰化物/千克 Cyanide（kg）
2011		1 436.773	39.385	12.946	0.104	6.768	39.502
2012		1 472.135	36.052	12.657	0.285	1.289	25.564
北　京	Beijing	2.700	2.480	0.000	0.000	0.000	0.000
天　津	Tianjin	7.016	0.085	0.007	0.003	0.000	0.004
河　北	Hebei	0.052	0.005	0.000	0.001	0.000	0.000
山　西	Shanxi	0.000	0.000	0.000	0.000	0.000	0.000
内蒙古	Inner Mongolia	0.545	0.360	0.010	0.000	0.000	0.000
辽　宁	Liaoning	946.361	10.791	10.004	0.001	0.001	0.001
吉　林	Jilin	0.130	0.040	0.001	0.000	0.027	0.043
黑龙江	Heilongjiang	0.767	0.105	0.003	0.000	0.000	0.000
上　海	Shanghai	21.785	2.025	0.063	0.140	0.000	2.770
江　苏	Jiangsu	29.381	7.205	0.561	0.047	0.015	0.066
浙　江	Zhejiang	130.056	3.066	0.033	0.051	0.000	0.000
安　徽	Anhui	0.826	0.178	0.000	0.000	0.000	0.000
福　建	Fujian	7.722	0.031	0.000	0.000	0.001	4.711
江　西	Jiangxi	0.549	0.341	0.023	0.000	0.010	0.020
山　东	Shandong	4.795	0.386	0.417	0.000	0.000	0.050
河　南	Henan	0.000	0.000	0.000	0.000	0.000	0.000
湖　北	Hubei	25.190	2.424	0.023	0.000	0.000	0.000
湖　南	Hunan	1.003	0.074	0.522	0.000	0.030	0.123
广　东	Guangdong	290.049	5.517	0.861	0.040	0.291	14.606
广　西	Guangxi	0.000	0.000	0.001	0.000	0.000	0.000
海　南	Hainan	0.000	0.000	0.000	0.000	0.000	0.000
重　庆	Chongqing	1.289	0.026	0.005	0.002	0.070	0.000
四　川	Sichuan	0.003	0.000	0.079	0.000	0.000	0.000
贵　州	Guizhou	0.680	0.320	0.004	0.000	0.000	0.000
云　南	Yunnan	0.000	0.000	0.000	0.000	0.000	0.000
西　藏	Tibet	0.000	0.000	0.000	0.000	0.000	0.000
陕　西	Shaanxi	0.024	0.000	0.002	0.000	0.004	0.010
甘　肃	Gansu	1.210	0.592	0.038	0.001	0.840	3.160
青　海	Qinghai	0.000	0.000	0.000	0.000	0.000	0.000
宁　夏	Ningxia	0.001	0.000	0.000	0.000	0.000	0.000
新　疆	Xinjiang	0.002	0.001	0.000	0.000	0.000	0.000

各地区危险（医疗）废物集中处置情况（四）
Centralized Treatment of Hazardous（Medical）Wastes by Region（4）
（2012）

单位：千克   （kg）

年 份 地 区 Year Region	渗滤液中污染物排放量 Amount of Pollutants Discharged in Landfill Leachate					
	汞 Mercury	镉 Cadmium	六价铬 Hexavalent Chromium	总铬 Total Chromium	铅 Plumbum	砷 Arsenic
2011	7.953	10.303	—	215.437	111.454	42.445
2012	5.987	6.448	7.889	37.012	33.620	14.837
北 京 Beijing	0.000	0.000	0.000	0.000	0.000	0.000
天 津 Tianjin	0.000	0.003	0.000	0.010	0.050	0.010
河 北 Hebei	0.000	0.000	0.003	0.003	0.000	0.000
山 西 Shanxi	0.000	0.000	0.000	0.000	0.000	0.000
内蒙古 Inner Mongolia	0.000	0.000	0.000	0.000	0.000	0.000
辽 宁 Liaoning	0.003	0.750	0.650	0.650	0.015	0.100
吉 林 Jilin	0.000	0.027	0.038	0.437	0.006	0.164
黑龙江 Heilongjiang	0.000	0.000	0.000	0.000	0.000	0.000
上 海 Shanghai	0.845	0.000	0.000	0.000	0.400	0.000
江 苏 Jiangsu	0.016	0.121	0.483	2.522	9.788	0.431
浙 江 Zhejiang	0.000	0.000	0.000	1.240	4.816	2.185
安 徽 Anhui	0.021	0.001	0.000	0.000	0.381	0.000
福 建 Fujian	4.880	0.341	0.006	2.266	2.449	1.859
江 西 Jiangxi	0.000	3.887	1.000	4.000	7.759	7.251
山 东 Shandong	0.000	0.010	0.000	0.160	0.100	0.030
河 南 Henan	0.000	0.000	0.000	0.000	0.000	0.000
湖 北 Hubei	0.000	0.000	0.000	0.000	0.000	0.000
湖 南 Hunan	0.010	0.027	0.054	0.087	0.319	0.139
广 东 Guangdong	0.002	0.162	5.316	14.316	0.474	0.084
广 西 Guangxi	0.000	0.000	0.000	0.000	0.000	0.000
海 南 Hainan	0.000	0.000	0.000	0.000	0.000	0.000
重 庆 Chongqing	0.000	0.154	0.339	0.779	0.216	0.055
四 川 Sichuan	0.000	0.000	0.000	0.000	0.000	0.000
贵 州 Guizhou	0.000	0.335	0.000	0.000	0.521	0.419
云 南 Yunnan	0.000	0.000	0.000	0.000	0.000	0.000
西 藏 Tibet	0.000	0.000	0.000	0.000	0.000	0.000
陕 西 Shaanxi	0.000	0.000	0.000	0.000	0.005	0.000
甘 肃 Gansu	0.210	0.630	0.000	10.540	6.320	2.110
青 海 Qinghai	0.000	0.000	0.000	0.000	0.000	0.000
宁 夏 Ningxia	0.000	0.000	0.000	0.000	0.000	0.000
新 疆 Xinjiang	0.000	0.000	0.000	0.002	0.001	0.000

注：2011 年未统计危险（医疗）废物处置厂渗滤液中六价铬的排放量。

各地区危险（医疗）废物集中处置情况（五）
Centralized Treatment of Hazardous（Medical）Wastes by Region（5）
（2012）

单位：吨 　　　　　　　　　　　　　　　　　　　　　　　　　　　　　　　　　　（ton）

年　份　Year 地　区　Region	焚烧废气中污染物排放量 Volume of Pollutants Emission in the Incineration Gas		
	二氧化硫 Sulfur Dioxide	氮氧化物 Nitrogen Oxide	烟尘 Soot
2011	1 491.408	1 406.524	1 011.210
2012	1 033.306	1 287.103	845.676
北　京　Beijing	0.592	20.486	2.296
天　津　Tianjin	62.728	22.594	29.104
河　北　Hebei	19.634	32.176	58.805
山　西　Shanxi	2.878	16.545	3.115
内蒙古　Inner Mongolia	4.150	7.763	4.201
辽　宁　Liaoning	10.770	24.230	12.474
吉　林　Jilin	1.482	13.882	21.100
黑龙江　Heilongjiang	1.998	11.065	3.439
上　海　Shanghai	59.165	235.549	52.513
江　苏　Jiangsu	126.749	281.366	164.395
浙　江　Zhejiang	144.513	110.523	59.548
安　徽　Anhui	13.201	17.124	120.072
福　建　Fujian	7.814	31.962	6.250
江　西　Jiangxi	9.345	22.957	1.862
山　东　Shandong	131.947	68.706	20.227
河　南　Henan	20.523	20.090	6.567
湖　北　Hubei	28.854	84.793	59.127
湖　南　Hunan	2.530	6.300	54.820
广　东　Guangdong	176.988	128.102	51.341
广　西　Guangxi	1.790	9.301	22.144
海　南　Hainan	0.154	0.175	0.123
重　庆　Chongqing	4.401	19.238	4.696
四　川　Sichuan	6.795	18.127	28.010
贵　州　Guizhou	2.568	2.097	0.991
云　南　Yunnan	10.949	11.068	4.083
西　藏　Tibet	0.000	0.000	0.000
陕　西　Shaanxi	4.636	12.424	2.208
甘　肃　Gansu	171.302	49.995	9.020
青　海　Qinghai	0.830	2.430	0.263
宁　夏　Ningxia	2.681	1.936	0.612
新　疆　Xinjiang	1.339	4.099	42.270

各地区环境污染治理投资情况

Investment in the Treatment of Environmental Pollution by Region

（2012）

单位：亿元 （100 million yuan）

年　份 地　区	Year Region	环境污染治理投资总额 Total Investment in Treatment of Environmental Pollution	城市环境基础设施 建设投资 Investment in Urban Environment Infrastructure Facilities	工业污染源 治理投资 Investment in Treatment of Industrial Pollution Sources	建设项目"三同时"环 保投资 Investments in Environment Components for New Construction Projects
	2011	6 026. 2	3 469. 4	444. 4	2 112. 4
	2012	8 253. 6	5 062. 7	500. 5	2 690. 4
国家级	State Level	326. 7	—	—	326. 7
北　京	Beijing	342. 6	312. 9	3. 3	26. 4
天　津	Tianjin	157. 5	105. 0	12. 6	39. 9
河　北	Hebei	486. 1	369. 5	23. 6	93. 0
山　西	Shanxi	328. 1	240. 0	32. 3	55. 8
内蒙古	Inner Mongolia	445. 1	343. 9	19. 0	82. 2
辽　宁	Liaoning	683. 4	339. 6	11. 9	331. 9
吉　林	Jilin	103. 4	60. 6	5. 7	37. 1
黑龙江	Heilongjiang	218. 1	163. 2	3. 9	51. 0
上　海	Shanghai	134. 1	65. 2	11. 6	57. 3
江　苏	Jiangsu	657. 2	380. 3	39. 0	237. 9
浙　江	Zhejiang	375. 0	163. 3	28. 3	183. 4
安　徽	Anhui	330. 2	251. 0	12. 7	66. 5
福　建	Fujian	222. 6	122. 3	23. 8	76. 5
江　西	Jiangxi	316. 0	273. 5	3. 9	38. 6
山　东	Shandong	739. 2	446. 5	67. 1	225. 6
河　南	Henan	209. 4	139. 8	14. 8	54. 8
湖　北	Hubei	285. 5	189. 8	14. 9	80. 8
湖　南	Hunan	190. 3	130. 2	18. 0	42. 1
广　东	Guangdong	260. 2	103. 0	28. 1	129. 1
广　西	Guangxi	190. 6	109. 7	8. 6	72. 3
海　南	Hainan	44. 7	23. 5	4. 8	16. 4
重　庆	Chongqing	186. 9	126. 5	3. 8	56. 6
四　川	Sichuan	178. 3	100. 3	11. 1	66. 9
贵　州	Guizhou	68. 9	35. 7	12. 5	20. 7
云　南	Yunnan	132. 4	48. 7	19. 7	64. 0
西　藏	Tibet	3. 9	0. 7	0. 2	3. 0
陕　西	Shaanxi	180. 6	123. 3	27. 1	30. 2
甘　肃	Gansu	121. 4	70. 0	21. 1	30. 3
青　海	Qinghai	24. 1	12. 3	2. 2	9. 6
宁　夏	Ningxia	55. 6	16. 8	6. 9	31. 9
新　疆	Xinjiang	255. 0	195. 1	7. 9	52. 0

各地区城市环境基础设施建设投资情况
Investment in Urban Environment Infrastructure by Region
（2012）

单位：亿元 （100 million yuan）

年 份 地 区 Year Region	投资总额 Total Investment	燃气 Gas Supply	集中供热 Gerneral Heating	排水 Sewerage Projects	园林绿化 Gardening & Greening	市容环境卫生 Sanitation
2011	3 469. 4	331. 4	437. 6	770. 1	1 546. 2	384. 1
2012	5 062. 7	551. 8	798. 1	934. 1	2 380. 0	398. 6
北 京 Beijing	312. 9	25. 3	69. 1	40. 1	153. 8	24. 5
天 津 Tianjin	105. 0	16. 3	5. 9	9. 8	64. 6	8. 5
河 北 Hebei	369. 5	59. 1	77. 7	49. 0	163. 1	20. 6
山 西 Shanxi	240. 0	40. 3	104. 6	12. 0	71. 3	11. 8
内蒙古 Inner Mongolia	343. 9	14. 5	89. 2	44. 4	184. 5	11. 3
辽 宁 Liaoning	339. 6	21. 6	97. 9	71. 2	126. 5	22. 4
吉 林 Jilin	60. 6	10. 2	23. 2	7. 0	15. 6	4. 6
黑龙江 Heilongjiang	163. 2	15. 4	84. 5	21. 7	30. 4	11. 2
上 海 Shanghai	65. 2	13. 9	0. 0	17. 1	19. 3	15. 0
江 苏 Jiangsu	380. 3	27. 5	1. 0	81. 3	247. 6	22. 9
浙 江 Zhejiang	163. 6	19. 3	1. 4	36. 3	88. 2	18. 5
安 徽 Anhui	251. 0	20. 5	2. 0	53. 9	155. 4	19. 1
福 建 Fujian	122. 3	6. 3	0. 0	27. 1	66. 5	22. 4
江 西 Jiangxi	273. 5	82. 0	0. 0	31. 7	147. 1	12. 7
山 东 Shandong	446. 5	51. 5	81. 7	69. 6	217. 8	25. 9
河 南 Henan	139. 8	18. 7	11. 8	34. 0	67. 5	7. 7
湖 北 Hubei	189. 8	10. 1	0. 7	98. 0	60. 1	20. 9
湖 南 Hunan	130. 2	8. 7	0. 0	27. 8	57. 0	36. 8
广 东 Guangdong	103. 0	15. 7	0. 0	27. 1	44. 6	15. 6
广 西 Guangxi	109. 7	6. 2	0. 0	22. 5	70. 7	10. 2
海 南 Hainan	23. 5	0. 2	0. 0	12. 6	7. 5	3. 1
重 庆 Chongqing	126. 5	4. 7	0. 0	19. 4	98. 1	4. 2
四 川 Sichuan	100. 3	7. 7	0. 7	31. 3	50. 7	9. 9
贵 州 Guizhou	35. 7	5. 7	0. 0	7. 4	17. 7	4. 9
云 南 Yunnan	48. 7	1. 2	1. 0	19. 0	17. 7	9. 9
西 藏 Tibet	0. 7	0. 0	0. 0	0. 1	0. 2	0. 4
陕 西 Shaanxi	123. 3	19. 4	17. 4	24. 4	50. 2	12. 0
甘 肃 Gansu	70. 0	2. 6	16. 5	12. 8	35. 9	2. 2
青 海 Qinghai	12. 3	1. 5	3. 7	4. 7	2. 2	0. 2
宁 夏 Ningxia	16. 8	2. 7	7. 3	1. 0	5. 3	0. 6
新 疆 Xinjiang	195. 1	22. 8	100. 7	19. 8	43. 2	8. 5

9

重点城市环境统计

ANNUAL STATISTIC REPORT ON ENVIRONMENT IN CHINA
2012

重点城市工业废水排放及处理情况（一）

（2012）

城 市 名 称	工业废水 排放量/ 万吨	直接排入 环境的	排入污水处理 厂的	工业废水 处理量/ 万吨	废水治理 设施数/套	废水治理设施 治理能力/ （万吨/日）	废水治理设施 运行费用/ 万元
总　　计	1 198 520	840 399	358 121	2 934 632	43 392	14 615	4 124 726.7
北　京	9 190	2 232	6 958	11 081	521	65	51 338.7
天　津	19 117	8 027	11 090	40 464	967	171	127 294.3
石 家 庄	31 058	15 813	15 245	70 658	809	233	159 852.4
唐　山	19 396	16 001	3 395	403 183	906	1 846	153 498.0
秦 皇 岛	6 055	2 236	3˙819	16 704	258	87	12 631.1
邯　郸	5 906	5 446	460	62 587	366	243	68 412.7
保　定	15 774	9 590	6 184	24 973	684	292	27 199.6
太　原	3 161	2 316	845	36 319	273	119	55 500.2
大　同	5 985	5 280	705	4 755	292	28	7 064.1
阳　泉	501	179	322	3 180	48	14	4 976.8
长　治	8 659	8 427	232	20 360	475	93	28 986.0
临　汾	6 151	6 151	0	15 954	353	80	36 073.3
呼和浩特	2 187	1 687	500	5 007	87	22	14 603.6
包　头	7 704	1 629	6 074	66 849	109	305	38 805.0
赤　峰	2 253	1 936	317	3 463	131	24	6 706.1
沈　阳	7 705	2 489	5 216	11 672	304	72	24 141.2
大　连	30 795	28 519	2 276	26 483	523	108	17 211.9
鞍　山	5 517	4 573	944	49 216	186	267	40 222.3
抚　顺	2 836	1 525	1 311	14 947	153	88	21 572.8
本　溪	4 461	4 322	138	66 966	195	321	36 670.0
锦　州	6 534	5 578	955	5 576	96	23	25 586.5
长　春	5 309	4 160	1 150	6 002	127	31	34 608.0
吉　林	10 615	6 791	3 824	9 890	106	42	13 111.5
哈 尔 滨	6 497	2 206	4 290	6 587	191	465	30 277.2
齐齐哈尔	14 934	14 644	290	7 486	105	182	6 493.1
牡 丹 江	2 435	1 999	437	2 657	56	19	3 024.1
上　海	46 359	19 728	26 631	78 777	1 802	321	328 406.1
南　京	24 223	14 543	9 680	116 498	658	342	108 924.7
无　锡	22 987	6 188	16 799	40 048	995	216	84 744.5
徐　州	16 050	13 789	2 261	24 911	321	103	35 685.9
常　州	14 630	7 962	6 668	22 414	604	82	34 371.7
苏　州	70 754	45 257	25 497	73 790	1 754	464	446 727.7
南　通	18 200	11 547	6 653	14 297	849	89	62 597.4
连 云 港	6 390	4 925	1 466	11 440	262	104	23 471.3
扬　州	9 387	7 339	2 048	5 707	314	79	25 519.6
镇　江	9 982	8 637	1 345	11 364	420	55	25 904.3
杭　州	42 724	15 208	27 515	118 473	1 216	556	101 665.0

重点城市工业废水排放及处理情况（一）（续表）

（2012）

城 市名 称	工业废水排放量/万吨	直接排入环境的	排入污水处理厂的	工业废水处理量/万吨	废水治理设施数/套	废水治理设施治理能力/（万吨/日）	废水治理设施运行费用/万元
宁　波	20 125	12 936	7 189	22 025	929	110	82 163.5
温　州	7 913	5 035	2 878	7 458	1 020	40	31 528.0
湖　州	11 069	5 524	5 545	9 512	535	60	18 750.9
绍　兴	30 418	6 970	23 448	30 687	1 032	186	104 471.5
合　肥	5 971	3 809	2 162	25 993	267	157	19 037.6
芜　湖	3 148	1 940	1 208	4 763	223	37	12 766.7
马 鞍 山	6 911	6 327	584	93 597	202	318	68 077.8
福　州	5 333	4 015	1 318	23 452	410	131	24 657.6
厦　门	26 948	23 809	3 139	28 435	321	31	16 381.6
泉　州	20 535	13 805	6 730	32 771	743	132	54 518.7
南　昌	10 929	9 972	956	16 230	228	67	15 597.4
九　江	9 952	9 892	60	17 258	203	67	25 462.0
济　南	6 653	5 691	962	28 118	263	115	24 542.2
青　岛	11 146	3 046	8 100	49 663	482	172	29 561.5
淄　博	16 621	8 999	7 622	27 245	721	174	77 871.6
枣　庄	10 412	8 569	1 843	11 904	152	58	15 206.5
烟　台	9 359	5 873	3 486	15 173	501	114	25 898.4
潍　坊	26 955	8 688	18 267	37 898	646	346	65 909.7
济　宁	16 854	14 033	2 821	22 630	274	118	36 571.7
泰　安	8 423	6 330	2 093	20 050	214	82	23 038.6
日　照	7 868	6 273	1 595	8 360	141	43	16 739.0
郑　州	12 538	8 586	3 952	12 457	319	130	10 009.7
开　封	8 204	8 062	142	3 686	78	16	5 158.1
洛　阳	8 410	7 400	1 010	12 281	760	97	32 399.0
平 顶 山	7 167	7 064	103	10 672	138	81	11 999.0
安　阳	6 547	6 542	4	37 947	310	78	31 378.1
焦　作	11 572	10 007	1 565	10 748	187	67	13 376.3
三 门 峡	6 687	6 687	0	7 192	253	43	11 188.9
武　汉	20 851	17 324	3 527	122 133	283	379	40 802.8
宜　昌	20 635	19 100	1 535	16 088	343	87	25 454.9
荆　州	10 579	10 210	369	4 685	105	29	10 993.1
长　沙	3 777	3 480	297	3 163	230	26	5 053.7
株　洲	6 492	5 911	582	11 210	446	65	15 884.0
湘　潭	6 053	5 694	360	9 985	187	38	6 736.6
岳　阳	13 634	13 323	311	10 885	184	45	17 618.5
常　德	12 855	12 814	42	10 762	166	56	8 858.8
张 家 界	579	579	0	104	20	1	102.3
广　州	22 677	12 794	9 882	23 456	994	201	59 390.2

重点城市工业废水排放及处理情况（一）（续表）

（2012）

城 市 名 称	工业废水排放量/ 万吨	直接排入 环境的	排入污水处理 厂的	工业废水 处理量/ 万吨	废水治理 设施数/套	废水治理设施 治理能力/ （万吨/日）	废水治理设施 运行费用/ 万元
韶　关	10 182	10 112	70	69 218	313	227	32 637.7
深　圳	13 831	7 810	6 021	12 253	1 611	54	66 337.5
珠　海	5 524	3 147	2 377	4 697	371	27	20 687.7
汕　头	5 097	4 327	770	4 470	364	50	12 043.3
湛　江	8 263	8 202	61	7 313	185	62	11 149.4
南　宁	12 496	12 068	428	11 297	282	127	19 625.7
柳　州	13 303	12 148	1 155	64 498	291	343	66 913.0
桂　林	4 999	4 502	497	8 293	293	47	5 651.7
北　海	2 291	1 889	402	13 779	47	67	8 019.1
海　口	879	131	748	746	66	8	1 926.3
重　庆	30 611	29 466	1 145	44 824	1 578	395	79 899.4
成　都	11 780	8 287	3 493	30 237	922	169	39 704.4
自　贡	2 335	2 231	103	1 638	116	17	2 444.7
攀枝花	4 012	3 948	63	82 776	257	259	52 177.7
泸　州	3 707	3 703	4	3 284	212	20	8 269.6
德　阳	6 420	5 998	422	7 822	278	41	12 886.7
绵　阳	6 798	6 212	586	9 940	303	53	7 247.3
南　充	3 361	2 355	1 006	1 009	44	3	2 551.6
宜　宾	6 120	6 106	14	5 256	170	44	13 469.1
贵　阳	2 008	2 008	0	23 444	213	87	15 342.2
遵　义	2 805	2 805	0	2 846	178	17	4 956.7
昆　明	5 211	4 868	343	25 513	507	118	24 711.4
曲　靖	3 224	3 119	105	14 895	386	77	16 230.7
玉　溪	4 996	4 667	329	23 572	262	158	16 976.9
拉　萨	333	333	0	273	23	5	440.3
西　安	10 224	7 923	2 301	9 089	312	41	12 323.5
铜　川	433	422	11	367	34	3	1 028.8
宝　鸡	5 498	5 028	470	4 883	333	52	9 203.1
咸　阳	6 033	4 055	1 978	5 259	208	28	12 796.8
渭　南	3 303	3 141	162	12 220	201	43	12 033.5
延　安	2 649	2 589	59	3 204	274	17	12 992.3
兰　州	4 625	3 318	1 307	15 117	109	68	10 220.3
金　昌	1 848	1 794	55	3 462	69	10	6 854.6
西　宁	3 185	3 185	0	16 193	84	52	9 912.0
银　川	5 963	4 678	1 285	6 631	81	30	17 108.6
石嘴山	1 230	1 170	61	7 469	174	68	6 447.6
乌鲁木齐	5 980	5 478	502	20 084	111	74	13 653.9
克拉玛依	1 749	1 187	561	5 348	104	5	38 816.4

重点城市工业废水排放及处理情况（二）

（2012）

单位：吨

城 市 名 称	工业废水中污染物产生量				
	化学需氧量	氨氮	石油类	挥发酚	氰化物
总　　计	12 528 711.5	734 164.4	182 638.9	41 802.2	3 231.1
北　　京	61 769.1	2 725.3	3 298.4	876.9	5.9
天　　津	130 525.3	7 368.7	1 135.9	2.6	0.5
石 家 庄	439 335.2	20 116.8	2 076.3	805.8	16.7
唐　　山	319 857.8	7 062.8	10 393.6	5 939.0	382.4
秦 皇 岛	62 173.6	1 506.8	78.2	—	0.4
邯　　郸	78 523.5	7 242.0	2 943.5	1 933.8	74.0
保　　定	220 210.5	4 618.7	577.8	4.4	1.5
太　　原	22 993.9	933.9	272.4	460.0	35.1
大　　同	43 127.9	4 178.4	118.6	67.8	—
阳　　泉	2 406.4	238.6	0.5	—	—
长　　治	110 828.5	8 349.3	3 324.2	2 898.3	90.6
临　　汾	45 907.5	4 402.1	1 139.6	920.2	54.7
呼和浩特	52 215.6	2 336.5	114.9	16.7	0.0
包　　头	46 618.1	22 681.6	2 445.5	972.6	18.7
赤　　峰	42 503.2	462.3	27.0	—	—
沈　　阳	46 846.5	1 720.9	327.0	16.2	2.3
大　　连	108 660.1	10 278.0	1 064.7	15.6	0.6
鞍　　山	31 721.1	1 014.1	299.9	0.4	14.5
抚　　顺	10 310.1	308.0	547.1	8.4	0.6
本　　溪	59 227.8	1 466.0	2 290.5	1 129.5	122.4
锦　　州	53 209.6	1 075.3	3 229.1	21.0	0.5
长　　春	258 179.8	11 520.0	365.4	1.9	0.2
吉　　林	58 637.1	1 315.8	106.2	188.8	10.9
哈 尔 滨	149 465.3	3 280.5	243.3	3 431.9	0.3
齐齐哈尔	60 913.9	4 258.4	43.8	127.7	0.2
牡 丹 江	45 752.3	4 795.8	22.3	16.0	0.9
上　　海	312 601.6	11 889.5	7 690.6	1 066.4	217.7
南　　京	178 950.2	13 555.9	5 298.9	2 898.3	154.8
无　　锡	166 243.4	4 615.8	953.8	94.1	2.7
徐　　州	144 396.1	5 873.4	451.1	164.8	16.4
常　　州	99 401.9	2 584.7	1 755.0	17.4	9.8
苏　　州	453 396.6	14 848.9	933.3	36.7	52.0
南　　通	173 884.6	5 570.4	1 482.4	22.2	45.0
连 云 港	338 797.5	3 432.5	513.6	137.0	0.0
扬　　州	120 195.2	2 828.5	368.4	19.6	28.1
镇　　江	83 583.5	10 146.4	104.1	85.4	59.7
杭　　州	1 017 406.3	10 022.9	787.7	147.9	2.2

重点城市工业废水排放及处理情况（二）（续表）

（2012）

单位：吨

城 市 名 称	工业废水中污染物产生量				
	化学需氧量	氨氮	石油类	挥发酚	氰化物
宁 波	257 911.6	30 769.7	16 244.6	0.4	233.4
温 州	152 303.1	5 184.9	3 107.3	4.1	342.2
湖 州	81 115.6	2 567.4	236.6	0.0	0.0
绍 兴	300 828.0	18 365.5	30.6	1.7	0.4
合 肥	76 298.9	4 964.3	1 018.8	0.1	3.0
芜 湖	39 619.8	812.1	441.6	0.2	0.3
马 鞍 山	127 355.3	1 563.1	3 278.1	1 520.7	155.7
福 州	95 757.6	2 246.6	1 250.3	209.7	5.2
厦 门	75 491.9	924.3	5 155.9	0.6	26.5
泉 州	353 418.4	17 153.7	2 107.8	0.2	47.9
南 昌	93 085.5	9 188.9	356.6	243.1	48.2
九 江	60 010.7	1 859.6	1 408.6	184.7	0.4
济 南	80 753.2	14 549.3	3 482.0	1 380.6	30.3
青 岛	94 141.0	5 587.9	1 152.1	4.1	24.3
淄 博	291 413.3	21 382.5	2 269.3	615.9	16.3
枣 庄	109 944.4	7 296.8	139.6	679.5	3.3
烟 台	133 606.6	3 629.2	13 325.1	4.6	6.9
潍 坊	383 681.3	9 586.7	1 003.7	70.5	3.3
济 宁	208 824.5	5 298.3	166.1	286.5	6.5
泰 安	53 525.9	1 962.1	827.2	319.5	6.2
日 照	90 183.8	2 945.1	168.6	108.0	0.0
郑 州	57 336.1	1 315.2	469.9	0.5	6.9
开 封	32 770.7	4 078.0	97.3	0.0	0.8
洛 阳	44 570.5	2 307.5	34 446.9	11.9	2.6
平 顶 山	81 267.5	2 770.3	1 103.3	568.9	28.6
安 阳	81 156.2	2 047.1	1 667.0	1 100.2	48.1
焦 作	241 280.9	5 986.5	293.0	0.9	—
三 门 峡	45 713.9	2 303.6	198.8	3.5	1.1
武 汉	135 415.9	3 726.9	829.1	616.6	13.4
宜 昌	63 970.2	8 027.9	440.0	16.1	1.3
荆 州	82 855.5	4 146.0	94.0	14.4	22.6
长 沙	36 263.2	950.5	23.8	—	1.1
株 洲	21 383.5	4 905.9	271.7	1.9	8.6
湘 潭	30 350.4	1 381.9	1 661.7	793.8	158.3
岳 阳	137 054.0	25 661.0	853.4	48.3	3.5
常 德	57 365.1	1 769.5	59.7	0.0	0.2
张 家 界	3 368.5	82.1	10.2	—	—
广 州	158 724.5	3 300.5	1 175.5	48.3	48.8

重点城市工业废水排放及处理情况（二）（续表）

（2012）

单位：吨

城 市 名 称	工业废水中污染物产生量				
	化学需氧量	氨氮	石油类	挥发酚	氰化物
韶 关	10 536.4	521.5	9.1	5.3	1.1
深 圳	82 116.2	5 676.3	62.6	—	138.5
珠 海	31 513.1	1 465.3	99.8	6.9	0.1
汕 头	46 355.6	974.1	1.7	0.0	0.2
湛 江	62 457.3	1 374.9	159.6	0.0	—
南 宁	271 271.3	3 240.1	43.3	26.2	1.6
柳 州	119 888.8	16 291.3	2 080.2	1 657.2	11.7
桂 林	32 759.9	3 108.4	67.1	0.3	0.3
北 海	82 873.6	1 125.0	303.7	56.2	0.1
海 口	2 458.6	102.0	9.9	—	—
重 庆	288 043.8	28 232.9	2 727.8	849.6	84.5
成 都	68 468.6	4 433.6	567.0	0.0	44.0
自 贡	7 523.6	882.4	76.4	0.1	0.0
攀 枝 花	30 525.5	850.7	1 224.5	139.3	11.6
泸 州	46 064.0	16 622.1	168.5	0.0	0.0
德 阳	116 239.6	1 347.6	177.7	0.0	1.1
绵 阳	32 614.1	889.6	38.9	0.0	0.0
南 充	8 603.7	750.5	242.7	0.0	0.0
宜 宾	109 429.9	1 543.1	122.7	46.2	1.4
贵 阳	19 106.2	1 416.2	587.5	216.7	4.7
遵 义	36 497.4	953.6	23.2	0.0	0.0
昆 明	51 403.0	1 644.4	688.4	603.9	29.4
曲 靖	47 398.3	21 497.1	1 120.1	1 078.8	68.2
玉 溪	35 886.6	3 919.3	164.7	61.0	3.2
拉 萨	5 093.4	214.7	3.9	—	—
西 安	110 848.3	6 761.7	887.1	5.5	14.3
铜 川	2 054.7	20.6	0.5	—	—
宝 鸡	127 377.9	15 482.3	509.1	1.3	3.6
咸 阳	43 562.3	2 575.0	423.5	19.3	1.6
渭 南	102 657.2	3 123.9	1 237.7	1 686.0	36.8
延 安	20 328.3	1 608.6	691.3	0.8	0.1
兰 州	21 585.1	5 301.4	1 616.6	109.3	1.8
金 昌	9 005.5	4 878.7	84.2	0.1	0.1
西 宁	24 019.1	1 003.3	499.4	2.3	—
银 川	195 277.2	61 740.3	1 341.3	20.2	0.8
石 嘴 山	32 579.1	3 102.6	1 273.7	509.7	20.7
乌鲁木齐	53 385.8	45 746.7	2 875.1	978.4	21.3
克拉玛依	20 008.6	518.9	2 737.9	316.4	—

重点城市工业废水排放及处理情况（三）

（2012）

单位：吨

城 市名 称	工业废水中污染物产生量					
	汞	镉	六价铬	总铬	铅	砷
总　　计	12.151	887.573	2 483.902	5 016.126	1 515.341	2 758.824
北　京	0.000	0.001	10.270	11.038	0.269	2.344
天　津	0.039	0.024	8.561	13.436	15.271	0.949
石 家 庄	0.001	—	0.004	26.581	0.037	0.031
唐　山	0.000	—	26.660	46.660	1.185	—
秦 皇 岛	0.000	0.000	9.653	9.674	0.000	0.001
邯　郸	0.000	—			—	0.010
保　定	0.000	0.000	3.424	3.654	6.537	—
太　原	0.005	0.013	1.297	1.380	0.089	0.032
大　同	0.000	0.011	0.031	0.031	0.620	0.976
阳　泉	0.001	—	—	—	—	—
长　治	0.356	—	2.368	2.676	—	—
临　汾	0.000	13.165	0.001	0.001	9.709	57.695
呼 和 浩 特	0.001	—	—	—	—	0.002
包　头	0.002	3.728	0.451	0.508	5.425	11.383
赤　峰	0.091	62.260	0.020	0.025	52.316	159.953
沈　阳	0.000	0.004	3.765	3.765	0.890	—
大　连	0.000	0.000	2.021	2.210	0.243	0.000
鞍　山	0.000	0.000	4.625	4.625	0.002	—
抚　顺	0.000	0.279	0.001	0.001	0.683	1.517
本　溪	0.044	0.000	0.243	0.243	0.001	0.001
锦　州	0.000	—	3.040	5.651	—	0.026
长　春	0.000	0.000	11.298	11.301	—	0.000
吉　林	0.000	0.002	0.004	0.004	0.021	0.024
哈 尔 滨	0.000	—	4.518	4.518	0.361	—
齐 齐 哈 尔	0.135	—	0.552	0.579	—	—
牡 丹 江	0.003	0.000	0.000	0.000	0.005	0.001
上　海	0.003	0.027	101.576	130.581	1.380	0.818
南　京	0.004	0.181	35.042	35.359	0.068	0.227
无　锡	0.000	2.251	40.896	59.360	1.266	0.000
徐　州	0.000	—	0.128	3.188	0.289	—
常　州	0.000	0.011	22.914	24.487	0.121	3.874
苏　州	0.000	0.000	30.315	75.605	2.215	—
南　通	0.388	0.032	60.324	82.708	3.182	0.370
连 云 港	0.000	—	—	—	0.245	—
扬　州	0.001	0.016	34.087	51.886	2.920	0.918
镇　江	0.000	—	38.068	95.667	—	0.894
杭　州	0.000	—	198.507	212.237	1.706	0.269

重点城市工业废水排放及处理情况（三）（续表）

（2012）

单位：吨

城 市 名 称	工业废水中污染物产生量					
	汞	镉	六价铬	总铬	铅	砷
宁　波	0.023	—	442.322	1 243.034	0.025	—
温　州	—	2.792	682.994	903.018	1.833	1.396
湖　州	—	—	2.043	17.225	1.516	—
绍　兴	—	—	1.414	2.935	0.013	0.027
合　肥	—	0.001	0.061	0.061	0.023	0.040
芜　湖	2.240	0.058	3.007	3.022	0.946	0.438
马 鞍 山	—	—	1.677	1.879	0.056	—
福　州	0.055	—	40.176	40.176	2.022	—
厦　门	0.004	0.145	263.593	484.815	0.744	0.558
泉　州	0.004	0.045	95.348	188.868	1.935	0.256
南　昌	0.000	0.077	34.343	35.590	0.175	—
九　江	0.011	10.275	5.531	7.267	12.059	9.443
济　南	—	—	5.905	5.954	—	0.806
青　岛	—	—	12.954	23.900	0.032	—
淄　博	0.001	0.010	0.767	15.866	0.995	1.049
枣　庄	0.001	—	0.060	0.072	0.001	0.015
烟　台	0.730	23.767	11.773	15.808	20.195	209.074
潍　坊	—	—	0.138	0.366	—	—
济　宁	0.000	—	0.006	0.010	0.065	—
泰　安	0.082	0.001	4.955	6.574	0.542	0.206
日　照	0.000	0.111	2.978	6.537	0.205	—
郑　州	—	—	4.415	4.424	—	0.021
开　封	0.000	0.468	0.529	15.061	1.043	0.063
洛　阳	0.003	0.655	1.103	1.356	29.967	1.421
平 顶 山	0.167	—	1.240	1.240	0.002	0.574
安　阳	—	53.809	0.275	0.275	19.401	4.465
焦　作	0.342	27.316	0.930	52.499	24.173	4.529
三 门 峡	0.477	77.357	0.097	0.097	113.631	35.202
武　汉	0.125	0.004	18.528	18.607	1.078	58.108
宜　昌	0.010	—	4.012	5.062	0.065	—
荆　州	0.001	—	3.056	3.514	—	0.098
长　沙	0.000	0.011	0.284	0.377	0.178	0.001
株　洲	2.539	44.980	5.107	7.120	70.367	41.497
湘　潭	—	1.701	4.030	8.086	5.752	0.682
岳　阳	0.088	10.173	0.098	0.131	16.748	4.130
常　德	0.000	0.001	0.049	0.378	0.000	0.011
张 家 界	—	0.009	0.018	0.018	0.046	0.024
广　州	0.000	0.048	11.870	124.187	0.576	0.159

重点城市工业废水排放及处理情况（三）（续表）

（2012）
单位：吨

城 市 名 称	工业废水中污染物产生量					
	汞	镉	六价铬	总铬	铅	砷
韶 关	0.016	1.039	0.382	3.452	5.547	2.812
深 圳	0.113	0.023	13.140	532.109	0.099	1.802
珠 海	0.000	0.009	0.186	2.745	0.013	—
汕 头	0.000	0.002	2.748	6.665	0.488	0.003
湛 江	—	0.000	0.022	0.162	0.001	1.332
南 宁	0.010	0.008	1.179	1.420	0.061	0.153
柳 州	0.313	10.904	4.811	4.811	20.159	5.519
桂 林	0.001	0.041	17.625	27.888	10.144	0.123
北 海	—	—	0.047	8.690	—	—
海 口	0.000	0.000	0.000	16.896	0.000	0.000
重 庆	0.000	0.001	85.731	163.953	6.990	2.998
成 都	0.000	0.008	8.175	10.197	0.134	44.100
自 贡	—	—	0.000	0.000	—	0.015
攀 枝 花	0.000	0.024	7.421	8.317	0.006	3.919
泸 州	—	—	1.347	2.018	—	—
德 阳	0.495	5.957	0.230	1.470	6.008	55.677
绵 阳	0.116	0.026	4.965	43.559	0.094	213.039
南 充	—	—	—	—	—	—
宜 宾	0.211	0.075	0.003	0.026	0.090	46.025
贵 阳	—	0.002	0.008	0.008	0.005	2.522
遵 义	0.003	0.002	0.049	0.207	0.009	0.000
昆 明	0.657	130.050	0.034	0.781	227.385	512.657
曲 靖	0.028	27.223	3.024	3.033	34.787	104.630
玉 溪	0.044	70.803	0.722	0.722	90.457	157.459
拉 萨	0.059	0.022	—	0.003	0.309	14.894
西 安	0.042	0.019	3.855	4.699	0.887	0.000
铜 川	—	—	—	—	—	—
宝 鸡	0.003	3.583	0.258	0.263	5.337	26.551
咸 阳	—	0.026	1.556	2.747	—	—
渭 南	—	0.000	—	—	0.001	—
延 安	0.000	0.004	0.020	0.020	0.006	0.040
兰 州	0.000	—	0.022	0.042	—	0.000
金 昌	0.157	290.580	—	—	661.544	826.106
西 宁	0.361	11.226	1.192	1.302	10.005	66.984
银 川	—	0.086	0.650	0.724	1.296	44.307
石 嘴 山	0.410	—	—	—	—	8.473
乌鲁木齐	1.134	0.007	2.152	2.152	0.016	0.076
克拉玛依	—	—	—	—	—	—

重点城市工业废水排放及处理情况（四）

（2012）

单位：吨

	工业废水中污染物排放量				
	化学需氧量	氨氮	石油类	挥发酚	氰化物
总　　计	1 419 680.7	132 300.8	8 616.8	644.3	97.0
北　　京	6 265.8	352.0	49.1	0.2	0.1
天　　津	26 533.3	3 294.5	138.2	1.2	0.0
石 家 庄	45 236.0	6 597.9	147.2	4.1	1.6
唐　　山	19 834.9	1 297.4	273.7	4.8	7.3
秦 皇 岛	15 030.1	846.5	21.5	0.0	—
邯　　郸	9 367.8	536.4	62.0	1.5	0.6
保　　定	22 187.7	1 300.9	27.9	4.0	0.0
太　　原	3 790.8	272.2	27.7	0.1	0.6
大　　同	9 289.7	1 312.5	53.7	67.8	0.0
阳　　泉	286.8	59.7	0.0	—	—
长　　治	16 406.1	1 576.2	430.5	287.0	9.5
临　　汾	9 427.2	913.8	186.9	6.2	1.2
呼和浩特	10 348.0	633.8	1.2	0.0	0.0
包　　头	5 400.1	5 367.8	589.3	1.0	1.3
赤　　峰	5 891.0	187.8	5.3	—	0.0
沈　　阳	9 342.0	830.4	50.3	13.6	0.0
大　　连	24 438.7	2 460.2	169.1	2.3	0.1
鞍　　山	5 096.5	462.1	43.7	0.4	1.8
抚　　顺	1 985.1	60.8	30.6	0.7	0.2
本　　溪	8 858.5	373.4	61.0	0.8	0.9
锦　　州	8 658.7	280.9	56.7	1.9	0.1
长　　春	11 126.2	1 334.5	22.9	0.0	0.0
吉　　林	14 795.4	879.5	66.5	0.1	0.0
哈 尔 滨	6 637.0	969.7	38.1	0.3	0.0
齐齐哈尔	21 219.0	2 077.0	25.6	4.0	0.2
牡 丹 江	6 857.4	100.3	8.1	0.6	0.9
上　　海	26 159.5	2 281.0	646.2	2.7	2.2
南　　京	22 439.4	1 203.1	203.7	8.6	1.0
无　　锡	11 383.1	413.1	20.2	0.1	0.3
徐　　州	17 167.4	911.0	92.6	0.1	0.6
常　　州	8 157.7	398.4	74.6	2.5	1.3
苏　　州	46 741.2	3 622.0	102.3	3.7	3.2
南　　通	27 611.3	1 654.7	243.4	6.9	1.2
连 云 港	15 299.7	1 729.1	15.8	5.5	0.0
扬　　州	12 103.0	977.3	87.5	5.8	2.0
镇　　江	5 974.3	376.3	20.2	3.1	1.9
杭　　州	36 039.6	1 556.9	40.3	18.0	0.3

重点城市工业废水排放及处理情况（四）（续表）

（2012）   单位：吨

城　市 名　称	工业废水中污染物排放量				
	化学需氧量	氨氮	石油类	挥发酚	氰化物
宁　波	20 066.1	1 019.0	41.8	0.1	1.1
温　州	14 491.3	1 326.0	141.2	0.7	3.4
湖　州	9 441.7	530.4	13.8	0.0	0.0
绍　兴	30 880.8	2 412.3	13.2	1.6	0.0
合　肥	8 440.0	516.2	24.8	0.0	0.0
芜　湖	6 797.4	105.1	30.6	0.1	0.0
马 鞍 山	7 135.8	286.3	68.8	2.4	1.7
福　州	6 084.1	492.0	25.3	0.0	0.1
厦　门	4 080.1	331.2	3.8	0.1	0.8
泉　州	23 853.4	1 682.0	110.0	0.0	0.7
南　昌	11 299.9	1 519.2	75.7	3.1	5.7
九　江	11 419.3	840.1	66.3	1.2	0.1
济　南	5 497.0	373.4	77.7	6.7	3.6
青　岛	8 280.9	749.6	32.6	0.2	0.1
淄　博	13 989.3	1 127.1	53.0	3.0	0.2
枣　庄	6 482.7	312.0	33.9	0.5	0.2
烟　台	6 787.5	393.0	497.0	2.3	0.1
潍　坊	19 814.7	2 641.5	18.0	1.6	0.7
济　宁	9 117.8	448.5	73.1	0.2	0.0
泰　安	6 710.7	254.3	45.7	1.3	0.1
日　照	5 928.6	397.4	9.6	18.0	0.0
郑　州	11 739.6	534.8	176.7	0.4	3.6
开　封	10 051.5	1 514.8	62.6	0.0	0.7
洛　阳	10 425.9	658.1	70.2	2.1	0.8
平 顶 山	12 947.3	934.0	189.0	89.0	3.4
安　阳	9 494.4	517.7	48.0	0.1	3.0
焦　作	13 699.5	978.9	39.2	0.3	—
三 门 峡	7 929.9	536.9	61.8	0.9	0.1
武　汉	16 425.1	1 412.9	111.5	0.6	1.9
宜　昌	17 722.2	2 871.0	58.4	0.2	0.3
荆　州	24 810.0	2 893.6	65.1	0.9	0.4
长　沙	13 799.0	449.1	14.6	—	0.0
株　洲	5 168.2	1 370.1	64.6	0.4	1.7
湘　潭	3 663.1	511.6	110.3	1.7	2.6
岳　阳	27 185.2	13 347.1	66.0	0.4	0.4
常　德	10 398.0	965.0	17.5	0.0	0.0
张 家 界	3 127.2	81.9	9.3	—	—
广　州	21 674.8	1 352.9	93.1	1.1	2.0

重点城市工业废水排放及处理情况（四）（续表）

（2012）

单位：吨

城　市名　称	工业废水中污染物排放量				
	化学需氧量	氨氮	石油类	挥发酚	氰化物
韶　关	5 596.9	239.8	5.1	0.5	1.1
深　圳	13 356.7	1 153.3	4.1	—	0.8
珠　海	6 308.9	605.2	27.7	0.2	0.0
汕　头	9 939.0	533.8	0.8	0.0	0.0
湛　江	12 174.6	592.1	6.6	0.0	—
南　宁	26 793.6	1 463.1	5.7	10.6	1.5
柳　州	12 463.6	1 674.0	35.7	0.9	1.8
桂　林	9 806.4	644.3	17.8	0.3	0.0
北　海	9 908.0	129.0	4.1	0.0	0.1
海　口	808.4	48.3	3.3	—	—
重　庆	49 157.2	3 077.2	353.1	9.5	1.5
成　都	13 744.4	842.2	33.1	0.0	0.1
自　贡	3 191.0	552.8	29.4	0.1	0.0
攀 枝 花	1 701.4	125.6	4.7	0.0	0.0
泸　州	12 083.5	305.4	14.4	0.0	0.0
德　阳	6 418.3	494.1	40.1	0.0	0.0
绵　阳	5 352.3	167.4	10.0	0.0	0.0
南　充	5 427.6	380.5	10.4	0.0	0.0
宜　宾	19 968.1	699.5	54.9	0.1	0.7
贵　阳	6 978.3	260.9	60.9	0.0	0.0
遵　义	13 376.7	406.0	22.2	0.0	0.0
昆　明	8 261.1	257.5	72.1	0.4	0.4
曲　靖	8 027.5	1 004.2	56.4	0.2	0.6
玉　溪	12 269.7	127.5	4.4	0.5	0.1
拉　萨	628.1	65.0	0.2	—	—
西　安	24 381.0	1 695.9	271.7	0.1	0.0
铜　川	1 463.4	7.5	0.4	—	—
宝　鸡	14 584.8	1 071.1	117.1	0.4	1.6
咸　阳	14 542.9	1 330.2	52.4	0.5	1.5
渭　南	20 427.2	1 096.6	45.2	1.2	2.2
延　安	4 879.3	245.5	59.2	0.3	0.1
兰　州	4 348.5	2 642.7	68.8	0.4	0.0
金　昌	4 418.5	4 719.0	39.5	0.1	0.1
西　宁	15 606.6	555.7	38.5	0.6	—
银　川	16 676.9	2 779.9	53.4	4.9	0.7
石 嘴 山	5 465.1	1 284.7	15.6	0.3	0.6
乌鲁木齐	6 812.2	723.7	53.3	3.7	1.2
克拉玛依	2 087.1	112.0	77.6	7.5	—

重点城市工业废水排放及处理情况（五）

（2012）

单位：吨

城 市 名 称	工业废水中污染物排放量					
	汞	镉	六价铬	总铬	铅	砷
总 计	0.401	4.371	41.835	95.278	24.073	16.364
北 京	—	0.001	0.326	0.429	0.054	0.012
天 津	0.004	0.010	0.169	0.454	1.004	0.019
石 家 庄	0.001	—	0.004	2.033	0.011	0.000
唐 山	0.000	—	0.092	0.186	0.004	—
秦 皇 岛	—	—	0.003	0.028	—	—
邯 郸	—	—	—	—	—	—
保 定	0.000	—	0.057	0.067	0.197	—
太 原	0.003	0.013	0.155	0.170	0.089	0.032
大 同	0.000	0.004	0.005	0.005	0.219	0.363
阳 泉	0.000	—	—	—	—	—
长 治	0.001	—	0.003	0.008	—	—
临 汾	—	0.027	0.001	0.001	0.068	0.173
呼 和 浩 特	0.001	—	—	—	—	0.002
包 头	0.001	—	0.001	0.002	—	—
赤 峰	0.034	0.102	0.000	0.005	1.215	1.900
沈 阳	—	0.000	0.086	0.086	0.037	—
大 连	0.000	0.000	0.047	0.070	0.169	0.000
鞍 山	—	—	0.000	0.000	—	—
抚 顺	0.000	0.019	0.001	0.001	0.177	0.085
本 溪	0.000	0.000	—	—	0.000	0.001
锦 州	—	—	0.159	0.241	—	0.026
长 春	—	0.000	0.088	0.091	—	0.000
吉 林	0.000	0.000	0.000	0.000	0.001	0.000
哈 尔 滨	—	—	0.041	0.041	0.017	—
齐 齐 哈 尔	—	—	0.005	0.005	—	—
牡 丹 江	—	—	—	—	—	—
上 海	0.000	0.001	1.011	2.773	0.243	0.073
南 京	0.002	0.006	0.326	0.398	0.003	0.046
无 锡	—	—	0.146	0.660	0.080	—
徐 州	—	—	0.089	0.679	0.023	—
常 州	—	0.005	0.196	0.217	0.020	0.007
苏 州	—	0.000	0.898	2.717	0.060	—
南 通	0.091	0.007	1.925	4.181	1.084	0.349
连 云 港	—	—	—	—	0.022	—
扬 州	0.001	0.000	0.388	1.066	0.166	0.002
镇 江	0.000	—	0.089	0.246	—	0.092
杭 州	—	—	2.317	2.779	0.029	0.000

重点城市工业废水排放及处理情况（五）（续表）

（2012）

单位：吨

城　市 名　称	工业废水中污染物排放量					
	汞	镉	六价铬	总铬	铅	砷
宁　波	0.004	—	0.906	3.617	0.003	—
温　州	—	0.174	3.766	5.414	0.082	0.031
湖　州	—	—	0.067	0.369	0.049	—
绍　兴	—	—	0.110	0.307	0.008	0.011
合　肥	—	0.001	0.007	0.007	0.013	0.001
芜　湖	0.000	0.003	0.084	0.099	0.254	0.158
马 鞍 山	—	—	0.006	0.009	—	—
福　州	0.008	—	0.454	0.454	0.015	—
厦　门	0.001	0.009	1.053	2.762	0.181	0.134
泉　州	0.002	0.002	0.675	7.432	0.139	0.101
南　昌	0.000	0.007	16.870	16.890	0.024	—
九　江	0.001	0.051	0.068	0.107	0.235	0.529
济　南	—	—	0.086	0.125	—	0.042
青　岛	—	—	0.074	0.327	0.002	—
淄　博	0.001	0.001	0.008	0.336	0.092	0.031
枣　庄	0.001	—	0.060	0.072	0.001	0.015
烟　台	0.000	0.001	0.156	0.217	0.033	0.037
潍　坊	—	—	0.030	0.247	—	—
济　宁	0.000	—	—	0.004	—	—
泰　安	—	—	0.006	0.062	0.009	0.135
日　照	0.000	0.001	0.058	0.124	0.002	—
郑　州	—	—	0.044	0.044	—	0.021
开　封	0.000	0.401	0.529	1.095	1.043	0.061
洛　阳	0.003	0.287	0.029	0.206	1.200	0.690
平 顶 山	0.002	—	0.014	0.014	0.001	0.061
安　阳	0.000	0.000	0.002	0.002	0.140	—
焦　作	0.003	0.010	0.076	22.615	0.201	0.012
三 门 峡	0.003	0.082	0.003	0.003	0.445	0.319
武　汉	0.002	0.004	1.194	1.270	0.123	0.209
宜　昌	0.000	—	0.008	0.062	0.002	—
荆　州	0.001	—	2.759	2.983	—	0.098
长　沙	0.000	0.011	0.133	0.168	0.062	0.001
株　洲	0.145	0.387	0.068	0.071	2.834	0.953
湘　潭	—	0.125	0.702	0.749	0.519	0.030
岳　阳	0.001	0.007	0.005	0.007	0.255	0.350
常　德	0.000	0.001	0.004	0.014	0.000	0.005
张 家 界	—	0.006	0.012	0.012	0.021	0.012
广　州	0.000	0.015	1.670	2.502	0.092	0.029

重点城市工业废水排放及处理情况（五）（续表）

（2012） 单位：吨

城 市 名 称	工业废水中污染物排放量					
	汞	镉	六价铬	总铬	铅	砷
韶 关	0.007	0.607	0.123	0.329	3.605	0.507
深 圳	0.008	0.002	0.136	0.332	0.010	0.001
珠 海	0.000	0.009	0.119	0.485	0.005	—
汕 头	0.000	0.002	0.002	0.028	0.058	0.003
湛 江	—	0.000	0.003	0.015	0.001	0.005
南 宁	0.005	0.001	0.010	0.035	0.008	0.021
柳 州	0.003	0.091	0.059	0.059	0.593	0.162
桂 林	0.000	0.006	0.090	0.159	0.048	0.041
北 海	—	—	0.006	0.051	—	—
海 口	0.000	0.000	0.000	0.131	0.000	0.000
重 庆	0.000	0.000	0.204	0.507	0.073	1.358
成 都	0.000	0.001	0.100	0.221	0.009	0.096
自 贡	—	—	0.000	0.000	—	0.015
攀 枝 花	—	0.000	0.001	0.012	0.000	0.017
泸 州	—	—	0.009	0.019	—	—
德 阳	0.002	—	0.002	0.033	—	0.062
绵 阳	0.005	0.002	0.161	0.531	0.002	0.236
南 充	—	—	—	—	—	—
宜 宾	0.007	0.000	0.000	0.006	0.001	0.113
贵 阳	—	0.001	0.004	0.004	0.001	—
遵 义	0.003	0.000	0.006	0.071	0.001	0.000
昆 明	0.000	1.060	0.000	0.007	4.914	3.636
曲 靖	0.000	0.121	0.008	0.008	0.309	0.465
玉 溪	0.001	0.004	0.000	0.000	0.011	0.209
拉 萨	—	—	—	0.000	—	—
西 安	0.000	0.001	0.084	0.198	0.024	—
铜 川	—	—	—	—	—	—
宝 鸡	0.000	0.073	0.017	0.018	0.461	0.178
咸 阳	—	0.001	0.101	1.334	—	—
渭 南	—	0.000	—	—	0.001	—
延 安	0.000	0.004	0.020	0.020	0.006	0.040
兰 州	0.000	—	0.002	0.013	—	0.000
金 昌	0.023	0.519	—	—	0.548	1.619
西 宁	0.002	0.077	0.007	0.009	0.283	0.207
银 川	—	—	0.006	0.080	0.010	0.037
石 嘴 山	0.002	—	—	—	—	—
乌鲁木齐	0.012	0.007	0.130	0.130	0.016	0.076
克拉玛依	—	—	—	—	—	—

重点城市工业废气排放及处理情况（一）

（2012）

城 市 名 称	工业废气排放 总量（标态）/ （亿米³）	工业废气中污染物产生量/万吨			工业废气中污染物排放量/万吨		
		二氧化硫	氮氧化物	烟（粉）尘	二氧化硫	氮氧化物	烟（粉）尘
总　　计	358 691	3 108.7	963.8	41 225.9	970.4	878.6	462.5
北　京	3 264	15.9	9.8	422.0	5.9	8.5	3.1
天　津	9 032	62.4	29.7	775.9	21.5	27.6	5.9
石 家 庄	6 747	70.5	22.3	1 021.7	18.0	21.0	9.8
唐　山	27 439	66.6	32.3	2 238.7	31.3	31.2	41.0
秦 皇 岛	2 794	15.6	6.0	222.9	7.2	5.7	7.9
邯　郸	10 459	54.0	17.2	1 358.9	20.3	17.1	19.6
保　定	2 350	21.2	7.1	211.2	7.5	7.0	3.7
太　原	5 504	37.9	11.1	588.3	10.2	10.5	4.2
大　同	3 532	44.6	12.7	735.1	13.5	12.0	6.2
阳　泉	1 403	22.4	7.4	165.7	9.9	7.4	2.6
长　治	4 392	42.0	13.6	700.6	13.7	11.2	20.3
临　汾	4 582	39.7	7.8	526.5	9.7	6.3	9.6
呼和浩特	2 328	39.2	16.3	354.8	9.9	16.3	1.8
包　头	6 825	54.8	13.3	730.7	21.0	13.3	9.1
赤　峰	1 632	69.9	6.0	298.1	12.6	6.0	2.2
沈　阳	1 949	20.2	8.2	213.3	9.7	8.1	5.3
大　连	2 576	22.0	12.1	549.8	11.5	11.6	5.2
鞍　山	4 652	15.1	6.8	220.9	12.3	6.8	8.5
抚　顺	2 159	13.5	6.0	253.3	5.6	5.1	4.9
本　溪	5 518	8.2	5.4	258.2	7.6	5.4	6.1
锦　州	2 149	9.5	3.0	110.2	4.7	2.8	3.4
长　春	3 250	13.4	10.1	66.1	6.9	9.5	4.1
吉　林	1 883	11.2	9.3	492.3	8.2	9.1	3.6
哈 尔 滨	2 027	14.8	10.5	408.4	8.1	9.8	5.2
齐齐哈尔	1 677	8.0	8.1	111.6	6.3	8.0	5.9
牡 丹 江	660	4.1	4.4	185.9	2.3	4.4	3.9
上　海	13 361	53.5	32.5	725.4	19.3	28.5	6.4
南　京	6 828	97.3	16.6	572.2	11.9	12.2	3.9
无　锡	5 635	27.9	14.8	477.5	8.7	13.4	4.8
徐　州	5 144	43.2	23.2	864.1	14.3	17.9	4.5
常　州	2 974	10.7	8.3	244.7	3.6	8.2	2.8
苏　州	14 821	58.3	25.4	630.6	18.3	21.1	5.6
南　通	2 823	22.0	6.6	166.1	7.0	6.5	3.5
连 云 港	970	6.7	3.5	97.4	4.1	3.4	1.7
扬　州	1 297	18.7	7.3	156.7	4.6	7.0	1.2
镇　江	2 649	49.9	11.1	254.2	7.0	8.2	1.8
杭　州	4 295	16.2	7.6	409.7	8.6	7.5	3.3

重点城市工业废气排放及处理情况（一）（续表）

（2012）

城 市 名 称	工业废气排放总量（标态）/亿米³	工业废气中污染物产生量/万吨			工业废气中污染物排放量/万吨		
		二氧化硫	氮氧化物	烟（粉）尘	二氧化硫	氮氧化物	烟（粉）尘
宁 波	6 218	99.2	27.3	608.8	14.4	23.6	2.6
温 州	1 701	12.6	4.7	110.3	3.7	4.3	1.9
湖 州	1 907	9.3	5.8	418.6	3.8	5.8	2.5
绍 兴	1 431	12.9	4.8	150.9	5.9	4.7	1.6
合 肥	2 769	9.8	8.6	454.4	4.6	7.9	4.1
芜 湖	2 490	10.9	7.2	726.8	3.7	7.0	4.8
马 鞍 山	8 656	16.1	11.1	380.9	6.8	9.8	3.2
福 州	3 206	22.4	11.2	173.5	7.6	8.7	3.7
厦 门	1 178	5.0	2.5	49.1	1.9	1.3	0.3
泉 州	2 900	33.5	8.6	163.1	10.1	7.9	5.2
南 昌	1 199	8.8	2.5	76.5	4.3	2.5	1.1
九 江	2 498	22.6	6.2	221.3	9.2	6.1	3.4
济 南	4 040	28.4	8.5	470.8	10.3	8.1	5.2
青 岛	2 266	26.9	9.5	269.1	7.3	7.5	2.7
淄 博	5 030	60.3	13.8	783.8	21.9	13.2	5.7
枣 庄	2 201	21.9	7.0	549.2	7.5	7.0	2.4
烟 台	2 745	39.2	8.2	445.7	8.7	8.0	3.2
潍 坊	2 509	37.7	8.2	497.1	12.7	8.1	3.3
济 宁	3 790	48.4	17.0	639.1	13.1	15.9	4.6
泰 安	1 960	27.0	6.0	346.7	7.1	6.0	2.0
日 照	4 571	16.2	6.2	322.7	5.6	6.2	3.1
郑 州	3 916	29.5	18.7	823.8	11.0	15.3	2.4
开 封	831	10.6	3.2	397.3	4.2	3.2	2.2
洛 阳	6 564	50.1	14.3	1 268.7	13.6	13.7	4.6
平 顶 山	2 798	24.2	8.1	586.9	8.8	7.7	6.4
安 阳	3 443	31.0	6.4	326.3	11.6	6.1	8.0
焦 作	1 524	12.0	6.4	343.6	6.9	6.4	3.0
三 门 峡	1 528	23.2	7.2	325.4	10.8	7.2	2.4
武 汉	6 030	25.1	11.2	323.1	10.0	10.6	2.1
宜 昌	1 434	40.5	3.6	173.2	6.5	3.6	1.5
荆 州	810	11.2	1.6	54.8	4.8	1.2	1.5
长 沙	547	5.6	4.8	132.9	2.1	2.1	1.2
株 洲	1 258	28.8	3.0	182.7	4.0	3.0	0.8
湘 潭	2 408	11.5	4.3	666.4	4.3	4.2	2.4
岳 阳	1 068	13.2	6.5	203.6	6.0	5.9	2.8
常 德	1 145	10.6	4.8	167.2	3.9	4.6	1.9
张 家 界	147	2.2	0.6	25.1	2.0	0.6	0.3
广 州	3 217	43.8	9.6	306.1	6.8	5.6	1.4

重点城市工业废气排放及处理情况（一）（续表）

（2012）

城　市 名　称	工业废气排放 总量（标态）/ 亿米³	工业废气中污染物产生量/万吨			工业废气中污染物排放量/万吨		
		二氧化硫	氮氧化物	烟（粉）尘	二氧化硫	氮氧化物	烟（粉）尘
韶　关	1 373	9.8	3.9	29.5	5.0	3.5	0.7
深　圳	2 137	3.7	3.4	39.7	1.0	3.0	0.5
珠　海	1 373	9.0	5.0	74.5	3.0	5.0	1.1
汕　头	673	10.4	3.1	52.5	2.8	2.6	0.5
湛　江	1 009	9.5	3.7	116.6	2.5	2.2	1.0
南　宁	1 049	11.1	3.5	253.5	3.1	3.1	2.5
柳　州	3 419	18.7	5.1	321.2	5.4	5.0	5.2
桂　林	974	12.7	3.6	39.2	3.9	3.6	1.6
北　海	722	4.3	1.8	60.0	1.3	1.6	0.7
海　口	27	0.2	0.0	0.2	0.2	0.0	0.1
重　庆	8 360	128.5	27.3	1 746.8	51.0	27.2	16.6
成　都	2 971	12.2	6.1	350.9	5.7	5.7	2.5
自　贡	256	3.2	0.8	26.8	3.1	0.8	0.6
攀枝花	2 782	15.2	3.3	259.1	11.1	3.0	4.1
泸　州	938	17.6	2.6	122.0	4.7	2.5	0.8
德　阳	824	10.8	1.3	95.0	1.9	1.3	1.4
绵　阳	1 106	8.9	3.3	239.4	3.6	3.2	0.8
南　充	169	1.1	0.2	1.5	0.9	0.2	0.5
宜　宾	3 187	38.0	6.5	253.6	13.8	4.5	1.5
贵　阳	2 242	15.7	2.6	106.6	6.5	2.3	2.0
遵　义	3 415	32.0	5.0	228.7	8.6	4.2	0.9
昆　明	3 637	32.9	7.0	365.9	11.3	7.0	5.9
曲　靖	3 662	68.2	11.9	792.0	17.5	11.3	4.2
玉　溪	940	11.9	1.6	106.1	3.0	1.6	12.2
拉　萨	55	0.1	0.2	19.9	0.1	0.2	0.1
西　安	1 043	17.3	4.7	135.9	8.3	4.2	1.7
铜　川	1 185	9.2	9.6	84.6	1.8	5.3	4.2
宝　鸡	1 349	23.0	6.6	109.5	3.1	5.8	1.4
咸　阳	1 380	20.9	8.8	273.4	6.7	8.2	1.6
渭　南	3 887	91.3	17.5	513.8	26.3	16.5	3.6
延　安	248	2.0	0.5	13.6	1.8	0.5	2.1
兰　州	3 954	15.8	9.6	384.8	6.9	8.4	3.4
金　昌	833	150.6	2.4	66.3	10.5	2.4	1.2
西　宁	3 673	13.4	4.9	228.8	7.1	4.9	4.9
银　川	2 063	36.2	8.9	428.6	10.6	7.5	2.6
石嘴山	1 584	34.1	10.3	894.9	9.1	9.6	7.1
乌鲁木齐	3 629	24.0	11.4	336.5	11.8	11.4	5.7
克拉玛依	1 055	41.8	3.4	68.9	4.8	3.4	1.2

重点城市工业废气排放及处理情况（二）

（2012）

城 市 名 称	废气治理设施数/ 套	废气治理设施处理能力（标态）/ （万米³/时）	废气治理设施运行费用/ 万元
总　　计	126 733	994 289	8 336 730
北　京	3 004	10 901	87 848
天　津	3 911	22 382	257 709
石 家 庄	2 102	12 092	106 414
唐　山	3 628	91 651	538 541
秦 皇 岛	3 770	7 133	52 617
邯　郸	1 497	24 075	289 945
保　定	1 415	4 732	45 164
太　原	1 563	6 862	124 120
大　同	1 590	5 525	89 781
阳　泉	529	3 473	25 121
长　治	980	7 894	89 974
临　汾	962	6 912	98 832
呼 和 浩 特	1 101	8 975	35 379
包　头	865	11 903	161 180
赤　峰	724	3 678	39 637
沈　阳	2 062	12 361	37 406
大　连	1 960	6 883	46 698
鞍　山	1 547	12 353	92 873
抚　顺	458	4 523	30 275
本　溪	744	7 857	153 984
锦　州	568	2 853	19 101
长　春	1 109	6 743	45 064
吉　林	734	11 876	42 051
哈 尔 滨	1 206	16 524	30 415
齐 齐 哈 尔	776	4 986	15 910
牡 丹 江	225	963	6 961
上　海	4 795	32 703	405 020
南　京	1 495	10 829	210 557
无　锡	2 613	13 175	95 226
徐　州	1 478	9 387	145 933
常　州	1 009	10 596	52 226
苏　州	4 783	25 609	299 262
南　通	1 506	5 098	52 805
连 云 港	481	3 246	23 000
扬　州	542	4 631	54 733
镇　江	622	5 419	80 053
杭　州	2 336	9 281	96 431

重点城市工业废气排放及处理情况（二）（续表）

（2012）

城　市 名　称	废气治理设施数/ 套	废气治理设施处理能力（标态）/ （万米^3/时）	废气治理设施运行费用/ 万元
宁　波	2 953	107 329	362 335
温　州	2 368	5 836	65 182
湖　州	1 334	10 806	107 801
绍　兴	2 338	3 774	40 084
合　肥	626	2 186	37 295
芜　湖	688	4 353	41 349
马　鞍　山	611	12 991	155 705
福　州	670	10 511	96 625
厦　门	937	2 663	27 394
泉　州	1 107	5 844	61 232
南　昌	604	2 405	32 431
九　江	614	3 254	51 730
济　南	996	14 968	107 991
青　岛	1 221	8 378	81 247
淄　博	3 201	12 433	128 450
枣　庄	877	5 052	37 387
烟　台	1 219	8 375	60 973
潍　坊	1 536	9 751	64 035
济　宁	1 112	12 538	104 786
泰　安	664	3 861	36 161
日　照	925	7 922	132 898
郑　州	1 242	8 041	71 233
开　封	200	756	6 040
洛　阳	1 036	7 945	61 175
平　顶　山	478	3 938	19 890
安　阳	966	10 870	51 958
焦　作	936	4 843	33 912
三　门　峡	572	2 924	29 818
武　汉	897	9 049	110 161
宜　昌	919	66 025	29 186
荆　州	402	702	13 651
长　沙	419	727	4 370
株　洲	745	2 556	48 895
湘　潭	213	1 892	21 497
岳　阳	369	4 237	19 706
常　德	704	3 282	30 251
张　家　界	46	88	798
广　州	2 549	10 971	237 511

重点城市工业废气排放及处理情况（二）（续表）
（2012）

城 市 名 称	废气治理设施数/ 套	废气治理设施处理能力（标态）/ （万米³/时）	废气治理设施运行费用/ 万元
韶 关	672	4 927	54 572
深 圳	1 083	2 440	39 935
珠 海	1 104	4 129	32 075
汕 头	513	2 392	14 278
湛 江	353	2 346	39 193
南 宁	874	2 530	33 462
柳 州	946	7 923	77 466
桂 林	584	1 360	11 842
北 海	131	1 140	12 822
海 口	42	62	417
重 庆	4 319	19 990	202 828
成 都	1 988	5 350	63 337
自 贡	143	477	5 075
攀 枝 花	698	6 165	77 247
泸 州	284	8 817	33 869
德 阳	659	2 592	11 947
绵 阳	637	2 956	29 289
南 充	32	353	1 133
宜 宾	557	4 492	37 229
贵 阳	556	2 407	39 085
遵 义	472	4 041	59 219
昆 明	1 535	7 298	117 588
曲 靖	875	7 358	62 813
玉 溪	973	2 322	24 113
拉 萨	150	117	870
西 安	649	3 124	28 076
铜 川	350	1 407	5 669
宝 鸡	808	3 356	37 787
咸 阳	635	2 537	27 903
渭 南	708	11 434	92 125
延 安	197	318	5 357
兰 州	910	6 641	56 599
金 昌	192	1 066	51 772
西 宁	574	5 370	72 574
银 川	419	3 968	66 061
石 嘴 山	545	3 709	45 343
乌 鲁 木 齐	459	6 496	44 880
克 拉 玛 依	123	715	19 493

重点城市工业固体废物产生及处置利用情况

（2012）

单位：万吨

城 市 名 称	一般工业固体废物产生量	一般工业固体废物综合利用量	一般工业固体废物处置量	一般工业固体废物贮存量	一般工业固体废物倾倒丢弃量	一般工业固体废弃物综合利用率/%
总　　计	141 869	95 018	37 475	11 089	33	66.3
北　京	1 104	872	219	13	0	79.0
天　津	1 820	1 816	7	0	0	99.6
石 家 庄	759	743	9	7	0	97.9
唐　山	10 024	7 494	2 535	198	0	73.6
秦 皇 岛	2 130	1 253	679	198	0	58.8
邯　郸	3 162	3 164	16	161	0	94.7
保　定	994	923	70	0	0	92.9
太　原	2 787	1 499	1 259	20	10	53.8
大　同	4 001	3 148	144	720	0	78.5
阳　泉	1 908	417	1 425	64	2	21.8
长　治	2 440	1 636	516	293	2	66.8
临　汾	2 475	1 815	409	338	1	70.8
呼和浩特	1 122	401	716	5	0	35.7
包　头	2 945	1 405	1 200	340	0	47.7
赤　峰	2 196	397	1 590	214	0	18.0
沈　阳	704	668	138	40	0	92.6
大　连	707	675	32	0	0	95.5
鞍　山	5 109	1 266	1 329	2 513	0	24.8
抚　顺	2 839	663	1 833	343	0	23.4
本　溪	6 488	979	5 210	289	9	15.1
锦　州	283	206	75	2	0	72.6
长　春	470	469	1	0	0	99.8
吉　林	1 427	685	67	675	0	48.0
哈 尔 滨	571	573	3	0	0	99.5
齐齐哈尔	463	330	74	78	0	68.6
牡 丹 江	312	305	0	7	0	97.8
上　海	2 199	2 140	56	10	0	97.0
南　京	1 616	1 127	363	128	0	69.7
无　锡	979	889	66	24	0	90.8
徐　州	1 651	1 538	47	93	0	92.6
常　州	584	571	13	0	0	97.8
苏　州	2 214	2 179	35	0	0	98.4
南　通	447	439	8	1	0	98.1
连 云 港	482	444	3	48	0	89.7
扬　州	292	289	3	1	0	98.7
镇　江	740	728	4	8	0	98.3
杭　州	707	656	51	0	0	92.8

重点城市工业固体废物产生及处置利用情况（续表）

（2012）

单位：万吨

城 市 名 称	一般工业固体废物产生量	一般工业固体废物综合利用量	一般工业固体废物处置量	一般工业固体废物贮存量	一般工业固体废物倾倒丢弃量	一般工业固体废弃物综合利用率/%
宁　波	1 247	1 147	93	7	0	92.0
温　州	243	234	9	0	0	96.2
湖　州	178	171	7	0	0	96.0
绍　兴	380	349	20	12	0	91.6
合　肥	1 077	1 012	7	59	0	93.8
芜　湖	373	359	12	4	0	95.7
马 鞍 山	2 367	1 639	488	240	0	69.2
福　州	728	655	66	12	0	89.9
厦　门	114	110	11	0	0	90.9
泉　州	719	685	20	14	0	95.3
南　昌	186	184	2	0	0	98.7
九　江	913	439	89	389	0	48.0
济　南	1 012	1 011	0	1	0	99.8
青　岛	853	841	20	15	0	96.0
淄　博	1 824	1 771	10	52	0	96.6
枣　庄	750	759	1	0	0	99.9
烟　台	2 565	2 075	446	44	0	80.9
潍　坊	825	749	23	71	0	90.3
济　宁	2 183	1 838	217	137	0	83.9
泰　安	1 155	1 312	0	2	0	99.8
日　照	993	988	5	0	0	99.5
郑　州	1 500	1 140	330	32	0	75.9
开　封	150	142	10	2	0	93.7
洛　阳	2 993	1 686	1 294	12	0	56.3
平 顶 山	2 504	2 293	1	245	0	90.3
安　阳	1 192	1 088	104	0	0	91.3
焦　作	782	484	266	62	0	59.6
三 门 峡	1 667	578	1 003	111	0	34.2
武　汉	1 381	1 364	62	6	0	95.2
宜　昌	1 447	675	611	164	0	46.6
荆　州	147	147	0	0	0	100.0
长　沙	104	95	3	6	0	90.6
株　洲	358	319	32	24	0	85.1
湘　潭	693	696	6	25	0	95.7
岳　阳	397	368	6	39	0	89.2
常　德	306	296	2	9	0	96.5
张 家 界	44	42	0	2	0	96.5
广　州	615	589	22	4	0	95.7

重点城市工业固体废物产生及处置利用情况（续表）

（2012）

单位：万吨

城 市名 称	一般工业固体废物产生量	一般工业固体废物综合利用量	一般工业固体废物处置量	一般工业固体废物贮存量	一般工业固体废物倾倒丢弃量	一般工业固体废弃物综合利用率/%
韶 关	791	757	16	235	1	75.1
深 圳	105	94	10	1	0	89.6
珠 海	274	264	10	0	0	96.4
汕 头	108	106	1	1	0	97.7
湛 江	253	243	12	1	0	95.1
南 宁	356	326	140	13	0	91.2
柳 州	1 184	1 144	44	3	0	96.0
桂 林	189	166	6	17	0	87.7
北 海	267	220	48	0	0	82.1
海 口	6	6	1	0	0	91.6
重 庆	3 115	2 569	475	97	5	81.6
成 都	585	577	8	0	0	98.6
自 贡	85	83	2	0	0	97.2
攀 枝 花	5 949	1 354	4 488	112	0	22.8
泸 州	291	259	11	21	0	89.1
德 阳	362	232	150	0	0	60.8
绵 阳	265	269	3	1	0	98.2
南 充	15	15	0	0	0	98.6
宜 宾	601	524	57	20	0	87.2
贵 阳	1 122	664	444	17	0	59.0
遵 义	665	479	134	53	0	71.9
昆 明	3 003	1 356	1 604	106	0	44.4
曲 靖	2 265	1 422	207	638	0	62.7
玉 溪	2 445	852	1 303	296	0	34.8
拉 萨	315	6	26	298	0	1.9
西 安	258	249	8	2	0	96.2
铜 川	174	166	4	5	0	95.0
宝 鸡	412	251	161	0	0	60.9
咸 阳	501	460	21	22	0	91.6
渭 南	2 575	1 122	1 248	205	0	43.6
延 安	81	73	7	0	1	90.3
兰 州	628	624	23	2	0	96.2
金 昌	1 237	210	881	146	0	17.0
西 宁	505	511	2	10	0	97.7
银 川	729	592	29	117	0	80.3
石 嘴 山	640	413	220	7	0	64.5
乌鲁木齐	1 300	1 157	135	7	1	89.1
克拉玛依	96	68	28	0	0	70.7

重点城市工业危险废物产生及处置利用情况

（2012）

城 市 名 称	危险废物 产生量/万吨	危险废物综合 利用量/万吨	危险废物 处置量/万吨	危险废物 贮存量/万吨	危险废物倾倒 丢弃量/吨	危险废物处置 利用率/%
总　　　计	1 495.10	1 043.05	463.36	35.98	3.85	97.7
北　　京	13.41	4.50	8.91	0.00	0	100.0
天　　津	11.47	3.96	7.51	0.00	0	100.0
石 家 庄	19.69	3.99	15.51	0.26	0	98.7
唐　　山	19.96	19.10	0.86	0.00	0	100.0
秦 皇 岛	1.57	0.19	1.37	0.00	0	100.0
邯　　郸	0.67	0.59	0.08	0.00	0	100.0
保　　定	1.91	0.71	1.20	0.00	0	100.0
太　　原	4.85	2.07	2.78	0.01	0	99.8
大　　同	0.25	0.21	0.04	0.00	0	99.6
阳　　泉	0.21	0.11	0.00	0.10	0	54.1
长　　治	2.57	2.45	0.11	0.02	0	99.3
临　　汾	4.18	4.01	0.17	0.01	0	99.8
呼和浩特	0.17	0.14	0.03	0.00	0	100.0
包　　头	6.27	2.70	3.37	0.21	0	96.7
赤　　峰	1.49	0.54	1.19	0.00	0	100.0
沈　　阳	6.91	4.25	2.62	0.04	0	99.4
大　　连	15.58	4.74	10.85	0.00	0	100.0
鞍　　山	2.10	1.76	0.34	0.00	0	100.0
抚　　顺	6.56	3.75	2.68	0.28	0	95.9
本　　溪	10.39	10.26	0.12	0.00	2.00	100.0
锦　　州	5.56	12.19	0.29	0.00	0	100.0
长　　春	3.03	0.18	2.85	0.00	0	99.9
吉　　林	64.97	63.88	1.09	0.00	0	100.0
哈 尔 滨	2.71	1.23	1.49	0.00	0	100.0
齐齐哈尔	0.51	0.26	0.13	0.13	0	75.4
牡 丹 江	0.39	0.16	0.22	0.01	0	98.6
上　　海	54.96	30.34	24.60	0.14	0	99.7
南　　京	32.52	20.46	11.85	1.12	0	96.7
无　　锡	56.97	28.50	28.40	0.08	0	99.9
徐　　州	0.60	0.08	0.53	0.01	0	98.6
常　　州	9.78	5.94	3.84	0.00	0	100.0
苏　　州	62.78	28.54	34.46	0.10	0	99.8
南　　通	12.63	8.14	4.28	0.23	0	98.2
连 云 港	1.00	0.63	0.37	0.00	0	100.0
扬　　州	9.42	6.59	2.59	0.68	0	93.1
镇　　江	9.41	2.22	7.17	0.02	0	99.7
杭　　州	12.65	4.58	8.04	0.09	0	99.3

重点城市工业危险废物产生及处置利用情况（续表）

（2012）

单位：万吨

城 市 名 称	危险废物 产生量	危险废物 综合利用量	危险废物 处置量	危险废物 贮存量	危险废物倾倒 丢弃量	危险废物处置 利用率/%
宁 波	20.96	14.40	6.41	0.21	0	99.0
温 州	5.10	1.91	3.19	0.03	0	99.5
湖 州	7.76	0.38	7.21	0.23	0	97.1
绍 兴	17.07	1.92	15.23	0.20	0	98.9
合 肥	2.86	1.21	1.65	0.01	0	99.8
芜 湖	1.30	0.17	1.13	0.00	0	99.8
马 鞍 山	2.17	1.86	0.29	0.02	0	99.0
福 州	2.20	1.16	1.12	0.01	0	99.5
厦 门	2.51	0.58	1.95	0.02	0	99.3
泉 州	1.70	0.14	1.56	0.00	0	99.9
南 昌	2.47	2.00	0.47	0.00	0	100.0
九 江	5.43	5.00	0.44	0.04	0	99.3
济 南	6.14	4.16	6.19	0.03	0	99.7
青 岛	3.55	1.90	1.64	0.05	0	98.7
淄 博	172.94	158.51	14.36	0.17	0	99.9
枣 庄	1.36	1.30	0.06	0.00	0	100.0
烟 台	120.61	115.54	7.99	5.81	0	95.5
潍 坊	7.62	6.13	1.46	0.08	0	98.9
济 宁	76.98	76.54	0.44	0.01	0	100.0
泰 安	6.42	4.06	2.37	0.04	0	99.4
日 照	0.31	0.07	0.25	0.00	0	100.0
郑 州	0.58	0.23	0.35	0.01	0	98.6
开 封	0.19	0.14	0.02	0.02	0	87.2
洛 阳	1.28	0.93	0.31	0.04	0	96.8
平 顶 山	0.27	0.27	0.00	0.00	0	100.0
安 阳	2.42	2.40	0.01	0.01	0	99.7
焦 作	3.50	3.17	0.32	0.01	0	99.8
三 门 峡	18.38	16.88	1.30	0.20	0	98.9
武 汉	20.74	16.51	4.23	0.00	0	100.0
宜 昌	0.83	0.58	0.17	0.09	0	88.9
荆 州	1.00	0.93	0.08	0.00	0	99.8
长 沙	0.30	0.11	0.16	0.07	0	79.8
株 洲	14.63	13.90	0.73	0.00	0	100.0
湘 潭	8.25	7.78	0.42	0.05	0	99.4
岳 阳	87.34	51.32	35.88	0.14	0	99.8
常 德	1.10	0.03	1.06	0.00	0	99.6
张 家 界	0.25	0.25	0.00	0.00	0	100.0
广 州	29.78	12.11	17.68	0.00	0	100.0

重点城市工业固体废物产生及处置利用情况（续表）

（2012）

单位：万吨

城 市 名 称	危险废物 产生量	危险废物 综合利用量	危险废物 处置量	危险废物 贮存量	危险废物倾倒 丢弃量	危险废物处置 利用率/%
韶 关	20.28	17.86	2.17	0.29	0	98.6
深 圳	31.27	15.84	15.43	0.00	0	100.0
珠 海	9.12	3.93	5.20	0.00	0	100.0
汕 头	0.75	0.64	0.12	0.00	0	99.8
湛 江	0.24	0.03	0.21	0.00	0	100.0
南 宁	0.94	0.01	0.16	0.77	0	18.5
柳 州	2.08	1.22	0.68	0.18	0	91.2
桂 林	0.09	0.09	0.00	0.00	0	100.0
北 海	0.53	0.26	0.14	0.13	0	75.8
海 口	0.21	0.01	0.20	0.00	0	100.0
重 庆	49.03	37.18	11.80	0.05	0	99.9
成 都	13.24	1.96	11.26	0.03	0	99.8
自 贡	0.75	0.05	0.70	0.00	0	99.4
攀 枝 花	78.82	47.27	31.54	0.01	0	100.0
泸 州	0.03	0.00	0.03	0.00	0	100.0
德 阳	0.73	0.00	0.73	0.00	0	100.0
绵 阳	10.69	9.93	0.76	0.00	0	100.0
南 充	0.62	0.55	0.07	0.00	0	100.0
宜 宾	0.15	0.03	0.12	0.00	0	100.0
贵 阳	0.80	0.77	0.03	0.00	0	99.9
遵 义	0.42	0.28	0.14	0.00	0	99.8
昆 明	65.74	50.03	15.68	0.03	0	99.9
曲 靖	12.12	3.21	18.06	8.80	0	70.7
玉 溪	1.66	1.57	0.44	0.12	1.85	94.5
拉 萨	0	0	0	0	0	—
西 安	0.92	0.05	0.86	0.00	0	99.6
铜 川	0	0	0	0	0	—
宝 鸡	1.22	0.48	0.74	0.00	0	100.0
咸 阳	0.35	0.05	0.32	0.00	0	99.8
渭 南	9.76	9.21	0.53	0.04	0	99.6
延 安	7.45	0.27	7.34	0.12	0	98.4
兰 州	6.14	1.14	5.06	0.00	0	100.0
金 昌	1.42	1.84	0.17	0.04	0	98.1
西 宁	14.90	15.02	0.12	4.35	0	77.7
银 川	4.07	2.24	0.66	1.17	0	71.3
石 嘴 山	1.36	1.35	0.01	0.00	0	100.0
乌鲁木齐	16.41	7.27	0.97	8.17	0	50.3
克拉玛依	2.37	0.97	0.84	0.56	0	76.4

重点城市汇总工业企业概况（一）

（2012）

城 市 名 称	汇总工业企业数/个	工业总产值（现价）/亿元	工业炉窑数/台	工业锅炉数/台	35 蒸吨及以上的	20（含）～35 蒸吨之间的	10（含）～20 蒸吨之间的	10 蒸吨以下的
总 计	68 744	243 451.8	46 196	54 332	7 910	4 114	7 598	34 710
北 京	896	6 174.9	317	1 760	308	268	342	842
天 津	2 021	11 723.4	702	2 202	419	307	345	1 131
石 家 庄	1 450	2 244.9	791	1 124	139	38	106	841
唐 山	1 551	6 460.4	1 438	731	141	71	163	356
秦 皇 岛	315	1 247.2	111	393	52	25	60	256
邯 郸	612	2 872.9	703	383	106	26	48	203
保 定	1 541	2 104.7	571	1 069	32	33	169	835
太 原	310	1 785.1	400	463	79	51	96	237
大 同	659	971.5	333	973	56	60	98	759
阳 泉	315	446.9	471	232	25	9	17	181
长 治	608	1 570.4	497	636	83	59	71	423
临 汾	487	1 465.6	340	434	69	59	55	251
呼和浩特	446	548.5	69	780	133	82	239	326
包 头	132	1 841.3	381	234	51	25	42	116
赤 峰	490	392.5	255	481	81	49	108	243
沈 阳	935	2 681.9	283	1 660	266	193	404	797
大 连	909	3 833.8	445	1 258	206	224	206	622
鞍 山	627	2 828.5	2 259	592	127	74	124	267
抚 顺	255	1 106.9	391	342	46	49	64	183
本 溪	379	1 337.6	359	371	58	22	52	239
锦 州	214	899.4	353	378	67	73	59	179
长 春	304	1 961.8	143	705	214	85	123	283
吉 林	248	1 348.2	164	373	104	49	33	187
哈 尔 滨	365	1 379.0	114	698	176	95	139	288
齐齐哈尔	371	469.2	346	715	76	85	130	424
牡 丹 江	102	171.8	32	205	43	23	47	92
上 海	2 158	18 166.1	1 285	2 122	160	70	160	1 732
南 京	736	7 283.1	501	467	74	20	41	332
无 锡	1 363	6 945.3	809	912	153	16	71	672
徐 州	623	1 892.6	254	356	86	27	44	199
常 州	1 104	4 224.0	477	566	68	8	31	459
苏 州	1 842	9 991.9	661	1 701	216	92	238	1 155
南 通	1 377	2 168.3	520	694	72	14	59	549
连 云 港	255	970.4	36	223	35	11	51	126
扬 州	559	1 699.4	157	259	43	5	16	195
镇 江	608	1 724.2	148	304	40	8	16	240
杭 州	1 475	4 306.3	489	994	169	44	157	624

重点城市汇总工业企业概况（一）（续表）

（2012）

城 市 名 称	汇总工业企业数/个	工业总产值（现价）/亿元	工业炉窑数/台	工业锅炉数/台	35 蒸吨及以上的	20（含）～35蒸吨之间的	10（含）～20蒸吨之间的	10 蒸吨以下的
宁　波	1 351	6 455.8	960	858	105	45	65	643
温　州	1 940	1 269.4	684	1 085	11	17	94	963
湖　州	1 152	1 235.2	375	595	47	15	47	486
绍　兴	1 024	3 025.7	284	1 082	122	46	129	785
合　肥	873	2 049.5	406	354	42	12	37	263
芜　湖	587	1 483.8	269	204	12	14	21	157
马 鞍 山	593	1 560.7	402	143	49	1	22	71
福　州	467	1 623.5	279	393	16	16	83	278
厦　门	370	2 387.2	91	182	17	7	34	124
泉　州	1 294	1 974.5	928	827	40	43	140	604
南　昌	325	821.5	158	282	7	24	47	204
九　江	579	915.9	465	211	26	7	34	144
济　南	629	1 946.5	619	359	132	30	81	116
青　岛	743	4 632.3	198	608	160	75	64	309
淄　博	1 038	3 436.0	1 199	516	197	55	64	200
枣　庄	219	518.2	113	233	75	6	20	132
烟　台	591	3 835.2	239	608	132	43	78	355
潍　坊	878	4 483.8	227	839	227	67	105	440
济　宁	456	2 443.9	392	467	153	24	36	254
泰　安	262	879.0	101	310	88	12	45	165
日　照	136	1 385.0	119	140	50	8	9	73
郑　州	626	1 575.9	593	468	55	32	53	328
开　封	392	311.8	274	147	11	8	28	100
洛　阳	674	1 849.9	602	265	50	22	31	162
平 顶 山	237	828.9	211	147	45	16	16	70
安　阳	463	1 154.7	479	261	28	22	20	191
焦　作	429	812.8	364	455	57	26	23	349
三 门 峡	378	523.4	529	182	44	5	30	103
武　汉	446	4 668.0	279	389	54	17	47	271
宜　昌	358	922.2	359	302	34	27	47	194
荆　州	480	556.6	167	315	17	7	29	262
长　沙	419	1 152.5	116	259	6	5	12	236
株　洲	542	925.9	609	225	15	9	21	180
湘　潭	195	744.2	162	129	16	0	17	96
岳　阳	278	1 310.1	115	232	25	7	36	164
常　德	218	834.1	77	219	69	13	21	116
张 家 界	70	13.8	43	22	2	0	1	19
广　州	1 204	7 095.3	287	860	86	29	138	607

重点城市汇总工业企业概况（一）（续表）

（2012）

城市名称	汇总工业企业数/个	工业总产值（现价）/亿元	工业炉窑数/台	工业锅炉数/台	35蒸吨及以上的	20（含）～35蒸吨之间的	10（含）～20蒸吨之间的	10蒸吨以下的
韶　关	339	626.7	173	134	16	7	20	91
深　圳	1 847	8 392.8	39	443	24	5	48	366
珠　海	520	2 797.8	74	346	20	8	37	281
汕　头	547	460.9	22	512	11	9	98	394
湛　江	320	739.6	103	234	50	40	19	125
南　宁	383	441.9	108	325	45	34	33	213
柳　州	357	1 771.0	309	229	56	23	13	137
桂　林	470	356.0	292	192	4	7	32	149
北　海	83	520.8	59	62	15	5	23	19
海　口	99	289.4	29	76	2	1	6	67
重　庆	3 107	6 814.8	1 947	1 456	112	53	125	1 166
成　都	1 303	3 545.5	2 315	960	29	32	100	799
自　贡	345	355.2	178	124	11	7	5	101
攀枝花	223	899.1	366	100	28	8	12	52
泸　州	598	606.1	252	360	21	3	11	325
德　阳	369	745.6	584	231	11	19	34	167
绵　阳	574	1 090.5	414	261	10	7	33	211
南　充	479	138.8	258	161	3	6	8	144
宜　宾	312	786.9	182	225	26	31	15	153
贵　阳	200	1 003.0	1 112	126	36	21	20	49
遵　义	240	870.5	1 198	230	10	26	25	169
昆　明	591	2 026.9	355	332	44	22	40	226
曲　靖	492	1 045.6	297	206	55	13	32	106
玉　溪	270	1 598.5	176	147	17	11	23	96
拉　萨	41	33.5	15	19	1	0	0	18
西　安	483	2 040.9	748	567	87	93	105	282
铜　川	109	163.3	41	69	3	8	3	55
宝　鸡	211	1 266.5	90	264	38	34	93	99
咸　阳	558	963.4	398	346	40	14	52	240
渭　南	404	1 051.7	257	322	53	12	49	208
延　安	173	532.8	125	322	20	8	63	231
兰　州	251	901.6	897	413	35	65	158	155
金　昌	38	789.4	80	80	29	10	1	40
西　宁	139	586.7	282	183	18	17	30	118
银　川	141	713.6	106	338	84	66	69	119
石嘴山	271	489.4	607	371	43	12	57	259
乌鲁木齐	244	1 389.8	184	379	170	42	22	145
克拉玛依	55	1 722.8	145	224	38	74	35	77

重点城市汇总工业企业概况（二）

（2012）

单位：万吨

城 市名 称	工业用水总量	取水量	重复用水量	工业煤炭消耗量	燃料煤
总　　计	24 682 947	3 018 819	21 664 129	202 591	150 103
北　京	325 063	18 462	306 601	1 612	1 513
天　津	1 004 487	40 473	964 013	4 583	4 319
石 家 庄	676 402	43 557	632 845	4 038	3 707
唐　山	1 510 219	62 453	1 447 767	8 504	3 607
秦 皇 岛	45 847	9 851	35 997	1 012	874
邯　郸	597 020	26 478	570 541	5 650	2 391
保　定	385 877	25 153	360 724	1 669	1 323
太　原	320 319	12 916	307 403	4 437	1 791
大　同	188 541	13 822	174 719	2 803	2 530
阳　泉	505 997	4 544	501 453	928	928
长　治	195 876	17 900	177 976	6 041	2 238
临　汾	195 022	20 794	174 229	5 732	1 736
呼和浩特	92 515	8 085	84 430	2 888	2 807
包　头	635 858	26 006	609 851	3 525	2 442
赤　峰	151 681	6 441	145 239	1 504	1 402
沈　阳	118 200	13 373	104 827	1 793	1 783
大　连	100 279	88 550	11 729	2 177	2 136
鞍　山	395 059	37 907	357 153	2 097	1 013
抚　顺	285 609	13 266	272 343	1 104	925
本　溪	228 698	20 845	207 853	1 562	426
锦　州	102 380	10 981	91 400	681	673
长　春	178 546	16 935	161 611	2 659	2 023
吉　林	315 633	37 993	277 640	2 310	1 653
哈 尔 滨	217 956	12 360	205 595	2 115	1 976
齐齐哈尔	165 992	75 446	90 546	1 320	1 167
牡 丹 江	55 251	7 834	47 417	772	747
上　海	860 514	81 673	778 841	5 419	4 087
南　京	718 936	181 184	537 752	3 561	3 361
无　锡	150 977	37 822	113 154	2 561	2 484
徐　州	561 763	32 497	529 267	3 880	3 407
常　州	165 640	21 324	144 316	1 141	1 132
苏　州	882 091	96 934	785 157	5 926	4 660
南　通	154 193	30 125	124 067	1 791	1 778
连 云 港	98 184	10 611	87 573	485	420
扬　州	133 772	16 325	117 447	1 194	1 193
镇　江	114 123	13 928	100 195	2 094	1 975
杭　州	313 086	61 179	251 906	1 471	1 319

重点城市汇总工业企业概况（二）（续表）

（2012）

单位：万吨

城 市 名 称	工业用水 总量	取水量	重复用水量	工业煤炭 消耗量	燃料煤
宁 波	961 111	466 658	494 453	4 499	4 348
温 州	33 398	11 030	22 368	1 031	1 019
湖 州	53 409	15 835	37 574	778	569
绍 兴	177 771	44 532	133 238	1 169	1 102
合 肥	206 094	12 550	193 544	1 377	1 322
芜 湖	41 243	7 814	33 429	1 457	830
马 鞍 山	301 075	13 132	287 943	2 101	1 243
福 州	57 726	14 853	42 873	1 845	1 719
厦 门	89 959	7 233	82 726	415	415
泉 州	189 601	28 663	160 938	1 420	1 331
南 昌	61 576	14 627	46 950	554	371
九 江	123 894	50 179	73 715	667	499
济 南	276 020	13 292	262 728	1 882	1 360
青 岛	151 174	27 301	123 873	1 436	1 319
淄 博	541 589	32 212	509 377	3 271	2 742
枣 庄	157 656	13 591	144 065	1 867	956
烟 台	55 264	15 664	39 599	1 646	1 514
潍 坊	182 441	34 853	147 587	2 323	1 889
济 宁	561 517	27 174	534 343	3 999	3 425
泰 安	93 497	14 562	78 934	1 305	979
日 照	227 103	13 854	213 249	1 268	698
郑 州	264 284	24 774	239 510	2 689	2 657
开 封	194 113	11 081	183 032	613	506
洛 阳	277 384	20 413	256 971	2 543	2 446
平 顶 山	246 999	15 483	231 516	5 487	2 128
安 阳	226 323	16 530	209 794	1 438	768
焦 作	101 540	21 323	80 217	913	658
三 门 峡	159 306	7 807	151 499	1 483	975
武 汉	655 052	125 644	529 407	2 297	1 009
宜 昌	182 972	27 907	155 065	1 239	529
荆 州	18 179	13 868	4 311	357	328
长 沙	8 861	5 518	3 343	479	453
株 洲	60 324	7 597	52 727	474	343
湘 潭	170 698	7 129	163 569	944	391
岳 阳	165 355	16 838	148 517	888	841
常 德	81 650	14 983	66 666	708	656
张 家 界	282	210	72	68	54
广 州	86 794	32 730	54 065	1 671	1 632

重点城市汇总工业企业概况（二）（续表）

（2012）   单位：万吨

城 市 名 称	工业用水 总量	取水量	重复用水量	工业煤炭 消耗量	燃料煤
韶 关	172 455	13 787	158 667	1 034	667
深 圳	24 378	17 496	6 883	435	435
珠 海	42 229	10 588	31 641	655	629
汕 头	9 942	6 812	3 130	1 037	999
湛 江	49 485	10 303	39 182	648	618
南 宁	97 380	14 195	83 185	513	496
柳 州	366 814	37 385	329 429	1 297	406
桂 林	57 353	7 491	49 862	368	277
北 海	40 817	3 166	37 651	248	209
海 口	4 120	1 160	2 960	1	1
重 庆	556 557	136 872	419 685	4 140	3 390
成 都	174 755	18 691	156 064	866	824
自 贡	37 841	2 987	34 854	156	141
攀 枝 花	123 279	16 025	107 254	1 102	618
泸 州	141 798	7 767	134 031	469	439
德 阳	50 799	8 431	42 368	216	186
绵 阳	80 821	10 060	70 762	499	373
南 充	8 922	3 756	5 166	62	40
宜 宾	42 632	10 272	32 360	1 157	950
贵 阳	141 883	4 791	137 092	724	404
遵 义	77 783	4 146	73 637	698	682
昆 明	215 198	15 925	199 273	1 834	1 314
曲 靖	276 335	11 544	264 792	3 240	2 639
玉 溪	36 754	7 478	29 277	344	205
拉 萨	1 279	584	695	20	19
西 安	48 643	15 425	33 218	838	801
铜 川	20 181	625	19 557	485	398
宝 鸡	46 127	13 905	32 222	913	886
咸 阳	17 409	10 582	6 827	1 339	1 214
渭 南	281 926	13 384	268 542	3 396	2 201
延 安	6 541	4 968	1 573	156	150
兰 州	55 037	17 391	37 645	1 362	1 197
金 昌	75 833	6 713	69 120	472	470
西 宁	119 640	8 671	110 970	799	572
银 川	97 979	9 027	88 953	1 982	1 965
石 嘴 山	74 834	8 435	66 399	2 806	1 241
乌鲁木齐	222 487	14 864	207 623	2 241	1 681
克拉玛依	193 862	7 648	186 214	396	396

重点城市汇总工业企业概况（三）

（2012）

城 市 名 称	燃料油消耗量/ 万吨	焦炭消耗量/ 万吨	天然气消耗量/ 亿米³	其他燃料消耗量/ 万吨标煤
总　　计	1 292	18 460	1 358	8 269
北　京	40	1	37	141
天　津	9	520	19	175
石 家 庄	5	472	1	107
唐　山	1	3 271	26	478
秦 皇 岛	0	282	3	1
邯　郸	1	1 439	2	96
保　定	0	75	25	26
太　原	1	265	3	103
大　同	0	58	0	30
阳　泉	0	1	0	6
长　治	0	239	0	188
临　汾	0	369	0	105
呼 和 浩 特	3	12	0	23
包　头	0	580	2	241
赤　峰	0	10	0	12
沈　阳	6	1	3	7
大　连	58	5	3	147
鞍　山	9	661	1	329
抚　顺	15	155	21	3
本　溪	0	836	0	314
锦　州	5	17	2	40
长　春	3	2	1	12
吉　林	14	214	5	44
哈 尔 滨	3	47	3	15
齐 齐 哈 尔	0	29	1	18
牡 丹 江	0	0	0	2
上　海	43	119	37	642
南　京	22	466	22	370
无　锡	7	258	4	20
徐　州	591	135	245	6
常　州	1	57	12	11
苏　州	56	873	34	35
南　通	4	5	0	61
连 云 港	0	114	1	21
扬　州	0	3	84	49
镇　江	2	103	156	554
杭　州	3	119	17	22

重点城市汇总工业企业概况（三）（续表）

（2012）

城市名称	燃料油消耗量/万吨	焦炭消耗量/万吨	天然气消耗量/亿米³	其他燃料消耗量/万吨标煤
宁　波	80	138	28	31
温　州	4	2	0	3
湖　州	7	4	2	43
绍　兴	11	0	10	1
合　肥	1	60	1	45
芜　湖	1	63	3	7
马鞍山	0	631	1	378
福　州	5	184	3	5
厦　门	13	0	5	4
泉　州	15	66	12	12
南　昌	1	133	0	110
九　江	1	165	1	40
济　南	2	407	2	50
青　岛	5	100	2	69
淄　博	21	330	15	118
枣　庄	0	6	0	10
烟　台	2	19	3	22
潍　坊	2	150	1	60
济　宁	0	0	57	37
泰　安	0	119	1	15
日　照	7	504	3	0
郑　州	2	1	10	0
开　封	0	1	0	1
洛　阳	16	16	2	17
平顶山	0	36	1	19
安　阳	0	458	1	6
焦　作	2	5	2	43
三门峡	0	6	2	0
武　汉	9	0	10	24
宜　昌	0	18	5	7
荆　州	0	0	0	47
长　沙	0	1	1	3
株　洲	2	14	3	11
湘　潭	0	3	114	20
岳　阳	1	1	3	21
常　德	0	1	0	18
张家界	0	0	0	0
广　州	70	102	8	92

重点城市汇总工业企业概况（三）（续表）

（2012）

城 市 名 称	燃料油消耗量/ 万吨	焦炭消耗量/ 万吨	天然气消耗量/ 亿米³	其他燃料消耗量/ 万吨标煤
韶 关	0	239	0	46
深 圳	25	0	26	20
珠 海	20	72	3	4
汕 头	1	0	0	1
湛 江	2	0	0	77
南 宁	5	1	0	54
柳 州	1	435	0	362
桂 林	0	29	0	23
北 海	0	39	2	15
海 口	0	0	8	2
重 庆	2	245	25	197
成 都	1	75	18	70
自 贡	0	1	2	2
攀 枝 花	0	296	0	910
泸 州	0	0	10	15
德 阳	0	2	4	1
绵 阳	0	9	2	4
南 充	1	1	1	3
宜 宾	0	3	1	19
贵 阳	5	14	2	18
遵 义	0	31	3	0
昆 明	6	180	32	22
曲 靖	1	215	1	15
玉 溪	0	332	0	2
拉 萨	11	0	0	0
西 安	2	1	2	0
铜 川	0	0	0	0
宝 鸡	0	8	0	5
咸 阳	2	0	3	1
渭 南	1	48	0	81
延 安	9	0	1	0
兰 州	2	89	70	109
金 昌	7	25	0	0
西 宁	0	93	11	27
银 川	1	2	3	0
石 嘴 山	0	44	1	9
乌鲁木齐	0	375	11	231
克拉玛依	1	0	29	79

重点城市工业污染防治投资情况（一）

（2012）

城市名称	汇总工业企业数	本年施工项目总数	工业废水治理项目	工业废气治理项目	脱硫治理项目	脱硝治理项目	工业固体废物治理项目	噪声治理项目	其他治理项目
总　　计	3 309	2 899	959	1 178	287	161	151	63	548
北　京	41	51	10	27	11	1	2	1	11
天　津	110	135	50	50	9	3	2	8	25
石 家 庄	26	20	5	10	6	2	1	0	4
唐　山	44	47	6	25	10	1	0	0	16
秦 皇 岛	9	7	1	6	4	1	0	0	0
邯　郸	17	20	3	16	2	2	0	0	1
保　定	8	6	1	5	1	0	0	0	0
太　原	27	28	2	14	7	0	8	0	4
大　同	46	46	6	19	12	2	15	2	4
阳　泉	15	24	3	7	3	2	1	0	13
长　治	20	27	3	8	2	3	1	2	13
临　汾	16	21	12	7	6	1	2	0	0
呼和浩特	31	71	31	32	6	4	2	0	6
包　头	12	17	3	13	4	1	0	0	1
赤　峰	9	4	1	2	0	0	0	0	1
沈　阳	8	2	1	0	0	0	0	0	1
大　连	19	4	3	0	0	0	0	0	1
鞍　山	2	0	0	0	0	0	0	0	0
抚　顺	9	15	3	10	1	1	0	0	2
本　溪	8	5	3	2	0	0	0	0	0
锦　州	5	4	1	2	2	0	0	1	0
长　春	16	10	2	5	1	1	0	0	3
吉　林	13	13	2	10	2	2	0	0	1
哈 尔 滨	7	7	0	4	2	1	3	0	0
齐齐哈尔	10	10	1	8	1	0	0	0	1
牡 丹 江	2	3	1	2	1	0	0	0	0
上　海	122	142	39	47	10	3	6	8	42
南　京	53	53	24	18	4	0	1	2	8
无　锡	25	24	5	17	0	5	0	0	2
徐　州	15	13	3	10	5	2	0	0	0
常　州	55	55	11	37	3	10	1	1	5
苏　州	57	63	20	28	4	5	3	3	9
南　通	65	48	24	16	3	3	2	0	6
连 云 港	24	10	4	5	1	1	1	0	0
扬　州	13	11	5	5	1	4	0	0	1
镇　江	21	10	7	1	0	0	0	0	2
杭　州	42	41	8	24	5	8	3	2	4

重点城市工业污染防治投资情况（一）（续表）

（2012）

单位：个

城 市 名 称	汇总工业 企业数	本年施工 项目总数	工业废水 治理项目	工业废气 治理项目	脱硫治理 项目	脱硝治理 项目	工业固体废 物治理项目	噪声治理 项目	其他治理 项目
宁 波	67	49	16	27	4	4	2	0	4
温 州	127	47	10	32	0	2	0	0	5
湖 州	51	41	14	12	0	1	0	1	14
绍 兴	110	97	38	36	7	8	3	1	19
合 肥	56	25	16	6	2	2	1	0	2
芜 湖	9	4	1	3	0	2	0	0	0
马 鞍 山	7	4	1	3	1	0	0	0	0
福 州	23	31	6	9	3	2	0	1	15
厦 门	51	36	11	12	1	1	0	0	13
泉 州	124	119	36	33	9	1	37	2	11
南 昌	6	10	8	0	0	0	0	2	0
九 江	11	9	4	3	0	1	0	0	2
济 南	23	24	9	7	3	1	2	0	6
青 岛	26	25	4	11	6	0	1	1	8
淄 博	72	93	23	58	14	6	2	0	10
枣 庄	31	25	8	5	2	1	0	0	12
烟 台	49	45	17	11	3	0	5	1	11
潍 坊	91	87	38	43	11	5	0	0	6
济 宁	29	37	5	21	5	3	0	0	11
泰 安	19	11	6	5	5	0	0	0	0
日 照	15	9	3	6	3	1	0	0	0
郑 州	19	21	5	12	3	0	2	1	1
开 封	3	2	1	1	1	0	0	0	0
洛 阳	14	11	4	6	1	0	0	0	1
平 顶 山	11	10	2	4	1	0	3	0	1
安 阳	11	9	4	5	4	0	0	0	0
焦 作	16	19	12	2	0	0	0	1	4
三 门 峡	8	4	2	0	0	0	2	0	0
武 汉	2	1	0	0	0	0	0	0	1
宜 昌	40	29	13	14	3	0	0	0	2
荆 州	5	7	0	3	1	0	1	2	1
长 沙	19	13	7	5	1	2	0	0	1
株 洲	21	22	8	11	3	1	0	0	3
湘 潭	7	7	2	4	0	2	0	0	1
岳 阳	8	2	2	0	0	0	0	0	0
常 德	16	6	3	3	2	1	0	0	0
张 家 界	1	0	0	0	0	0	0	0	0
广 州	50	40	9	16	2	5	0	2	13

重点城市工业污染防治投资情况（一）（续表）

（2012） 单位：个

城　市 名　称	汇总工业 企业数	本年施工 项目总数	工业废水 治理项目	工业废气 治理项目	脱硫治理 项目	脱硝治理 项目	工业固体废 物治理项目	噪声治理 项目	其他治理 项目
韶　关	36	36	16	14	4	2	3	0	3
深　圳	12	14	4	6	0	3	0	0	4
珠　海	55	13	5	2	0	0	1	0	5
汕　头	14	11	2	6	1	2	0	0	3
湛　江	20	18	8	7	2	1	0	3	0
南　宁	18	24	6	1	0	0	1	0	16
柳　州	45	42	19	9	1	0	2	0	12
桂　林	8	5	3	2	0	0	0	0	0
北　海	11	7	3	1	0	1	1	0	2
海　口	13	3	3	0	0	0	0	0	0
重　庆	168	111	49	36	5	8	3	4	19
成　都	59	47	9	14	1	2	0	2	22
自　贡	3	0	0	0	0	0	0	0	0
攀枝花	33	29	7	11	3	0	6	1	4
泸　州	36	28	28	0	0	0	0	0	0
德　阳	12	10	8	1	1	0	0	0	1
绵　阳	34	28	5	14	0	0	1	1	7
南　充	4	0	0	0	0	0	0	0	0
宜　宾	12	11	7	2	2	0	0	0	2
贵　阳	32	28	13	11	4	1	1	0	3
遵　义	71	17	9	5	0	1	2	0	1
昆　明	71	62	18	21	4	2	3	2	18
曲　靖	29	17	10	3	0	0	0	1	3
玉　溪	12	10	3	3	0	0	0	0	4
拉　萨	9	9	1	3	0	0	2	0	3
西　安	13	13	4	2	0	1	3	1	3
铜　川	0	0	0	0	0	0	0	0	0
宝　鸡	16	23	10	8	3	2	0	2	3
咸　阳	44	28	9	15	11	4	0	0	4
渭　南	53	57	29	15	5	4	1	0	12
延　安	14	25	9	0	0	0	2	1	13
兰　州	4	2	0	1	1	0	0	0	1
金　昌	5	5	1	3	0	0	0	0	1
西　宁	25	19	5	5	1	2	1	0	8
银　川	14	13	2	10	3	0	0	0	1
石嘴山	26	29	1	20	3	0	3	0	5
乌鲁木齐	8	7	1	6	2	4	0	0	0
克拉玛依	0	0	0	0	0	0	0	0	0

重点城市工业污染防治投资情况（二）

（2012）

单位：个

城 市 名 称	本年竣工 项目总数	工业废水 治理项目	工业废气 治理项目	脱硫治理 项目	脱硝治理 项目	工业固体废 物治理项目	噪声治理 项目	其他治理 项目
总　　计	2 990	1 100	1 155	286	114	129	70	536
北　京	52	13	25	8	1	2	1	11
天　津	73	20	26	5	3	2	6	19
石 家 庄	25	7	12	8	1	1	0	5
唐　山	54	10	28	10	2	1	0	15
秦 皇 岛	9	1	8	2	1	0	0	0
邯　郸	23	4	18	4	2	0	0	1
保　定	6	3	3	1	0	0	0	0
太　原	22	2	13	7	1	3	0	4
大　同	60	10	32	16	2	12	0	6
阳　泉	19	2	6	3	1	1	0	10
长　治	22	2	3	1	0	1	2	14
临　汾	11	7	3	3	0	1	0	0
呼 和 浩 特	56	29	17	4	1	2	1	7
包　头	15	2	12	4	1	0	0	1
赤　峰	13	9	2	0	0	0	0	2
沈　阳	5	4	0	0	0	0	0	1
大　连	5	5	0	0	0	0	0	0
鞍　山	2	0	2	0	0	0	0	0
抚　顺	8	2	6	1	0	0	0	0
本　溪	6	3	3	0	0	0	0	0
锦　州	4	2	1	1	0	0	1	0
长　春	8	1	4	1	1	0	0	3
吉　林	17	4	12	1	2	1	0	0
哈 尔 滨	8	0	4	2	1	3	0	1
齐 齐 哈 尔	13	2	9	1	0	0	0	2
牡 丹 江	2	1	1	0	0	0	0	0
上　海	158	41	54	10	2	6	8	49
南　京	53	21	18	4	0	1	2	11
无　锡	25	4	19	0	5	0	0	2
徐　州	14	4	10	5	2	0	0	0
常　州	52	13	33	1	10	1	1	4
苏　州	69	32	27	3	2	1	2	7
南　通	57	29	18	2	2	3	2	5
连 云 港	16	10	5	0	0	1	0	0
扬　州	12	4	7	3	3	0	0	1
镇　江	5	3	2	1	0	0	0	0
杭　州	37	6	24	5	7	2	2	3

重点城市工业污染防治投资情况（二）（续表）

（2012）

单位：个

城 市 名 称	本年竣工项目总数	工业废水治理项目	工业废气治理项目	脱硫治理项目	脱硝治理项目	工业固体废物治理项目	噪声治理项目	其他治理项目
宁　波	66	28	29	6	1	1	1	7
温　州	40	9	23	0	2	0	0	8
湖　州	50	19	14	0	0	0	1	16
绍　兴	88	39	30	5	5	2	1	16
合　肥	27	16	8	2	2	0	0	3
芜　湖	4	2	2	0	1	0	0	0
马 鞍 山	8	4	4	0	1	0	0	0
福　州	28	4	8	2	1	0	1	15
厦　门	39	15	13	1	0	0	0	11
泉　州	93	24	18	6	0	36	5	10
南　昌	11	8	1	0	0	0	2	0
九　江	11	5	3	0	0	0	0	3
济　南	29	14	9	4	0	2	0	4
青　岛	24	4	9	4	1	1	1	9
淄　博	68	23	38	9	1	0	0	7
枣　庄	30	10	6	4	0	0	0	14
烟　台	44	22	12	4	0	5	1	4
潍　坊	83	36	41	11	3	0	0	6
济　宁	31	3	18	4	2	0	1	9
泰　安	14	8	5	5	0	0	0	1
日　照	16	5	9	4	1	0	0	2
郑　州	17	3	11	2	0	2	1	0
开　封	2	1	1	1	0	0	0	0
洛　阳	12	3	8	3	0	0	0	1
平 顶 山	10	2	3	1	0	2	0	3
安　阳	9	5	4	3	0	0	0	0
焦　作	13	4	4	1	0	0	2	3
三 门 峡	7	3	3	2	1	0	0	1
武　汉	8	1	1	0	0	1	1	4
宜　昌	28	12	12	1	0	1	0	3
荆　州	9	0	5	3	0	1	2	1
长　沙	10	4	5	2	1	0	0	1
株　洲	17	3	11	4	1	0	0	3
湘　潭	7	2	4	0	1	0	0	1
岳　阳	4	1	3	1	1	0	0	0
常　德	8	5	3	2	1	0	0	0
张 家 界	1	1	0	0	0	0	0	0
广　州	50	16	19	1	5	0	3	12

重点城市工业污染防治投资情况（二）（续表）

（2012）

单位：个

城 市 名 称	本年竣工 项目总数	工业废水 治理项目	工业废气 治理项目	脱硫治理 项目	脱硝治理 项目	工业固体废 物治理项目	噪声治理 项目	其他治理 项目
韶　关	35	15	13	5	0	2	0	5
深　圳	13	4	6	0	3	0	0	3
珠　海	14	4	4	0	0	1	0	5
汕　头	18	7	8	1	2	0	0	3
湛　江	18	8	7	2	1	0	3	0
南　宁	27	7	2	0	0	2	0	16
柳　州	48	24	10	3	0	3	0	11
桂　林	3	2	1	0	0	0	0	0
北　海	10	7	0	0	0	1	0	2
海　口	18	9	9	0	0	0	0	0
重　庆	150	78	43	8	6	4	4	21
成　都	59	12	24	1	2	0	2	21
自　贡	1	1	0	0	0	0	0	0
攀枝花	21	5	9	2	0	3	2	2
泸　州	29	29	0	0	0	0	0	0
德　阳	11	7	3	2	0	0	0	1
绵　阳	32	10	14	0	0	0	1	7
南　充	0	0	0	0	0	0	0	0
宜　宾	13	9	2	2	0	0	0	2
贵　阳	40	23	13	6	1	1	1	2
遵　义	82	67	9	3	0	2	1	3
昆　明	58	17	22	3	2	2	1	16
曲　靖	16	7	4	0	0	0	0	5
玉　溪	13	3	6	0	0	0	0	4
拉　萨	11	2	3	0	0	3	0	3
西　安	17	4	6	1	3	3	1	3
铜　川	0	0	0	0	0	0	0	0
宝　鸡	16	6	5	2	2	0	2	3
咸　阳	36	15	17	12	3	0	0	4
渭　南	43	21	10	5	1	0	0	12
延　安	19	7	1	1	0	0	1	10
兰　州	3	1	1	1	0	0	0	1
金　昌	2	1	1	0	0	0	0	0
西　宁	13	4	3	2	0	1	0	5
银　川	19	6	12	4	2	0	0	1
石嘴山	27	2	20	3	0	2	0	3
乌鲁木齐	3	0	3	2	1	0	0	0
克拉玛依	0	0	0	0	0	0	0	0

重点城市工业污染治理项目建设情况

（2012）

城 市 名 称	本年竣工项目新增设计处理能力		
	治理废水/ （万吨/日）	治理废气（标态）/ （万米^3/时）	治理固废/ （万吨/日）
总　　计	189	27 614	10
北　京	1	266	0
天　津	4	721	0
石 家 庄	2	209	0
唐　山	4	1 536	0
秦 皇 岛	0	27	0
邯　郸	3	1 038	0
保　定	0	0	0
太　原	0	255	1
大　同	1	49	0
阳　泉	3	479	0
长　治	0	392	0
临　汾	3	321	0
呼和浩特	0	932	0
包　头	1	999	0
赤　峰	5	37	0
沈　阳	0	0	0
大　连	1	0	0
鞍　山	0	61	2
抚　顺	0	24	0
本　溪	1	7	0
锦　州	0	12	0
长　春	0	23	0
吉　林	0	159	0
哈 尔 滨	0	37	0
齐齐哈尔	0	22	0
牡 丹 江	0	42	0
上　海	1	655	0
南　京	3	136	0
无　锡	1	1 054	0
徐　州	1	2	0
常　州	1	1 332	0
苏　州	3	283	0
南　通	2	111	0
连 云 港	0	9	0
扬　州	0	658	0
镇　江	2	1	0
杭　州	1	1 291	0

重点城市工业污染治理项目建设情况（续表）

（2012）

城　市 名　称	本年竣工项目新增设计处理能力		
	治理废水/ （万吨/日）	治理废气（标态）/ （万米³/时）	治理固废/ （万吨/日）
宁　波	3	652	0
温　州	5	313	0
湖　州	0	202	0
绍　兴	5	272	0
合　肥	0	523	0
芜　湖	0	200	0
马 鞍 山	0	143	0
福　州	11	1 165	0
厦　门	1	48	0
泉　州	6	89	0
南　昌	1	6	0
九　江	0	140	0
济　南	2	65	0
青　岛	1	34	0
淄　博	2	1 959	0
枣　庄	3	176	0
烟　台	2	617	0
潍　坊	6	483	0
济　宁	2	191	0
泰　安	2	18	0
日　照	4	1 660	0
郑　州	1	127	0
开　封	0	0	0
洛　阳	0	26	0
平 顶 山	0	206	0
安　阳	4	172	0
焦　作	2	2	0
三 门 峡	2	120	0
武　汉	1	24	0
宜　昌	2	5	0
荆　州	0	132	0
长　沙	0	475	0
株　洲	0	90	1
湘　潭	0	232	0
岳　阳	0	300	0
常　德	5	2	0
张 家 界	0	0	0
广　州	2	536	0

重点城市工业污染治理项目建设情况（续表）

（2012）

城 市 名 称	本年竣工项目新增设计处理能力		
	治理废水/ （万吨/日）	治理废气（标态）/ （万米³/时）	治理固废/ （万吨/日）
韶　关	1	201	0
深　圳	0	11	0
珠　海	1	2	0
汕　头	1	29	0
湛　江	4	2	0
南　宁	3	200	0
柳　州	5	327	0
桂　林	4	0	0
北　海	0	0	0
海　口	3	9	0
重　庆	4	282	0
成　都	1	116	0
自　贡	0	0	0
攀 枝 花	3	85	0
泸　州	1	0	0
德　阳	0	22	0
绵　阳	1	33	0
南　充	0	0	0
宜　宾	0	6	0
贵　阳	4	15	0
遵　义	1	304	0
昆　明	1	111	0
曲　靖	2	10	3
玉　溪	3	20	0
拉　萨	0	0	0
西　安	0	338	0
铜　川	0	0	0
宝　鸡	3	121	0
咸　阳	1	1	0
渭　南	10	52	0
延　安	0	0	0
兰　州	0	3	0
金　昌	0	310	0
西　宁	0	113	0
银　川	1	294	0
石 嘴 山	0	10	0
乌鲁木齐	5	1	0
克拉玛依	0	0	0

重点城市农业污染排放情况

(2012)　　　　　　　　　　　　　　　　　　　　　　　　　　　单位：吨

城 市 名 称	农业污染物排放（流失）总量			
	化学需氧量	氨氮	总氮	总磷
总　　计	4 989 694	325 350	1 709 283	195 430
北　　京	78 292	4 732	32 580	4 362
天　　津	113 889	5 946	32 924	3 616
石 家 庄	176 663	6 384	45 527	4 617
唐　　山	131 740	6 542	49 907	6 400
秦 皇 岛	39 225	2 539	16 222	2 291
邯　　郸	74 508	4 471	30 379	3 166
保　　定	100 701	5 759	46 531	4 490
太　　原	9 130	859	3 981	429
大　　同	15 633	839	6 331	456
阳　　泉	3 307	294	1 216	218
长　　治	14 231	1 191	6 019	582
临　　汾	20 302	1 516	10 175	853
呼和浩特	104 673	1 230	23 330	1 698
包　　头	63 387	1 023	14 731	1 079
赤　　峰	91 299	2 041	19 293	1 849
沈　　阳	213 448	6 736	51 778	6 286
大　　连	101 840	3 905	18 812	3 359
鞍　　山	50 773	1 941	10 750	1 709
抚　　顺	22 623	830	3 787	557
本　　溪	16 438	758	5 055	776
锦　　州	81 897	4 486	25 455	3 375
长　　春	135 389	5 020	33 292	4 715
吉　　林	89 334	2 762	20 936	2 415
哈 尔 滨	207 772	6 239	48 662	4 875
齐齐哈尔	176 835	4 551	28 832	2 565
牡 丹 江	23 415	987	5 377	524
上　　海	33 090	3 469	15 460	1 804
南　　京	19 325	1 896	8 532	938
无　　锡	8 916	1 161	5 295	577
徐　　州	50 508	3 499	24 616	2 493
常　　州	15 859	1 503	6 767	859
苏　　州	22 329	2 205	8 764	1 092
南　　通	52 344	6 585	23 652	2 279
连 云 港	48 360	2 891	15 093	1 684
扬　　州	11 438	1 798	7 810	612
镇　　江	7 962	1 102	4 871	547
杭　　州	29 448	3 718	10 016	1 210

重点城市农业污染排放情况（续表）

（2012） 单位：吨

城 市 名 称	农业污染物排放（流失）总量			
	化学需氧量	氨氮	总氮	总磷
宁　波	16 947	2 857	10 157	1 207
温　州	14 490	1 892	7 319	884
湖　州	21 738	2 063	7 677	977
绍　兴	15 385	2 319	8 214	796
合　肥	66 782	3 477	15 891	2 240
芜　湖	10 192	1 515	6 023	689
马 鞍 山	5 366	817	3 827	425
福　州	33 840	6 024	14 517	2 066
厦　门	13 527	1 849	4 392	554
泉　州	15 005	2 201	8 001	919
南　昌	35 602	3 295	13 144	1 636
九　江	15 150	2 134	9 564	1 008
济　南	73 827	3 083	27 418	2 765
青　岛	115 659	5 606	21 393	2 869
淄　博	35 540	1 835	15 564	1 885
枣　庄	29 834	1 886	13 141	1 421
烟　台	110 771	5 763	31 402	4 464
潍　坊	138 730	9 226	55 916	6 586
济　宁	98 110	7 388	40 967	4 577
泰　安	81 011	3 864	31 132	3 889
日　照	27 281	1 686	8 227	1 238
郑　州	54 427	3 188	19 096	2 505
开　封	47 652	3 386	21 657	2 165
洛　阳	40 590	2 296	15 379	1 526
平 顶 山	28 165	2 246	19 488	2 416
安　阳	40 418	3 540	18 252	2 316
焦　作	37 646	1 833	12 015	1 403
三 门 峡	10 147	831	5 483	622
武　汉	42 647	3 662	15 187	1 903
宜　昌	33 197	4 749	17 873	2 176
荆　州	95 270	5 725	26 423	3 121
长　沙	47 304	4 579	16 152	1 929
株　洲	27 100	3 496	10 788	1 020
湘　潭	39 619	4 024	12 140	1 190
岳　阳	66 536	7 678	24 845	2 747
常　德	55 960	5 890	24 974	2 777
张 家 界	5 127	575	2 602	233
广　州	45 037	4 003	13 212	1 778

重点城市农业污染排放情况（续表）

（2012）

单位：吨

城 市 名 称	农业污染物排放（流失）总量			
	化学需氧量	氨氮	总氮	总磷
韶 关	19 904	2 519	8 365	940
深 圳	1 867	193	699	71
珠 海	11 104	876	2 906	346
汕 头	16 473	1 581	4 752	529
湛 江	46 659	5 642	21 245	2 651
南 宁	36 780	4 077	19 252	2 368
柳 州	19 428	2 426	9 619	1 037
桂 林	32 314	3 172	13 603	1 826
北 海	9 806	844	4 462	542
海 口	11 118	1 069	5 111	730
重 庆	123 595	12 962	54 139	6 400
成 都	75 558	8 229	31 208	3 483
自 贡	18 510	1 663	5 651	644
攀 枝 花	2 917	322	1 686	198
泸 州	24 238	2 695	9 665	966
德 阳	25 641	2 245	10 385	1 185
绵 阳	31 809	3 926	16 780	1 968
南 充	52 048	5 247	18 192	2 568
宜 宾	26 120	3 154	11 438	1 119
贵 阳	8 324	599	3 959	428
遵 义	11 126	1 968	9 118	871
昆 明	15 705	1 879	7 518	953
曲 靖	14 918	2 193	10 805	957
玉 溪	5 402	386	2 223	301
拉 萨	1 274	98	1 401	86
西 安	28 132	978	11 846	1 168
铜 川	7 598	340	622	66
宝 鸡	20 789	1 154	6 280	378
咸 阳	33 411	1 507	13 209	1 034
渭 南	31 414	2 428	16 715	1 376
延 安	7 422	792	2 718	190
兰 州	7 012	449	3 613	371
金 昌	7 619	174	2011	122
西 宁	10 059	391	1 866	145
银 川	29 704	625	8 422	720
石 嘴 山	7 156	164	2 425	165
乌鲁木齐	7 944	399	2 660	214
克拉玛依	845	91	506	29

重点城市城镇生活污染排放及处理情况（一）

（2012）

年　份 地　区	城镇人口/ 万人	生活煤炭消费量/ 万吨	生活天然气消费量/ 万米³	生活用水总量/ 万吨	居民家庭用水 总量	公共服务用水 总量
总　　计	40 458	8 224	2 205 197	3 218 234	2 624 737	593 497
北　　京	1 784	461	480 152	156 113	89 254	66 859
天　　津	1 152	231	28 109	68 329	54 329	14 000
石 家 庄	553	107	9 750	34 835	26 115	8 720
唐　　山	412	74	5 800	27 451	24 401	3 050
秦 皇 岛	155	55	2 475	12 387	6 974	5 413
邯　　郸	435	166	6 930	25 918	23 557	2 361
保　　定	469	110	11 734	29 689	22 267	7 422
太　　原	379	321	36 925	25 641	18 458	7 184
大　　同	190	74	12 564	11 109	8 377	2 731
阳　　泉	86	45	25 200	6 542	5 000	1 542
长　　治	156	227	698	10 405	7 556	2 849
临　　汾	189	193	23 834	10 959	10 132	827
呼和浩特	192	32	42 350	14 429	9 967	4 462
包　　头	222	154	29 580	11 149	9 366	1 783
赤　　峰	189	106	1	11 100	11 000	100
沈　　阳	657	256	11 771	44 693	36 648	8 045
大　　连	532	120	0	36 312	33 205	3 107
鞍　　山	254	65	471	20 064	14 825	5 239
抚　　顺	156	40	0	9 709	6 472	3 237
本　　溪	133	45	14	5 862	3 030	2 832
锦　　州	161	56	2 091	13 095	9 166	3 928
长　　春	387	80	7 202	24 240	7 117	17 123
吉　　林	286	75	3 063	16 706	15 247	1 459
哈 尔 滨	681	665	11 229	42 488	39 387	3 102
齐齐哈尔	240	300	18 364	12 565	10 000	2 565
牡 丹 江	141	101	443	11 209	9 040	2 169
上　　海	2 125	172	239 963	191 700	176 340	15 360
南　　京	655	22	22 175	57 929	34 333	23 596
无　　锡	478	4	3	46 668	34 122	12 545
徐　　州	482	17	3 370	36 525	34 055	2 470
常　　州	317	5	28 000	27 734	26 001	1 733
苏　　州	760	6	35 449	82 270	62 092	20 178
南　　通	429	12	3 909	34 417	32 224	2 193
连 云 港	238	39	8 623	19 286	17 809	1 477
扬　　州	265	23	8 738	18 490	13 248	5 242
镇　　江	205	16	4 699	16 599	15 403	1 196
杭　　州	657	4	17 692	61 832	56 832	5 000

重点城市城镇生活污染排放及处理情况（一）（续表）

（2012）

年　份 地　区	城镇人口/ 万人	生活煤炭消费量/ 万吨	生活天然气消费量/ 万米³	生活用水总量/ 万吨	居民家庭用水 总量	公共服务用水 总量
宁　波	536	40	7 301	42 615	35 423	7 193
温　州	611	5	548	50 287	46 041	4 246
湖　州	161	12	1 457	23 112	15 071	8 041
绍　兴	296	7	8 010	24 714	17 352	7 362
合　肥	446	24	10 551	41 727	37 512	4 215
芜　湖	217	10	3 604	17 173	15 901	1 272
马 鞍 山	125	10	4 706	9 589	8 400	1 188
福　州	455	7	4 059	36 835	34 186	2 649
厦　门	326	2	4 584	25 344	20 000	5 344
泉　州	501	22	427	40 573	37 649	2 924
南　昌	333	4	34	40 544	36 489	4 054
九　江	231	25	12	15 238	11 323	3 915
济　南	457	140	22 000	31 169	28 335	2 834
青　岛	596	122	19 025	48 968	44 612	4 356
淄　博	297	52	8 608	21 126	17 520	3 605
枣　庄	186	54	4 386	11 163	8 491	2 672
烟　台	397	114	7 008	25 182	22 897	2 284
潍　坊	458	152	3 613	25 232	20 186	5 046
济　宁	376	153	3 167	29 978	27 360	2 619
泰　安	291	93	5 300	20 698	18 563	2 135
日　照	141	57	0	8 150	7 190	960
郑　州	572	164	27 271	50 106	39 580	10 527
开　封	165	90	5 277	13 002	11 981	1 021
洛　阳	313	120	1 350	24 448	22 849	1 599
平 顶 山	188	4	2 100	17 237	10 342	6 895
安　阳	191	49	7 191	13 588	12 773	815
焦　作	177	60	496	13 833	12 928	905
三 门 峡	106	152	404	7 669	6 980	689
武　汉	802	130	2 500	67 340	62 655	4 684
宜　昌	219	61	3 223	14 298	12 060	2 238
荆　州	266	41	3 094	18 254	17 050	1 204
长　沙	529	28	20 100	49 328	45 875	3 453
株　洲	223	18	22 448	19 535	16 279	3 256
湘　潭	156	11	3 630	13 664	9 565	4 099
岳　阳	279	43	3 306	20 510	18 110	2 400
常　德	241	51	5 784	17 737	15 100	2 637
张 家 界	64	8	304	4 645	3 841	804
广　州	1 092	17	20 000	131 121	95 955	35 167

重点城市城镇生活污染排放及处理情况（一）（续表）

（2012）

年　份 地　区	城镇人口/ 万人	生活煤炭消费量/ 万吨	生活天然气消费量/ 万米³	生活用水总量/ 万吨	居民家庭用水 总量	公共服务用水 总量
韶　　关	153	11	0	10 629	9 885	744
深　　圳	1 061	3	20 000	116 340	116 220	120
珠　　海	175	0	5 100	19 096	13 558	5 538
汕　　头	379	1	400	22 400	20 137	2 262
湛　　江	272	12	5 042	26 295	23 603	2 691
南　　宁	382	51	5 517	28 754	23 106	5 648
柳　　州	223	20	907	22 417	18 607	3 810
桂　　林	207	22	1 640	16 324	14 812	1 511
北　　海	82	6	1 628	6 306	5 741	565
海　　口	162	1	1 822	13 302	9 236	4 067
重　　庆	1 678	147	194 906	120 134	106 853	13 281
成　　都	992	30	157 540	89 696	46 126	43 570
自　　贡	123	15	9 057	7 741	6 886	855
攀 枝 花	79	6	0	3 833	2 550	1 284
泸　　州	181	47	12 934	11 004	9 403	1 601
德　　阳	162	8	9 781	11 200	10 460	740
绵　　阳	202	16	58 903	14 538	13 538	1 000
南　　充	254	8	18 036	14 554	11 140	3 413
宜　　宾	187	172	13 643	11 247	9 264	1 983
贵　　阳	295	98	9 322	30 400	18 466	11 934
遵　　义	249	135	0	13 385	12 047	1 339
昆　　明	450	30	0	55 777	44 621	11 155
曲　　靖	233	63	284	14 810	13 022	1 787
玉　　溪	97	3	0	5 706	5 008	698
拉　　萨	36	2	4	2 359	1 976	382
西　　安	612	140	69 407	42 183	29 907	12 276
铜　　川	51	14	2 808	2 810	2 221	589
宝　　鸡	170	98	3 515	9 226	8 395	831
咸　　阳	223	100	8 632	12 155	7 946	4 208
渭　　南	194	24	12 054	10 383	9 623	760
延　　安	115	53	7 035	6 433	5 316	1 117
兰　　州	289	77	31 000	17 256	14 590	2 666
金　　昌	30	20	132	2 119	1 929	190
西　　宁	152	58	26 603	8 657	5 586	3 071
银　　川	149	33	69 758	14 852	12 289	2 564
石 嘴 山	52	21	3 000	3 913	2 312	1 601
乌鲁木齐	258	48	54 000	23 347	17 241	6 106
克拉玛依	28	2	542	4 470	3 342	1 128

重点城市城镇生活污染排放及处理情况（二）

（2012）

城 市 名 称	城镇生活污水 排放量/万吨	城镇生活化学需氧量 产生量/吨	城镇生活氨氮 产生量/吨	城镇生活化学需氧量 排放量/吨	城镇生活氨氮 排放量/吨
总　计	2 760 049	9 946 794.3	1 279 252.8	4 116 443.2	721 286.2
北　京	130 985	514 451.0	63 166.7	95 661.0	14 762.0
天　津	63 651	289 313.9	39 471.9	88 544.0	16 138.0
石 家 庄	28 631	121 166.3	16 815.3	6 892.5	2 645.1
唐　山	22 796	90 381.0	12 352.0	14 112.0	5 748.1
秦 皇 岛	11 148	24 363.3	3 423.8	4 660.8	1 711.9
邯　郸	24 951	95 949.0	13 046.0	44 428.0	8 579.0
保　定	26 717	104 300.6	13 468.8	37 125.1	6 642.0
太　原	18 033	86 680.9	11 512.3	12 733.8	3 474.7
大　同	8 548	35 551.2	5 111.9	16 726.2	3 008.5
阳　泉	5 651	19 915.3	2 663.9	7 590.0	1 372.8
长　治	8 792	36 452.4	4 841.3	12 031.7	2 805.2
临　汾	8 615	44 110.7	5 858.5	27 437.9	4 434.7
呼和浩特	11 543	44 287.0	5 000.1	17 040.5	2 828.2
包　头	7 961	53 502.0	6 841.2	29 530.3	4 423.5
赤　峰	11 000	44 089.7	5 511.2	14 211.9	5 039.3
沈　阳	34 415	153 428.5	20 856.7	37 608.0	14 879.0
大　连	30 875	124 275.2	17 981.1	57 290.8	8 571.0
鞍　山	12 526	58 418.7	7 941.3	47 326.9	7 667.0
抚　顺	8 848	36 324.8	4 937.9	11 429.5	4 025.7
本　溪	4 279	31 092.2	4 266.6	16 980.2	2 178.7
锦　州	11 131	37 492.8	5 096.7	19 314.8	3 734.2
长　春	20 781	89 673.5	11 554.0	31 721.9	7 015.9
吉　林	14 295	71 652.3	8 833.4	33 911.9	5 893.6
哈 尔 滨	33 543	145 428.7	19 902.3	100 258.1	15 590.3
齐齐哈尔	11 360	52 805.9	7 357.3	37 091.1	5 584.7
牡 丹 江	7 230	36 147.6	4 577.0	20 150.8	3 991.0
上　海	172 530	589 475.0	75 235.6	177 504.0	41 369.8
南　京	47 928	165 353.0	21 568.0	66 535.0	14 300.0
无　锡	42 001	121 258.6	15 603.0	20 147.1	2 430.4
徐　州	31 055	121 476.4	15 844.9	74 963.7	9 980.4
常　州	24 961	79 735.7	10 400.3	15 897.1	3 879.1
苏　州	61 189	191 882.4	25 028.1	24 640.6	10 586.7
南　通	29 518	108 231.1	14 117.1	38 692.5	9 376.1
连 云 港	16 332	59 942.8	7 818.6	46 537.5	6 704.7
扬　州	18 209	66 830.9	8 717.8	32 862.3	4 825.1
镇　江	14 095	52 482.6	6 747.8	29 146.5	3 774.0
杭　州	52 557	172 542.6	23 485.0	38 958.8	8 449.1

重点城市城镇生活污染排放及处理情况（二）（续表）

（2012）

城　市　名　称	城镇生活污水排放量/万吨	城镇生活化学需氧量产生量/吨	城镇生活氨氮产生量/吨	城镇生活化学需氧量排放量/吨	城镇生活氨氮排放量/吨
宁　波	36 223	154 682.5	18 992.7	30 296.0	10 081.0
温　州	43 359	154 617.8	20 932.0	106 748.2	16 655.2
湖　州	11 211	40 114.1	4 259.0	17 928.1	3 912.7
绍　兴	20 981	78 082.9	10 194.2	39 251.0	5 324.0
合　肥	37 547	105 833.6	12 537.2	49 112.6	6 837.3
芜　湖	14 628	51 540.3	6 105.6	31 989.0	5 015.5
马 鞍 山	8 767	29 738.3	3 522.8	16 489.0	2 700.1
福　州	31 363	111 204.2	14 440.0	66 097.0	9 276.0
厦　门	20 539	79 625.5	10 339.4	26 336.0	6 032.0
泉　州	34 542	122 450.1	15 900.2	88 071.3	12 974.9
南　昌	32 749	87 844.8	9 705.2	41 125.2	6 297.6
九　江	13 475	55 926.0	6 043.0	42 889.5	5 478.0
济　南	26 668	108 340.3	14 500.9	36 223.9	6 129.6
青　岛	40 151	141 840.7	18 984.8	25 206.0	6 187.0
淄　博	19 625	62 458.3	9 565.9	17 188.2	3 223.3
枣　庄	9 980	44 198.9	5 915.9	19 279.1	3 705.0
烟　台	22 684	92 786.1	12 421.0	31 203.7	6 656.8
潍　坊	25 186	108 776.8	14 559.4	22 008.0	5 637.0
济　宁	25 482	89 285.1	11 950.5	35 556.0	6 568.9
泰　安	14 850	68 939.9	9 003.2	34 001.0	5 967.0
日　照	7 373	33 515.5	4 485.9	15 401.7	2 867.5
郑　州	45 346	129 124.8	19 368.7	30 983.0	9 075.0
开　封	11 752	36 941.0	4 740.0	18 831.1	3 040.2
洛　阳	19 993	68 547.0	8 796.9	17 264.0	4 145.0
平 顶 山	13 790	41 147.9	5 280.6	20 414.3	4 117.7
安　阳	11 550	40 763.9	5 231.4	19 327.5	3 627.4
焦　作	13 267	38 784.9	4 977.4	8 440.9	2 323.7
三 门 峡	6 519	22 879.5	3 434.2	9 322.9	1 839.6
武　汉	61 484	187 379.9	23 422.5	96 353.6	13 298.0
宜　昌	13 606	48 407.3	6 050.9	26 374.2	4 842.4
荆　州	15 516	62 065.2	7 938.4	52 539.0	7 228.6
长　沙	40 548	122 625.3	14 863.7	61 358.0	8 786.0
株　洲	15 628	53 720.7	6 511.6	33 757.6	4 903.4
湘　潭	11 387	37 573.2	4 554.6	21 142.3	3 467.9
岳　阳	18 071	63 515.7	7 698.9	53 527.0	6 670.0
常　德	14 722	57 963.3	7 025.8	49 307.0	5 977.0
张 家 界	3 716	15 327.8	1 857.6	12 477.3	1 488.1
广　州	130 031	307 623.4	42 046.7	112 030.3	18 489.3

重点城市城镇生活污染排放及处理情况（二）（续表）

（2012）

城　市名　称	城镇生活污水排放量/万吨	城镇生活化学需氧量产生量/吨	城镇生活氨氮产生量/吨	城镇生活化学需氧量排放量/吨	城镇生活氨氮排放量/吨
韶　关	9 303	36 683.1	4 417.7	27 718.7	3 920.0
深　圳	116 220	298 109.7	37 554.1	68 395.4	11 506.5
珠　海	17 246	47 906.2	5 812.6	25 725.1	4 224.9
汕　头	19 962	89 304.1	10 918.1	56 549.9	9 080.7
湛　江	18 883	62 804.7	8 064.2	39 324.1	5 702.9
南　宁	25 809	96 023.0	9 735.0	59 593.0	7 616.0
柳　州	20 176	57 726.3	7 325.6	31 654.0	5 243.0
桂　林	13 905	47 509.0	6 265.9	28 973.0	4 477.0
北　海	5 355	19 747.7	2 304.0	12 528.0	1 267.0
海　口	10 642	37 756.8	4 660.6	4 945.0	3 457.4
重　庆	101 677	416 506.9	53 288.4	229 447.5	37 215.3
成　都	76 242	250 970.8	30 438.9	107 828.9	13 785.8
自　贡	6 571	30 982.3	3 784.9	18 620.0	2 526.4
攀 枝 花	3 765	19 723.0	2 405.4	13 871.7	1 813.7
泸　州	9 580	45 017.5	5 541.1	38 972.8	4 801.6
德　阳	9 367	40 619.3	4 957.6	23 230.4	3 317.6
绵　阳	13 799	53 709.9	6 503.4	32 197.4	4 489.6
南　充	12 680	63 502.4	7 775.6	46 542.4	5 834.3
宜　宾	9 190	44 929.6	5 361.7	34 277.8	4 375.9
贵　阳	20 997	61 540.9	7 858.7	25 718.1	3 941.2
遵　义	12 298	63 353.1	8 062.4	48 594.4	6 116.3
昆　明	47 410	122 791.8	13 721.0	2 340.3	4 828.1
曲　靖	11 848	60 451.8	7 234.6	49 931.2	6 156.2
玉　溪	4 689	21 649.7	2 611.5	15 361.1	2 268.8
拉　萨	2 003	8 696.5	1 067.3	7 886.5	995.3
西　安	29 824	141 026.5	18 264.1	62 360.9	10 550.3
铜　川	2 423	12 315.1	1 525.4	9 146.1	1 189.0
宝　鸡	8 066	35 450.9	4 862.2	14 357.1	3 090.6
咸　阳	10 975	49 610.9	6 425.0	15 513.5	3 351.5
渭　南	9 188	42 475.2	5 358.9	25 495.5	4 000.9
延　安	5 486	22 355.7	2 756.6	16 816.3	2 292.4
兰　州	13 688	62 242.6	8 399.7	37 838.1	5 453.1
金　昌	1 423	6 458.3	853.8	1 698.5	447.0
西　宁	7 374	33 332.3	4 333.2	16 430.3	3 604.7
银　川	13 707	33 616.9	4 229.2	4 413.7	3 029.2
石 嘴 山	3 130	11 844.5	1 490.1	3 846.3	1 012.7
乌鲁木齐	18 276	65 431.7	7 456.0	14 101.5	4 971.2
克拉玛依	3 890	5 959.5	515.4	653.9	66.8

重点城市城镇生活污染排放及处理情况（三）

（2012）

单位：吨

城 市 名 称	城镇生活二氧化硫 排放量	城镇生活氮氧化物 排放量	城镇生活烟尘 排放量
总 计	984 035	206 231	712 779
北 京	34 485	11 991	31 868
天 津	8 959	4 447	18 400
石 家 庄	8 955	2 144	6 635
唐 山	4 680	1 550	7 360
秦 皇 岛	4 338	965	55 498
邯 郸	13 022	2 557	10 000
保 定	18 648	2 047	11 017
太 原	31 296	8 260	30 650
大 同	8 384	1 698	6 754
阳 泉	1 424	1 092	4 005
长 治	2 686	4 554	11 939
临 汾	14 193	4 673	2 753
呼 和 浩 特	4 625	637	3 584
包 头	15 673	3 309	28 268
赤 峰	17 937	1 793	8 512
沈 阳	14 458	3 944	9 745
大 连	16 366	2 403	9 612
鞍 山	8 880	1 492	6 931
抚 顺	4 224	600	4 800
本 溪	4 629	939	2 149
锦 州	4 773	867	3 662
长 春	7 344	1 600	8 000
吉 林	6 885	1 525	6 750
哈 尔 滨	32 016	13 255	86 241
齐 齐 哈 尔	20 400	6 147	70 000
牡 丹 江	10 322	4 575	6 327
上 海	34 754	19 138	15 454
南 京	2 580	970	2 150
无 锡	512	300	387
徐 州	1 471	373	1 384
常 州	82	594	384
苏 州	799	401	353
南 通	2 496	335	978
连 云 港	6 588	844	3 178
扬 州	3 089	582	1 365
镇 江	2 649	497	2 100
杭 州	628	450	91

重点城市城镇生活污染排放及处理情况（三）（续表）

（2012）

単位：吨

城　市名　称	城镇生活二氧化硫排放量	城镇生活烟尘排放量	城镇生活氮氧化物排放量
宁　波	2 477	692	487
温　州	599	370	467
湖　州	2 057	208	239
绍　兴	845	194	240
合　肥	2 465	120	6 080
芜　湖	1 912	400	4 980
马　鞍　山	1 781	258	1 650
福　州	1 279	169	547
厦　门	486	181	190
泉　州	3 199	559	1 771
南　昌	719	70	280
九　江	6 509	1 781	2 129
济　南	11 306	3 629	8 355
青　岛	27 059	2 601	9 795
淄　博	4 973	1 223	3 988
枣　庄	11 940	1 506	5 374
烟　台	13 574	3 479	10 168
潍　坊	24 906	2 883	13 646
济　宁	13 956	2 752	10 809
泰　安	19 323	2 600	8 477
日　照	10 067	1 190	5 250
郑　州	9 395	1 305	8 900
开　封	12 878	1 397	1 141
洛　阳	25 500	2 400	1 320
平　顶　山	477	384	400
安　阳	7 840	980	1 069
焦　作	5 065	1 192	5 065
三　门　峡	26 551	3 808	2 768
武　汉	5 720	1 416	8 051
宜　昌	5 711	892	3 692
荆　州	5 569	596	2 496
长　沙	2 366	153	2 946
株　洲	2 592	540	1 080
湘　潭	271	310	919
岳　阳	7 014	851	2 700
常　德	6 088	1 069	3 581
张　家　界	1 920	172	800
广　州	2 363	1 963	738

重点城市城镇生活污染排放及处理情况（三）（续表）

（2012）

单位：吨

城 市 名 称	城镇生活二氧化硫排放量	城镇生活氮氧化物排放量	城镇生活烟尘排放量
韶 关	238	87	227
深 圳	488	885	300
珠 海	21	44	23
汕 头	455	117	55
湛 江	2 458	395	258
南 宁	8 748	1 123	4 631
柳 州	4 117	411	1 514
桂 林	4 515	456	1 438
北 海	1 240	141	395
海 口	178	32	174
重 庆	54 984	4 487	9 139
成 都	4 373	1 852	593
自 贡	5 256	382	309
攀 枝 花	576	113	113
泸 州	7 515	872	1 467
德 阳	1 362	291	132
绵 阳	5 565	1 090	159
南 充	4 291	293	134
宜 宾	10 848	646	2 149
贵 阳	33 543	1 820	5 380
遵 义	49 959	2 167	2 709
昆 明	5 310	951	322
曲 靖	8 626	1 142	1 269
玉 溪	964	208	43
拉 萨	671	39	197
西 安	20 570	11 401	12 624
铜 川	3 321	611	1 520
宝 鸡	8 621	1 979	6 967
咸 阳	21 097	2 937	10 860
渭 南	5 381	570	2 772
延 安	4 562	1 193	5 623
兰 州	11 753	2 576	1 341
金 昌	3 824	590	1 613
西 宁	6 855	1 365	4 608
银 川	5 613	1 218	2 972
石 嘴 山	5 375	446	1 897
乌鲁木齐	6 554	1 396	4 819
克拉玛依	205	63	160

重点城市机动车污染排放情况

（2012）

单位：万吨

城 市 名 称	机动车污染物排放总量			
	总颗粒物	氮氧化物	一氧化碳	碳氢化合物
总　　计	31.2	320.2	1 754.0	216.1
北　京	0.4	8.0	78.1	8.6
天　津	0.7	5.4	44.7	5.1
石 家 庄	0.7	6.8	24.5	3.2
唐　山	0.9	7.8	34.2	4.3
秦 皇 岛	0.5	3.9	18.2	2.4
邯　郸	1.1	7.3	23.6	3.8
保　定	0.1	7.2	45.9	5.3
太　原	0.3	2.9	15.9	1.9
大　同	0.2	2.9	15.1	1.8
阳　泉	0.0	2.4	22.6	2.5
长　治	0.2	1.6	6.3	0.8
临　汾	0.6	5.1	26.6	3.5
呼和浩特	0.4	2.8	9.4	1.4
包　头	0.5	3.3	11.1	1.6
赤　峰	0.4	2.1	7.3	1.1
沈　阳	0.5	4.6	25.1	3.2
大　连	0.7	4.9	21.0	2.8
鞍　山	0.2	2.5	13.4	1.6
抚　顺	0.2	1.8	8.4	1.2
本　溪	0.1	1.1	4.5	0.6
锦　州	0.2	2.3	7.3	1.2
长　春	0.6	5.5	37.0	4.6
吉　林	0.4	3.9	21.5	2.8
哈 尔 滨	0.5	3.9	25.3	3.1
齐齐哈尔	0.4	3.1	13.6	1.8
牡 丹 江	0.2	1.6	11.6	1.5
上　海	0.8	9.7	43.5	6.3
南　京	0.3	3.3	14.2	1.8
无　锡	0.3	3.1	14.6	1.6
徐　州	0.5	5.9	17.5	2.5
常　州	0.1	1.9	10.6	1.2
苏　州	0.5	6.5	51.6	6.0
南　通	0.1	1.8	14.9	1.7
连 云 港	0.1	1.4	5.9	0.7
扬　州	0.2	1.8	8.6	1.0
镇　江	0.1	1.1	5.8	0.7
杭　州	0.4	4.2	23.7	2.6

重点城市机动车污染排放情况（续表）

（2012） 单位：万吨

城 市 名 称	机动车污染物排放总量			
	总颗粒物	氮氧化物	一氧化碳	碳氢化合物
宁　波	0.2	2.6	17.0	2.0
温　州	0.1	1.6	13.8	1.5
湖　州	0.1	0.9	8.8	1.1
绍　兴	0.1	1.4	8.8	1.0
合　肥	0.2	2.3	11.0	1.3
芜　湖	0.1	1.2	7.1	0.8
马 鞍 山	0.1	0.6	2.7	0.3
福　州	0.2	2.1	10.3	1.2
厦　门	0.2	2.1	9.1	1.1
泉　州	0.2	1.8	18.0	2.1
南　昌	0.4	3.4	15.0	1.9
九　江	0.2	1.9	7.1	1.0
济　南	0.3	2.8	16.1	1.8
青　岛	0.5	4.2	22.4	2.7
淄　博	0.2	2.0	14.1	1.7
枣　庄	0.2	1.7	9.2	1.1
烟　台	0.4	3.4	22.4	2.8
潍　坊	0.6	6.1	32.0	4.1
济　宁	0.6	3.8	15.3	2.1
泰　安	0.1	1.6	9.4	1.2
日　照	0.1	1.1	6.1	0.7
郑　州	0.6	6.1	29.1	3.4
开　封	0.2	1.7	5.9	0.9
洛　阳	0.4	3.2	13.3	1.7
平 顶 山	0.2	1.8	8.5	1.1
安　阳	0.3	3.1	12.7	1.8
焦　作	0.3	3.4	15.7	1.9
三 门 峡	0.1	1.3	8.4	1.0
武　汉	0.5	4.8	21.6	2.6
宜　昌	0.2	1.9	10.3	1.3
荆　州	0.1	1.3	9.3	1.1
长　沙	0.2	2.7	15.9	1.8
株　洲	0.1	1.0	5.1	0.6
湘　潭	0.1	0.6	4.0	0.4
岳　阳	0.1	0.8	5.0	0.6
常　德	0.1	1.2	6.5	0.8
张 家 界	0.0	0.5	2.6	0.3
广　州	0.6	6.8	32.5	3.8

重点城市机动车污染排放情况（续表）

（2012）　　　　　　　　　　　　　　　　　　单位：万吨

城　市名　称	机动车污染物排放总量			
	总颗粒物	氮氧化物	一氧化碳	碳氢化合物
韶　关	0.1	0.6	4.3	0.5
深　圳	0.9	8.0	44.6	4.9
珠　海	0.3	1.9	11.1	1.3
汕　头	0.3	2.0	14.6	1.8
湛　江	0.1	1.2	6.8	0.8
南　宁	0.4	3.9	22.4	2.6
柳　州	0.1	1.2	6.3	0.7
桂　林	0.1	1.3	8.6	1.1
北　海	0.0	0.5	3.5	0.4
海　口	0.1	1.0	5.6	0.6
重　庆	0.7	10.6	61.6	7.8
成　都	0.3	4.7	30.4	3.4
自　贡	0.0	0.7	3.9	0.5
攀枝花	0.1	0.7	2.0	0.3
泸　州	0.0	1.0	8.7	0.9
德　阳	0.1	1.0	6.5	0.8
绵　阳	0.1	1.2	6.1	0.7
南　充	0.2	1.5	5.6	0.7
宜　宾	0.1	0.9	4.8	0.6
贵　阳	0.2	2.3	10.7	1.4
遵　义	0.1	1.2	7.9	1.0
昆　明	0.1	4.1	23.2	3.3
曲　靖	0.3	2.7	14.4	2.1
玉　溪	0.2	1.8	13.2	1.9
拉　萨	0.1	1.1	12.7	1.2
西　安	0.2	3.8	23.4	2.6
铜　川	0.0	0.5	3.5	0.5
宝　鸡	0.1	1.5	15.7	1.7
咸　阳	0.1	1.5	10.8	1.1
渭　南	0.2	2.3	8.4	1.1
延　安	1.0	1.5	6.7	0.7
兰　州	0.1	2.1	20.3	2.4
金　昌	0.0	0.4	4.2	0.5
西　宁	0.1	1.6	11.5	1.7
银　川	0.4	3.4	14.1	1.9
石嘴山	0.1	0.7	2.8	0.4
乌鲁木齐	0.5	5.7	21.1	2.9
克拉玛依	0.2	1.8	7.1	0.9

重点城市污水处理情况（一）

（2012）

城 市 名 称	污水处理厂数/ 座	污水处理厂 设计处理能力/ （万吨/日）	本年运行费用/ 万元	污水处理厂累计 完成投资/万元	新增固定资产/ 万元
总 计	2 519	10 216.1	2 353 225.2	21 401 036.2	1 022 532.0
北 京	130	431.0	109 625.0	942 199.9	7 641.5
天 津	57	268.6	70 103.3	652 905.2	72 965.1
石 家 庄	28	187.9	34 824.7	360 868.4	22 557.2
唐 山	24	115.7	21 521.3	260 165.6	14 582.5
秦 皇 岛	18	56.8	13 689.3	67 732.8	2 424.0
邯 郸	28	81.8	13 391.7	149 486.4	33 940.8
保 定	35	116.0	26 708.4	305 925.0	1 930.8
太 原	13	57.8	15 131.8	110 796.9	840.7
大 同	15	27.3	7 261.2	71 410.9	385.4
阳 泉	7	17.0	4 744.0	53 501.0	1 529.0
长 治	20	36.0	7 461.7	94 344.6	1 378.0
临 汾	19	31.6	5 022.2	92 036.0	4 355.9
呼和浩特	14	36.7	9 752.8	126 310.8	11 408.1
包 头	12	55.6	12 134.4	172 366.9	110.0
赤 峰	13	49.5	7 712.0	104 030.0	3 493.5
沈 阳	36	166.1	36 660.9	242 097.5	23 645.0
大 连	24	106.5	25 094.8	164 190.9	646.0
鞍 山	5	32.5	7 611.7	70 136.6	6 467.0
抚 顺	9	44.3	3 433.3	56 329.5	0.0
本 溪	5	31.5	3 420.5	28 187.7	6 915.0
锦 州	8	35.0	7 785.7	75 492.6	684.1
长 春	13	98.8	17 898.4	223 328.0	24 463.0
吉 林	8	65.0	14 874.4	106 601.0	1 084.0
哈 尔 滨	13	111.2	19 220.9	175 242.0	0.0
齐齐哈尔	6	36.0	4 259.4	80 571.6	8 239.5
牡 丹 江	6	20.0	5 194.4	42 460.6	80.0
上 海	55	681.7	145 061.5	1 350 103.0	10 278.6
南 京	68	212.1	54 787.6	570 047.9	30 171.4
无 锡	76	223.8	75 167.2	521 172.6	45 151.2
徐 州	36	99.0	18 432.9	158 847.4	4 050.4
常 州	44	111.9	35 972.7	402 387.2	939.6
苏 州	127	341.0	105 201.3	973 556.3	66 322.6
南 通	89	114.2	24 002.9	290 797.3	10 805.0
连 云 港	21	30.4	8 019.6	106 494.5	1 250.5
扬 州	39	72.8	17 491.9	115 690.0	19 207.6
镇 江	34	48.8	8 076.5	153 554.5	17 449.8
杭 州	39	349.0	99 338.8	763 294.7	29 627.5

重点城市污水处理情况（一）（续表）

（2012）

城 市 名 称	污水处理厂数/ 座	污水处理厂 设计处理能力/ （万吨/日）	本年运行费用/ 万元	污水处理厂累计 完成投资/万元	新增固定资产/ 万元
宁 波	33	166.9	38 281.9	359 469.2	3 320.5
温 州	30	87.8	29 772.0	231 185.0	38 973.8
湖 州	32	63.2	18 993.5	176 330.2	5 193.1
绍 兴	6	139.2	59 274.3	369 224.0	36 113.6
合 肥	15	109.8	23 033.6	219 242.8	26 113.3
芜 湖	8	49.3	10 246.9	86 113.1	4 037.0
马 鞍 山	9	33.5	7 646.0	87 425.0	11 362.0
福 州	17	86.8	12 824.3	131 081.0	1 140.4
厦 门	10	83.8	15 962.4	124 098.0	1 560.0
泉 州	37	99.8	19 666.7	185 131.1	13 181.0
南 昌	14	105.7	15 555.3	119 611.0	68.4
九 江	18	32.8	7 940.0	70 751.9	4 012.6
济 南	34	89.5	26 973.1	154 305.6	7 420.9
青 岛	31	242.5	53 908.0	412 487.6	11 492.3
淄 博	14	80.5	18 527.8	124 794.7	14 089.0
枣 庄	16	45.2	9 132.0	67 755.0	4 500.0
烟 台	32	80.8	19 293.4	186 757.7	6 360.0
潍 坊	32	135.1	52 517.3	230 995.8	45 682.4
济 宁	32	89.3	24 098.1	199 910.9	4 337.9
泰 安	11	45.2	8 865.3	81 213.0	1 842.0
日 照	12	25.7	7 728.2	73 177.0	1 314.0
郑 州	19	141.5	25 080.2	283 269.3	46 508.3
开 封	10	34.6	2 752.1	32 033.9	1 668.0
洛 阳	16	70.6	9 132.1	97 355.8	119.8
平 顶 山	7	36.0	6 963.4	75 116.0	160.0
安 阳	10	31.4	6 162.1	55 349.8	3 172.3
焦 作	17	46.9	11 738.1	90 400.8	1 056.3
三 门 峡	7	19.0	3 307.5	23 828.0	1 069.1
武 汉	19	212.3	35 616.2	502 104.7	34.3
宜 昌	26	50.0	8 329.4	77 015.2	865.0
荆 州	11	44.0	8 468.8	77 080.9	6 000.0
长 沙	14	152.9	30 335.8	452 779.8	943.8
株 洲	12	41.8	6 988.5	121 684.0	27 222.0
湘 潭	6	32.0	5 754.3	60 412.6	167.0
岳 阳	11	39.0	7 475.9	102 900.3	3 837.0
常 德	9	27.0	5 818.4	48 080.0	30.0
张 家 界	5	9.5	2 095.5	24 788.3	12.3
广 州	28	416.1	84 843.9	855 999.7	17 786.7

重点城市污水处理情况（一）（续表）

（2012）

城 市 名 称	污水处理厂数/ 座	污水处理厂 设计处理能力/ （万吨/日）	本年运行费用/ 万元	污水处理厂累计 完成投资/万元	新增固定资产/ 万元
韶 关	11	26.0	4 892.6	28 382.3	4 349.2
深 圳	42	470.7	65 606.0	492 178.4	14 039.0
珠 海	10	54.3	16 687.2	126 293.5	387.0
汕 头	8	58.8	13 961.4	131 007.0	107.6
湛 江	7	47.0	5 446.5	76 978.2	3 830.3
南 宁	11	92.4	27 388.0	205 733.8	1 725.3
柳 州	10	60.5	17 019.1	195 758.2	500.0
桂 林	21	45.5	10 696.4	77 074.2	2 435.2
北 海	2	15.0	3 450.4	50 951.8	2 642.5
海 口	9	55.8	9 727.3	104 450.2	1.4
重 庆	131	274.1	98 203.4	693 787.5	28 418.5
成 都	131	216.6	56 532.0	444 389.6	11 713.2
自 贡	5	14.5	3 958.9	35 814.3	170.3
攀 枝 花	8	14.0	3 036.1	30 043.0	2 389.0
泸 州	9	12.2	4 272.7	63 711.5	150.0
德 阳	8	23.5	6 089.7	38 361.1	1 034.5
绵 阳	11	39.1	10 110.8	91 411.2	16 919.5
南 充	10	27.6	8 537.8	50 875.1	5 291.1
宜 宾	10	14.1	4 344.1	39 547.5	124.0
贵 阳	15	67.0	10 578.2	101 506.0	7 012.8
遵 义	23	23.6	3 961.9	55 672.6	2 946.9
昆 明	20	128.5	28 829.4	266 655.7	435.0
曲 靖	9	26.0	4 350.0	55 796.6	481.9
玉 溪	9	16.8	2 446.7	40 677.7	1.5
拉 萨	1	5.0	2 000.0	12 210.0	0.0
西 安	24	137.4	27 586.9	327 132.4	21 834.5
铜 川	3	6.7	1 619.3	7 408.0	5.4
宝 鸡	13	34.2	4 684.8	60 975.0	8 296.8
咸 阳	15	39.4	7 955.6	68 415.7	3 946.9
渭 南	11	25.5	4 063.3	58 770.0	2 269.1
延 安	16	10.5	3 940.2	81 157.2	192.0
兰 州	7	54.0	10 123.9	51 856.3	35.6
金 昌	2	9.2	1 223.8	21 054.7	0.8
西 宁	10	28.4	5 185.5	71 643.6	298.0
银 川	7	36.5	7 562.0	59 120.2	0.0
石 嘴 山	3	11.5	3 079.5	32 046.0	6 150.3
乌鲁木齐	9	74.0	15 568.5	85 837.6	32 610.5
克拉玛依	4	23.0	1 311.7	60 430.0	0.0

重点城市污水处理情况（二）

（2012）

单位：万吨

城　市 名　称	污水实际 处理量	污水再生 利用量	工业用水量	市政用水量	景观用水量
总　　计	2 809 496.9	89 338.8	52 665.2	6 594.0	30 079.6
北　京	125 644.6	9 623.6	4 677.5	1 753.7	3 192.5
天　津	68 581.2	2 707.1	2 017.9	338.2	351.0
石 家 庄	51 619.1	2 043.7	1 926.7	117.0	0.0
唐　山	28 210.6	5 970.0	4 675.1	0.0	1 294.9
秦 皇 岛	16 476.2	1 536.6	306.0	366.0	864.6
邯　郸	20 551.0	4 308.0	2 715.5	60.8	1 531.7
保　定	32 637.6	2 222.0	434.3	1 006.0	781.7
太　原	17 791.2	2 918.1	2 766.4	57.4	94.3
大　同	5 325.6	1 436.9	1 382.9	20.0	34.0
阳　泉	4 844.8	213.5	213.5	0.0	0.0
长　治	8 530.1	199.8	199.8	0.0	0.0
临　汾	5 581.0	433.1	431.4	1.7	0.0
呼和浩特	10 152.3	1 143.8	600.1	150.5	393.3
包　头	13 042.1	5 283.1	4 943.8	339.3	0.0
赤　峰	11 558.4	259.9	259.9	0.0	0.0
沈　阳	47 482.2	3 018.8	3 018.8	0.0	0.0
大　连	31 271.7	1 206.1	473.6	0.5	732.0
鞍　山	6 736.7	0.0	0.0	0.0	0.0
抚　顺	10 317.4	447.6	447.6	0.0	0.0
本　溪	9 179.5	0.0	0.0	0.0	0.0
锦　州	10 282.1	361.9	361.9	0.0	0.0
长　春	21 805.0	629.9	629.9	0.0	0.0
吉　林	18 366.9	0.0	0.0	0.0	0.0
哈 尔 滨	22 330.9	0.0	0.0	0.0	0.0
齐齐哈尔	5 463.6	0.0	0.0	0.0	0.0
牡 丹 江	6 537.0	1.0	1.0	0.0	0.0
上　海	202 335.7	0.0	0.0	0.0	0.0
南　京	64 172.7	838.6	263.7	0.0	574.9
无　锡	58 770.3	2 381.6	398.1	183.5	1 800.0
徐　州	27 661.9	1 068.7	993.6	7.6	67.5
常　州	31 025.1	2 007.5	932.1	0.0	1 075.3
苏　州	80 585.4	757.4	716.3	36.9	4.2
南　通	31 350.1	0.0	0.0	0.0	0.0
连 云 港	8 335.3	0.0	0.0	0.0	0.0
扬　州	19 572.7	0.0	0.0	0.0	0.0
镇　江	13 330.3	185.1	56.9	13.3	114.9
杭　州	85 760.6	3 640.4	3 639.7	0.7	0.0

重点城市污水处理情况（二）（续表）

（2012）   单位：万吨

城　市 名　称	污水实际 处理量	污水再生 利用量	工业用水量	市政用水量	景观用水量
宁　波	44 232.0	2 758.8	2 502.2	36.7	219.8
温　州	26 415.0	30.4	0.0	30.0	0.4
湖　州	15 604.4	1 022.1	1 022.1	0.0	0.0
绍　兴	38 477.8	2.0	2.0	0.0	0.0
合　肥	36 408.2	0.0	0.0	0.0	0.0
芜　湖	13 910.5	0.0	0.0	0.0	0.0
马 鞍 山	9 266.4	22.0	20.1	0.0	1.8
福　州	27 036.8	0.0	0.0	0.0	0.0
厦　门	22 452.7	43.8	0.0	35.5	8.3
泉　州	25 438.9	154.8	154.8	0.0	0.0
南　昌	31 920.7	0.0	0.0	0.0	0.0
九　江	9 866.9	525.2	313.9	10.7	200.6
济　南	27 390.5	1 949.9	839.7	286.3	823.9
青　岛	45 929.4	0.0	0.0	0.0	0.0
淄　博	25 054.7	94.9	94.9	0.0	0.0
枣　庄	13 173.5	55.0	0.0	0.0	55.0
烟　台	24 686.9	304.5	46.1	120.0	138.4
潍　坊	44 033.9	821.3	0.0	0.0	821.3
济　宁	27 166.4	2 173.5	2 113.5	0.0	60.0
泰　安	12 312.6	559.9	559.9	0.0	0.0
日　照	8 333.2	633.0	0.0	633.0	0.0
郑　州	45 993.9	2 397.1	827.1	4.0	1 566.0
开　封	8 809.4	0.0	0.0	0.0	0.0
洛　阳	19 712.5	758.7	758.7	0.0	0.0
平 顶 山	12 331.5	0.0	0.0	0.0	0.0
安　阳	8 401.5	921.6	638.6	283.0	0.0
焦　作	14 885.8	219.0	0.0	0.0	219.0
三 门 峡	5 458.1	0.0	0.0	0.0	0.0
武　汉	62 228.0	48.4	6.0	41.4	1.0
宜　昌	12 514.9	0.0	0.0	0.0	0.0
荆　州	11 561.1	0.0	0.0	0.0	0.0
长　沙	40 514.2	0.0	0.0	0.0	0.0
株　洲	13 265.2	0.0	0.0	0.0	0.0
湘　潭	10 011.9	12.5	1.5	3.0	8.0
岳　阳	10 756.2	0.0	0.0	0.0	0.0
常　德	8 830.0	0.0	0.0	0.0	0.0
张 家 界	2 658.1	0.0	0.0	0.0	0.0
广　州	123 559.3	8 573.9	741.2	48.6	7 784.1

重点城市污水处理情况（二）（续表）

（2012）

单位：万吨

城 市名 称	污水实际处理量	污水再生利用量	工业用水量	市政用水量	景观用水量
韶 关	7 638.5	0.0	0.0	0.0	0.0
深 圳	·110 560.5	3 986.9	17.3	82.9	3 886.6
珠 海	17 797.9	45.1	0.0	27.8	17.3
汕 头	17 870.1	162.0	60.0	82.0	20.0
湛 江	12 992.6	0.0	0.0	0.0	0.0
南 宁	25 152.3	3.0	0.0	0.0	3.0
柳 州	17 968.2	0.0	0.0	0.0	0.0
桂 林	11 783.8	0.0	0.0	0.0	0.0
北 海	4 591.7	0.0	0.0	0.0	0.0
海 口	17 665.9	7.0	0.0	7.0	0.0
重 庆	83 104.5	0.0	0.0	0.0	0.0
成 都	73 719.5	369.0	0.0	41.0	328.0
自 贡	5 480.2	0.0	0.0	0.0	0.0
攀 枝 花	2 448.9	16.7	16.7	0.0	0.0
泸 州	3 162.3	12.9	0.0	10.4	2.6
德 阳	7 627.7	24.3	13.9	0.7	9.7
绵 阳	9 776.6	0.0	0.0	0.0	0.0
南 充	7 408.5	0.0	0.0	0.0	0.0
宜 宾	4 500.3	2.0	2.0	0.0	0.0
贵 阳	20 670.4	0.0	0.0	0.0	0.0
遵 义	7 381.3	168.0	0.0	168.0	0.0
昆 明	42 748.9	0.0	0.0	0.0	0.0
曲 靖	6 041.8	264.3	264.3	0.0	0.0
玉 溪	3 340.2	0.0	0.0	0.0	0.0
拉 萨	1 800.0	0.0	0.0	0.0	0.0
西 安	29 015.3	716.0	595.0	105.0	16.0
铜 川	1 308.3	0.0	0.0	0.0	0.0
宝 鸡	8 875.4	728.4	698.9	29.5	0.0
咸 阳	11 483.4	964.1	21.1	0.0	943.0
渭 南	6 073.3	36.2	0.0	0.0	36.2
延 安	2 446.0	0.0	0.0	0.0	0.0
兰 州	11 454.3	107.2	104.2	0.0	3.0
金 昌	1 379.3	0.0	0.0	0.0	0.0
西 宁	6 154.1	0.0	0.0	0.0	0.0
银 川	11 888.3	0.0	0.0	0.0	0.0
石 嘴 山	2 538.5	800.0	745.6	54.4	0.0
乌鲁木齐	16 347.0	0.0	0.0	0.0	0.0
克拉玛依	3 480.0	0.0	0.0	0.0	0.0

重点城市污水处理情况（三）

（2012）

单位：万吨

城 市 名 称	污泥产生量	污泥处置量	土地利用量	填埋处置量	建筑材料利用量	焚烧处置量	污泥倾倒丢弃量
总　　计	1 831.6	1 831.6	371.7	893.1	165.5	401.3	0.00
北　京	117.0	117.0	82.2	25.6	2.3	6.9	0.00
天　津	34.3	34.3	1.7	31.1	1.2	0.2	0.00
石 家 庄	42.5	42.5	6.9	34.9	0.0	0.7	0.00
唐　山	26.7	26.7	2.6	23.8	0.3	0.0	0.00
秦 皇 岛	9.1	9.1	5.3	2.3	0.0	1.5	0.00
邯　郸	12.0	12.0	0.0	11.0	0.2	0.8	0.00
保　定	18.3	18.3	0.5	17.5	0.0	0.3	0.00
太　原	13.1	13.1	11.8	1.2	0.0	0.1	0.00
大　同	0.8	0.8	0.0	0.7	0.0	0.0	0.00
阳　泉	2.7	2.7	2.5	0.2	0.0	0.0	0.00
长　治	4.4	4.4	1.0	3.2	0.0	0.2	0.00
临　汾	2.7	2.7	1.7	0.6	0.0	0.4	0.00
呼和浩特	3.8	3.8	0.2	3.6	0.0	0.1	0.00
包　头	1.7	1.7	0.0	1.7	0.0	0.0	0.00
赤　峰	1.4	1.4	0.0	1.4	0.0	0.0	0.00
沈　阳	32.1	32.1	0.0	32.1	0.0	0.0	0.00
大　连	12.2	12.2	0.0	1.6	6.8	3.9	0.00
鞍　山	6.4	6.4	0.0	6.4	0.0	0.0	0.00
抚　顺	3.5	3.5	0.0	3.5	0.0	0.0	0.00
本　溪	3.6	3.6	0.0	3.6	0.1	0.0	0.00
锦　州	4.4	4.4	0.3	4.0	0.0	0.1	0.00
长　春	10.7	10.7	0.0	10.7	0.0	0.0	0.00
吉　林	6.2	6.2	0.0	6.2	0.0	0.0	0.00
哈 尔 滨	13.5	13.5	0.0	13.5	0.0	0.0	0.00
齐齐哈尔	1.8	1.8	0.0	1.8	0.0	0.0	0.00
牡 丹 江	1.0	1.0	0.0	1.0	0.0	0.0	0.00
上　海	124.4	124.4	4.5	107.7	6.7	5.5	0.00
南　京	16.1	16.1	0.3	5.6	2.6	7.7	0.00
无　锡	67.8	67.8	0.6	8.9	19.1	39.2	0.00
徐　州	14.1	14.1	2.1	1.8	1.2	9.0	0.00
常　州	33.7	33.7	0.0	1.1	6.6	26.0	0.00
苏　州	68.4	68.4	7.8	11.9	1.1	47.6	0.00
南　通	19.4	19.4	0.7	0.1	4.6	14.0	0.00
连 云 港	0.9	0.9	0.0	0.1	0.1	0.7	0.00
扬　州	8.8	8.8	0.0	2.4	0.3	6.2	0.00
镇　江	6.8	6.8	0.0	0.2	0.7	5.8	0.00
杭　州	152.4	152.4	23.4	16.7	39.5	72.8	0.00

重点城市污水处理情况（三）（续表）

（2012）

单位：万吨

城 市 名 称	污泥产生量	污泥处置量	土地利用量	填埋处置量	建筑材料利用量	焚烧处置量	污泥倾倒丢弃量
宁　波	10.0	10.0	1.4	2.0	0.1	6.5	0.00
温　州	17.4	17.4	1.4	11.8	2.6	1.6	0.00
湖　州	9.8	9.8	0.0	3.2	2.3	4.3	0.00
绍　兴	65.9	65.9	0.0	4.3	1.3	60.3	0.00
合　肥	20.1	20.1	8.1	0.5	0.0	11.5	0.00
芜　湖	6.3	6.3	0.0	1.0	0.0	5.3	0.00
马 鞍 山	3.9	3.9	0.2	3.7	0.0	0.0	0.00
福　州	10.4	10.4	3.2	7.3	0.0	0.0	0.00
厦　门	12.4	12.4	4.3	6.1	1.3	0.7	0.00
泉　州	11.4	11.4	0.0	2.1	2.2	7.1	0.00
南　昌	6.7	6.7	0.0	0.3	4.4	1.9	0.00
九　江	3.2	3.2	0.2	2.7	0.2	0.2	0.00
济　南	18.5	18.5	17.1	0.5	0.0	0.9	0.00
青　岛	35.7	35.7	15.5	7.8	1.1	11.4	0.00
淄　博	29.2	29.2	0.0	25.4	0.6	3.1	0.00
枣　庄	5.2	5.2	0.2	0.0	0.0	5.0	0.00
烟　台	18.5	18.5	0.8	15.8	0.0	1.8	0.00
潍　坊	31.5	31.5	4.6	13.8	1.3	11.8	0.00
济　宁	12.1	12.1	1.5	5.8	1.1	3.7	0.00
泰　安	5.9	5.9	3.5	0.0	0.9	1.5	0.00
日　照	1.4	1.4	1.1	0.0	0.1	0.1	0.00
郑　州	49.3	49.3	20.3	29.1	0.0	0.0	0.00
开　封	4.8	4.8	0.0	4.8	0.0	0.0	0.00
洛　阳	12.7	12.7	0.0	12.7	0.0	0.0	0.00
平 顶 山	7.6	7.6	0.0	7.6	0.0	0.0	0.00
安　阳	5.9	5.9	0.3	5.6	0.0	0.0	0.00
焦　作	8.6	8.6	2.0	4.4	2.2	0.0	0.00
三 门 峡	2.5	2.5	0.1	2.4	0.0	0.0	0.00
武　汉	24.5	24.5	0.1	24.4	0.0	0.0	0.00
宜　昌	2.6	2.6	0.0	1.7	0.0	0.9	0.00
荆　州	4.2	4.2	1.2	2.7	0.3	0.0	0.00
长　沙	5.3	5.3	0.0	5.3	0.0	0.0	0.00
株　洲	6.6	6.6	0.0	3.2	3.4	0.0	0.00
湘　潭	4.9	4.9	0.0	0.7	3.7	0.5	0.00
岳　阳	2.3	2.3	0.1	2.2	0.0	0.0	0.00
常　德	4.4	4.4	0.0	4.0	0.4	0.0	0.00
张 家 界	1.3	1.3	0.0	1.3	0.0	0.0	0.00
广　州	77.4	77.4	56.0	0.6	18.0	2.8	0.00

重点城市污水处理情况（三）（续表）

（2012）   单位：万吨

城 市 名 称	污泥产生量	污泥处置量	土地利用量	填埋处置量	建筑材料利用量	焚烧处置量	污泥倾倒 丢弃量
韶 关	0.8	0.8	0.0	0.8	0.0	0.0	0.00
深 圳	109.9	109.9	25.1	84.7	0.0	0.0	0.00
珠 海	9.1	9.1	0.0	3.2	0.0	5.9	0.00
汕 头	9.5	9.5	0.0	9.5	0.0	0.0	0.00
湛 江	5.4	5.4	5.4	0.0	0.0	0.0	0.00
南 宁	8.8	8.8	8.8	0.0	0.0	0.0	0.00
柳 州	4.5	4.5	0.4	0.7	3.5	0.0	0.00
桂 林	1.4	1.4	1.1	0.3	0.1	0.0	0.00
北 海	1.9	1.9	0.0	1.9	0.0	0.0	0.00
海 口	4.2	4.2	4.2	0.0	0.0	0.0	0.00
重 庆	38.3	38.3	3.0	19.3	16.0	0.0	0.00
成 都	28.6	28.6	6.3	21.3	0.8	0.2	0.00
自 贡	1.1	1.1	0.0	1.1	0.0	0.0	0.00
攀 枝 花	9.0	9.0	0.0	9.0	0.0	0.0	0.00
泸 州	1.9	1.9	0.0	1.9	0.0	0.0	0.00
德 阳	1.6	1.6	0.0	1.0	0.0	0.5	0.00
绵 阳	4.6	4.6	0.0	0.5	4.1	0.0	0.00
南 充	1.3	1.3	0.0	1.3	0.0	0.0	0.00
宜 宾	1.6	1.6	0.0	1.5	0.1	0.0	0.00
贵 阳	3.5	3.5	0.0	3.4	0.0	0.0	0.00
遵 义	1.7	1.7	0.0	1.7	0.0	0.0	0.00
昆 明	24.1	24.1	6.6	17.6	0.0	0.0	0.00
曲 靖	2.7	2.7	0.3	2.5	0.0	0.0	0.00
玉 溪	1.1	1.1	1.1	0.1	0.0	0.0	0.00
拉 萨	36.0	36.0	0.0	36.0	0.0	0.0	0.00
西 安	19.9	19.9	6.0	13.8	0.0	0.0	0.00
铜 川	0.5	0.5	0.0	0.5	0.0	0.0	0.00
宝 鸡	1.4	1.4	0.1	1.2	0.0	0.0	0.00
咸 阳	6.9	6.9	0.4	5.9	0.0	0.6	0.00
渭 南	3.8	3.8	0.0	3.8	0.0	0.0	0.00
延 安	2.2	2.2	0.0	2.2	0.0	0.0	0.00
兰 州	9.3	9.3	0.0	9.3	0.0	0.0	0.00
金 昌	0.8	0.8	0.7	0.1	0.0	0.0	0.00
西 宁	4.9	4.9	0.0	4.9	0.0	0.0	0.00
银 川	7.2	7.2	2.5	3.8	0.0	0.9	0.00
石 嘴 山	0.8	0.8	0.0	0.0	0.0	0.8	0.00
乌鲁木齐	1.8	1.8	0.2	1.6	0.0	0.0	0.00
克拉玛依	0.2	0.2	0.0	0.2	0.0	0.0	0.00

重点城市污水处理情况（四）

（2012）

単位：吨

城 市 名 称	污染物去除量						
	化学需氧量	氨氮	油类	总氮	总磷	挥发酚	氰化物
总　　计	7 384 050.5	639 348.3	35 776.0	526 022.2	85 332.4	692.8	489.3
北　　京	430 303.5	52 609.6	7 298.9	56 603.4	5 393.6	16.3	2.4
天　　津	228 078.1	22 648.4	753.2	25 903.2	1 984.9	0.6	0.1
石 家 庄	155 763.8	16 728.9	53.0	2 893.6	494.4	0.0	2.2
唐　　山	82 799.0	6 845.7	109.4	5 649.3	844.0	0.0	0.0
秦 皇 岛	45 015.0	4 186.9	0.0	1 279.4	190.3	0.0	0.0
邯　　郸	56 192.3	5 666.6	39.3	2 060.5	328.4	0.0	0.0
保　　定	90 621.6	8 236.3	39.6	5 083.2	634.0	0.0	0.0
太　　原	79 963.7	7 029.4	1 325.0	7 624.0	1 105.4	3.3	0.3
大　　同	22 780.7	2 208.8	3.1	23.5	2.5	0.0	0.0
阳　　泉	12 325.3	1 291.1	0.0	465.1	22.0	0.0	0.0
长　　治	25 630.7	2 200.9	123.4	2 108.8	273.8	21.3	0.2
临　　汾	16 987.9	1 534.6	0.0	940.7	88.4	0.0	0.0
呼和浩特	27 246.5	2 171.8	0.0	459.6	156.5	0.0	0.0
包　　头	27 973.0	2 983.6	904.2	4 960.2	362.0	0.0	11.5
赤　　峰	35 190.1	1 398.1	0.0	4 717.2	0.0	0.0	0.0
沈　　阳	114 567.8	6 085.1	86.4	1 714.5	355.9	0.0	0.0
大　　连	71 123.0	8 273.5	191.1	5 355.1	733.9	0.0	0.0
鞍　　山	15 160.0	640.4	0.0	1 096.6	119.4	0.0	0.0
抚　　顺	23 402.5	1 429.8	0.0	1 573.5	227.5	0.0	0.0
本　　溪	16 655.2	1 903.4	0.0	2 516.7	61.9	0.0	0.0
锦　　州	20 232.5	1 753.9	32.1	1 786.1	246.3	0.0	0.0
长　　春	57 952.0	4 539.2	0.0	1 054.9	187.6	0.0	0.0
吉　　林	50 327.8	877.6	823.0	1 292.6	719.3	27.5	1.1
哈 尔 滨	51 899.9	4 968.6	229.3	5 359.2	852.3	0.0	0.0
齐齐哈尔	16 998.3	1 865.0	0.0	1 794.7	395.2	0.0	0.0
牡 丹 江	17 375.0	1 082.8	3.0	894.9	218.5	0.0	0.0
上　　海	569 078.9	43 450.9	1 332.0	43 062.4	5 990.7	163.3	14.8
南　　京	121 832.7	8 975.4	457.5	6 764.4	1 325.9	0.2	0.0
无　　锡	184 797.6	14 289.6	1 999.3	14 624.0	1 439.1	0.0	0.1
徐　　州	47 750.3	6 098.8	0.9	895.7	1 587.9	113.3	0.0
常　　州	88 820.8	7 024.7	132.1	4 463.9	943.3	30.3	22.9
苏　、州	231 598.6	19 025.5	315.3	10 746.7	2 112.0	54.2	0.0
南　　通	90 319.3	6 630.0	0.0	909.8	515.0	0.0	0.0
连 云 港	14 390.4	1 450.0	6.5	1 232.2	144.3	2.0	0.0
扬　　州	44 226.6	4 281.6	0.0	104.8	82.1	0.0	0.0
镇　　江	25 282.0	2 985.8	263.3	2 222.5	271.6	0.0	0.0
杭　　州	434 660.2	19 500.1	32.1	6 213.7	2 667.8	0.0	0.0

重点城市污水处理情况（四）（续表）

（2012）

单位：吨

城 市 名 称	污染物去除量						
	化学需氧量	氨氮	油类	总氮	总磷	挥发酚	氰化物
宁 波	118 260.3	9 543.8	640.7	4 323.9	854.8	0.0	83.3
温 州	71 351.2	5 319.4	145.2	1 053.1	283.5	0.0	223.7
湖 州	37 789.4	2 135.9	205.3	1 669.0	261.9	0.0	0.0
绍 兴	280 481.4	17 143.1	3 451.8	17 322.4	1 902.9	0.0	0.0
合 肥	48 319.2	4 923.5	352.5	4 049.7	1 813.7	34.7	0.0
芜 湖	20 716.2	1 507.0	13.8	1 693.4	230.3	0.0	0.0
马 鞍 山	13 826.2	885.4	1.7	1 557.3	150.4	0.0	0.0
福 州	50 788.4	5 234.0	360.3	6 273.7	533.6	0.0	0.0
厦 门	41 019.2	5 760.9	92.5	4 931.4	747.2	0.0	1.6
泉 州	111 392.3	3 570.4	366.4	3 593.7	310.6	0.0	98.2
南 昌	41 568.2	3 336.9	0.0	3 634.5	492.5	0.0	0.0
九 江	13 149.2	1 019.6	23.5	1 327.3	226.9	0.0	0.0
济 南	70 448.5	8 765.3	0.0	6 765.4	1 250.9	0.0	0.0
青 岛	189 166.4	16 259.7	0.0	18 137.4	2 064.7	0.0	22.9
淄 博	74 497.2	7 245.1	114.3	431.9	924.9	0.0	0.0
枣 庄	27 407.3	2 884.6	0.0	1 241.7	176.8	0.0	0.0
烟 台	70 786.8	7 248.8	12.7	4 454.1	756.5	0.0	0.0
潍 坊	148 360.7	12 091.7	49.1	5 953.1	735.7	0.5	0.0
济 宁	65 823.9	5 907.5	0.0	4 093.6	511.5	0.0	0.0
泰 安	33 575.2	2 542.7	0.0	1 089.8	48.6	0.0	0.0
日 照	23 206.0	1 841.5	0.0	1 499.4	160.8	0.0	0.6
郑 州	122 521.5	13 587.8	758.6	16 846.8	1 116.5	51.4	1.7
开 封	19 489.6	1 215.9	0.0	1 476.2	281.2	0.0	0.0
洛 阳	46 761.4	5 491.4	22.0	6 154.9	949.9	0.1	0.0
平 顶 山	27 863.3	1 909.6	0.0	3 138.8	251.2	0.0	0.0
安 阳	21 075.7	1 638.3	0.0	1 958.6	449.9	0.0	0.0
焦 作	60 138.9	3 218.1	0.0	3 539.1	371.0	0.0	0.0
三 门 峡	13 480.3	1 531.1	0.0	2 406.4	276.7	0.0	0.0
武 汉	83 970.4	9 040.9	3.0	653.3	872.7	0.0	0.0
宜 昌	20 671.6	1 549.3	13.3	1 911.3	200.9	0.0	0.0
荆 州	17 011.0	1 153.9	0.0	1 007.3	11 421.8	0.0	0.0
长 沙	66 512.2	6 917.6	0.0	5 997.7	772.2	0.0	0.0
株 洲	21 030.1	1 778.6	0.0	1 319.8	175.7	0.0	0.0
湘 潭	17 751.8	1 221.0	0.2	1 296.1	134.8	0.0	0.0
岳 阳	17 359.2	1 766.1	0.0	1 289.0	100.5	0.0	0.0
常 德	10 228.9	1 106.0	25.0	544.0	179.9	8.7	0.0
张 家 界	2 850.5	369.1	0.0	0.0	5.5	0.0	0.0
广 州	245 094.1	20 752.3	92.5	17 050.8	3 424.4	0.6	0.0

重点城市污水处理情况（四）（续表）

单位：吨

城市名称	污染物去除量						
	化学需氧量	氨氮	油类	总氮	总磷	挥发酚	氰化物
韶 关	5 964.4	497.7	97.8	406.2	72.7	0.0	0.0
深 圳	212 686.0	25 088.5	26.4	18 869.0	4 362.8	0.0	0.0
珠 海	30 618.2	2 364.8	180.0	1 797.2	231.8	0.4	0.0
汕 头	21 974.1	2 480.1	39.7	1 783.6	147.9	0.0	0.0
湛 江	23 880.7	2 398.3	0.0	351.5	347.6	0.0	0.0
南 宁	41 985.8	5 219.7	2 075.1	3 923.3	625.5	0.0	0.0
柳 州	26 453.4	2 500.3	70.8	1 790.3	227.8	0.0	0.0
桂 林	19 207.7	1 668.6	0.0	1 917.7	413.9	0.0	0.0
北 海	6 926.3	1 027.3	0.0	854.5	139.5	0.0	0.0
海 口	32 877.9	1 043.3	24.6	1 560.9	451.0	0.6	0.0
重 庆	201 319.2	16 180.9	1 078.7	24 113.9	2 701.6	0.2	0.0
成 都	143 141.8	16 653.2	1 302.7	21 813.5	1 536.9	70.4	1.3
自 贡	12 362.4	1 258.5	0.0	1 008.4	121.5	0.0	0.0
攀 枝 花	5 851.3	591.7	350.5	687.5	74.7	0.0	0.0
泸 州	6 044.7	739.5	5.3	581.0	70.7	0.0	0.0
德 阳	17 388.9	1 640.0	16.3	1 275.1	70.0	0.0	0.0
绵 阳	21 512.6	2 013.7	2.6	1 292.3	130.0	0.0	0.0
南 充	16 959.9	1 941.4	349.2	284.1	53.4	0.0	0.0
宜 宾	10 651.8	985.8	39.2	1 063.3	220.6	0.0	0.0
贵 阳	35 822.9	3 915.0	70.3	3 147.9	102.4	0.0	0.0
遵 义	14 808.4	1 960.0	101.0	504.7	80.8	0.0	0.0
昆 明	111 375.6	9 106.6	492.6	11 550.8	1 132.4	0.0	0.0
曲 靖	8 614.9	1 108.9	7.5	724.6	104.4	0.0	0.0
玉 溪	6 313.9	394.0	3.1	441.8	48.9	0.0	0.0
拉 萨	810.0	72.0	198.0	81.0	12.6	0.0	0.0
西 安	81 211.2	7 923.3	343.9	10 563.9	1 115.6	0.0	0.0
铜 川	3 948.2	354.0	0.0	532.7	50.3	0.0	0.0
宝 鸡	22 695.8	2 060.3	0.0	2 065.5	222.0	0.0	0.0
咸 阳	36 422.5	3 612.7	1 294.3	2 574.4	219.0	0.0	0.0
渭 南	16 153.0	1 426.6	106.8	1 568.2	156.6	0.0	0.0
延 安	7 124.9	650.1	0.0	107.4	15.3	0.0	0.0
兰 州	40 473.5	3 703.9	320.2	4 063.6	714.7	83.6	0.4
金 昌	4 824.5	415.4	54.6	461.7	38.1	0.3	0.0
西 宁	18 166.7	727.8	0.0	0.0	0.0	0.0	0.0
银 川	29 474.0	2 287.9	784.4	2 742.7	654.8	9.1	0.0
石 嘴 山	6 019.0	462.0	127.2	453.1	70.3	0.0	0.0
乌鲁木齐	61 585.7	7 564.1	0.0	0.0	0.0	0.0	0.0
克拉玛依	9 332.9	1 024.0	2 915.6	928.3	93.0	0.0	0.0

重点城市生活垃圾处理情况（一）

（2012）

城　市 名　称	生活垃圾处理厂数/ 座	实际处理量/ 万吨	本年运行费用/ 万元	生活垃圾处理厂累计 完成投资/万元	新增固定资产/ 万元
总　　　计	821	11 414.6	597 977.9	5 778 588.0	220 509.9
北　京	19	637.2	75 005.9	320 297.2	21 690.3
天　津	11	222.9	21 647.3	250 456.4	2 138.0
石 家 庄	16	66.9	2 266.1	47 195.2	322.4
唐　山	17	114.4	5 308.5	50 963.3	134.0
秦 皇 岛	4	16.3	1 307.5	8 245.0	1.7
邯　郸	16	119.4	2 090.8	68 986.6	7 800.0
保　定	8	81.0	3 263.0	28 469.2	32.0
太　原	5	165.1	2 329.0	22 802.0	189.4
大　同	4	15.0	883.0	6 383.0	0.0
阳　泉	2	17.4	553.0	8 344.0	2 300.0
长　治	9	31.8	2 059.1	32 093.1	20.0
临　汾	14	57.1	1 922.9	19 145.0	320.0
呼 和 浩 特	5	53.9	1 331.0	54 575.0	0.0
包　头	3	73.0	5 170.5	27 000.0	5 200.0
赤　峰	13	65.7	3 490.0	25 799.4	80.0
沈　阳	3	241.3	10 459.2	29 179.0	350.0
大　连	2	104.3	6 800.0	35 935.0	12 000.0
鞍　山	1	55.0	4 270.0	7 300.0	88.0
抚　顺	1	50.0	305.0	305.0	0.0
本　溪	2	38.5	540.0	14 751.5	0.0
锦　州	2	20.7	1 125.6	13 900.0	0.0
长　春	6	117.8	6 540.1	45 504.5	0.0
吉　林	6	54.1	2 777.6	37 781.2	9 761.1
哈 尔 滨	3	110.4	1 878.0	18 400.0	173.0
齐 齐 哈 尔	2	26.0	1 325.0	10 070.0	0.0
牡 丹 江	6	61.2	1 287.0	22 165.0	730.0
上　海	24	711.0	82 735.4	378 839.6	8 900.4
南　京	6	261.1	2 999.5	26 529.4	316.0
无　锡	4	77.5	16 535.5	87 060.9	22.3
徐　州	3	26.5	1 683.1	11 551.6	109.8
常　州	7	112.3	6 794.2	73 253.5	9 572.8
苏　州	7	207.4	12 953.8	200 820.6	425.8
南　通	5	93.8	3 164.0	57 175.0	1 200.0
连 云 港	7	60.8	4 693.6	37 269.0	280.0
扬　州	6	88.6	5 280.0	27 145.6	0.0
镇　江	5	84.3	871.9	67 610.0	559.0
杭　州	12	207.6	16 923.1	135 349.1	15 497.8

重点城市生活垃圾处理情况（一）（续表）

（2012）

城　市名　称	生活垃圾处理厂数/座	实际处理量/万吨	本年运行费用/万元	生活垃圾处理厂累计完成投资/万元	新增固定资产/万元
宁　波	9	246.9	19 472.9	199 123.8	2 320.1
温　州	20	179.8	17 876.3	112 709.4	4 413.7
湖　州	1	14.6	200.0	1 540.0	1 000.0
绍　兴	11	93.8	8 742.9	107 488.4	530.0
合　肥	25	86.7	3 925.0	62 582.6	44 657.5
芜　湖	20	42.8	5 864.8	33 589.5	840.0
马 鞍 山	18	35.2	3 209.6	13 523.0	286.0
福　州	4	52.3	1 901.5	54 108.0	6 254.9
厦　门	1	81.0	2 054.0	140 900.0	0.0
泉　州	28	52.5	1 514.0	22 684.6	0.0
南　昌	1	78.0	1 100.0	11 000.0	0.0
九　江	12	90.2	576.5	14 151.2	162.0
济　南	5	70.2	4 560.5	31 276.5	454.3
青　岛	6	151.6	11 928.2	37 675.0	2 175.0
淄　博	2	8.9	248.0	273 500.0	150.0
枣　庄	1	12.8	680.0	5 000.0	0.0
烟　台	8	100.8	5 364.5	44 334.0	3 845.0
潍　坊	8	125.2	3 370.0	48 202.5	1 703.5
济　宁	4	60.6	1 471.0	33 591.0	2 056.0
泰　安	4	23.5	1 412.0	11 721.0	2 791.0
日　照	4	32.5	1 960.5	25 660.0	633.0
郑　州	6	119.5	7 660.0	33 899.0	1 128.0
开　封	3	11.5	533.0	7 943.0	0.0
洛　阳	10	82.6	4 325.0	30 305.0	1 700.0
平 顶 山	7	52.0	2 277.0	30 171.0	1 217.0
安　阳	3	48.7	608.4	14 115.9	78.0
焦　作	7	53.8	3 314.0	29 563.7	575.1
三 门 峡	6	39.8	695.0	16 099.0	723.0
武　汉	8	432.3	11 373.8	503 671.6	0.0
宜　昌	20	64.3	1 696.0	51 196.9	967.2
荆　州	10	23.6	2 075.2	42 971.0	1 520.0
长　沙	4	190.0	23 178.0	40 845.0	0.0
株　洲	6	54.0	2 649.6	21 862.6	10.0
湘　潭	7	47.6	1 881.5	29 449.0	821.0
岳　阳	11	72.0	2 276.4	38 055.1	1 080.0
常　德	6	35.3	902.0	20 730.0	1 500.0
张 家 界	1	14.6	858.0	3 880.0	0.0
广　州	6	519.8	12 018.2	84 169.1	11 840.0

重点城市生活垃圾处理情况（一）（续表）

（2012）

城 市 名 称	生活垃圾处理厂数/ 座	实际处理量/ 万吨	本年运行费用/ 万元	生活垃圾处理厂累计 完成投资/万元	新增固定资产/ 万元
韶 关	9	48.1	1 715.3	28 021.6	1 550.0
深 圳	11	725.0	14 497.6	39 373.5	63.4
珠 海	2	62.3	1 329.9	60 414.0	9.6
汕 头	4	299.8	1 769.8	11 361.0	143.5
湛 江	4	39.5	949.5	14 373.0	8.0
南 宁	4	81.6	1 760.7	38 692.8	86.4
柳 州	1	40.0	824.5	16 110.5	2.7
桂 林	7	46.6	1 378.0	18 365.3	260.0
北 海	1	18.3	1 240.3	27 000.0	0.0
海 口	0	0.0	0.0	0.0	0.0
重 庆	44	512.8	23 813.8	282 435.2	3 772.1
成 都	11	226.0	5 339.0	166 644.2	4 206.9
自 贡	2	27.9	842.0	8 048.7	0.0
攀 枝 花	4	21.9	963.0	8 705.0	564.0
泸 州	4	40.2	1 147.1	11 734.0	42.0
德 阳	5	33.4	2 064.0	18 499.0	4 403.5
绵 阳	8	59.5	3 441.1	25 440.0	0.0
南 充	5	52.5	1 758.0	16 223.0	100.0
宜 宾	9	56.2	5 306.7	31 891.9	0.0
贵 阳	4	40.5	2 042.8	37 291.5	1 594.3
遵 义	9	87.5	2 556.0	39 046.0	710.0
昆 明	6	21.5	1 659.8	20 957.5	0.0
曲 靖	4	26.9	7 199.6	43 605.9	316.6
玉 溪	11	34.9	1 863.9	16 450.7	2 108.9
拉 萨	1	18.4	151.0	5 187.8	0.0
西 安	3	243.2	2 395.0	25 030.4	277.0
铜 川	4	18.2	206.0	5 643.0	150.0
宝 鸡	8	43.0	2 069.3	43 504.0	47.3
咸 阳	9	39.5	1 994.4	18 332.2	371.0
渭 南	5	32.3	857.0	10 949.0	670.0
延 安	12	34.1	895.8	24 611.9	19.7
兰 州	10	29.7	1 568.9	2 885.5	0.0
金 昌	2	18.6	248.0	2 206.4	0.0
西 宁	7	82.3	1 121.9	16 072.2	260.0
银 川	3	47.7	898.8	10 537.0	0.0
石 嘴 山	3	11.8	857.0	5 627.0	164.0
乌鲁木齐	3	127.1	1 349.4	24 482.5	964.1
克拉玛依	3	17.8	1 580.8	10 897.4	0.0

重点城市生活垃圾处理情况（二）

（2012）

单位：吨

城 市 名 称	渗滤液中污染物排放量					
	化学需氧量	氨氮	油类	总磷	挥发酚	氰化物
总　　计	83 956.5	8 923.7	59.3	115.9	8.2	0.3
北　京	6 279.5	634.8	2.4	11.6	0.3	—
天　津	460.3	37.6	—	0.1	—	—
石 家 庄	291.0	28.6	0.3	0.3	0.1	0.0
唐　山	1 045.4	122.1	1.2	2.4	0.2	0.0
秦 皇 岛	2.2	1.1	0.0	0.0	0.0	0.0
邯　郸	3 966.6	179.1	2.8	5.2	0.6	0.0
保　定	1 077.5	106.7	1.0	1.5	0.2	0.0
太　原	169.1	18.0	0.1	0.2	0.0	0.0
大　同	142.9	12.8	0.1	0.2	0.0	0.0
阳　泉	240.0	24.0	0.4	0.0	0.0	0.0
长　治	334.6	17.3	0.4	0.5	0.0	0.0
临　汾	285.3	24.3	2.0	2.2	0.1	0.0
呼和浩特	383.3	68.6	0.3	0.6	0.0	0.0
包　头	241.8	16.1	0.0	0.0	0.0	0.0
赤　峰	350.9	25.0	0.1	0.1	0.0	0.0
沈　阳	37.2	2.0	0.2	2.5	0.0	0.0
大　连	2.7	0.3	0.0	1.0	0.0	0.0
鞍　山	1 379.1	473.4	0.3	0.3	0.1	0.0
抚　顺	1 000.0	100.0	1.5	1.6	0.0	0.0
本　溪	157.8	14.5	0.0	0.0	0.0	0.0
锦　州	0.6	0.1	0.0	0.0	0.0	0.0
长　春	5 067.4	573.2	3.5	5.7	0.2	0.0
吉　林	944.1	94.4	1.2	1.7	0.2	0.0
哈 尔 滨	20.7	3.5	0.4	0.4	0.1	0.0
齐齐哈尔	274.0	31.3	0.0	0.4	0.0	0.0
牡 丹 江	795.0	89.8	0.3	0.2	0.1	0.0
上　海	5 805.3	294.1	3.4	6.3	1.3	—
南　京	598.7	88.5	0.6	1.0	0.3	0.0
无　锡	60.1	67.8	0.1	0.0	0.0	0.0
徐　州	591.6	7.5	0.0	0.0	0.0	0.0
常　州	10.4	0.4	0.0	0.0	0.0	0.0
苏　州	158.3	44.0	2.8	0.1	0.0	0.0
南　通	258.6	31.3	0.2	2.4	0.0	0.0
连 云 港	1 416.0	120.0	0.4	1.5	0.0	0.0
扬　州	887.1	129.8	0.1	0.3	0.0	0.0
镇　江	157.8	41.0	0.0	0.0	0.0	0.0
杭　州	247.4	26.1	0.3	0.1	0.0	0.0

重点城市生活垃圾处理情况（二）（续表）

（2012）

单位：吨

城 市 名 称	渗滤液中污染物排放量					
	化学需氧量	氨氮	油类	总磷	挥发酚	氰化物
宁 波	1 065.6	98.0	0.0	0.0	0.0	0.0
温 州	2 295.9	180.5	1.6	3.1	0.4	0.0
湖 州	40.9	6.8	0.2	0.1	0.0	0.0
绍 兴	818.9	69.0	1.0	3.0	0.0	0.0
合 肥	968.5	99.7	0.5	0.7	0.0	0.0
芜 湖	203.8	17.2	0.1	0.2	0.0	0.0
马 鞍 山	314.2	26.5	0.1	0.5	0.0	0.0
福 州	754.9	58.5	0.8	1.4	0.1	0.0
厦 门	89.0	13.4	0.0	0.0	0.0	0.0
泉 州	203.5	17.1	0.6	0.4	0.1	0.0
南 昌	710.0	62.0	1.0	1.0	0.0	0.0
九 江	1 006.1	89.8	1.2	2.0	0.1	0.0
济 南	259.1	27.4	0.4	0.6	0.1	0.0
青 岛	283.4	30.3	0.2	0.3	0.0	0.0
淄 博	0.2	0.0	0.0	0.0	0.0	0.0
枣 庄	0.6	0.0	0.0	0.0	0.0	0.0
烟 台	43.6	4.9	0.2	0.3	0.1	0.0
潍 坊	737.9	81.6	0.3	0.4	0.0	0.0
济 宁	217.2	38.4	0.4	0.5	0.0	0.0
泰 安	77.5	4.4	0.1	0.3	0.0	0.0
日 照	250.9	24.7	0.4	0.5	0.1	0.0
郑 州	981.2	156.3	1.2	2.8	0.2	0.0
开 封	303.7	40.6	0.3	0.4	0.1	0.0
洛 阳	843.4	87.9	0.5	0.4	0.1	0.0
平 顶 山	169.2	19.0	0.1	0.0	0.0	0.0
安 阳	7.8	0.9	0.0	0.0	0.0	0.0
焦 作	197.7	23.5	0.1	0.2	0.0	0.0
三 门 峡	51.7	5.9	0.0	0.0	0.0	0.0
武 汉	3 624.9	251.0	0.0	0.0	0.0	0.0
宜 昌	816.9	169.3	0.4	0.8	0.0	0.0
荆 州	388.3	83.2	0.4	0.6	0.1	0.0
长 沙	491.9	83.1	0.2	0.5	0.0	0.0
株 洲	524.8	43.5	0.4	0.9	0.1	0.0
湘 潭	1 314.8	120.5	0.7	2.4	0.1	0.0
岳 阳	2 655.0	205.5	1.2	2.9	0.2	0.0
常 德	1 731.1	131.9	0.5	1.4	0.1	0.0
张 家 界	5.6	1.5	0.0	0.0	0.0	0.0
广 州	368.2	25.7	0.9	0.6	0.0	0.0

重点城市生活垃圾处理情况（二）（续表）

（2012）

单位：吨

城 市 名 称	渗滤液中污染物排放量					
	化学需氧量	氨氮	油类	总磷	挥发酚	氰化物
韶 关	1 174.8	94.6	0.2	0.8	0.0	0.0
深 圳	1 513.4	330.2	0.1	0.7	0.0	0.0
珠 海	295.5	19.7	0.0	0.5	0.0	0.0
汕 头	939.2	93.2	0.2	0.0	0.0	0.0
湛 江	752.4	94.2	0.1	0.2	0.0	0.0
南 宁	541.3	46.6	0.7	0.7	0.0	0.0
柳 州	27.2	2.7	0.1	0.2	0.0	0.0
桂 林	702.4	58.9	0.4	0.6	0.0	0.0
北 海	9.4	0.9	0.0	0.0	0.0	0.0
海 口	0.0	0.0	0.0	0.0	0.0	0.0
重 庆	600.3	155.1	1.3	0.9	—	—
成 都	1 247.1	129.9	1.8	0.9	0.1	0.0
自 贡	357.5	44.7	0.4	0.6	0.1	0.0
攀 枝 花	142.6	16.3	0.0	0.1	0.0	0.0
泸 州	77.1	9.3	0.0	0.1	0.1	0.0
德 阳	45.7	1.0	0.1	0.2	0.0	0.0
绵 阳	83.6	17.2	0.3	0.2	0.0	0.0
南 充	766.3	123.3	0.5	1.5	0.0	0.0
宜 宾	161.9	18.1	0.3	1.2	0.0	0.0
贵 阳	2 591.7	298.0	0.1	0.1	0.0	0.0
遵 义	956.5	121.0	0.6	1.0	0.0	0.0
昆 明	1 935.8	213.0	4.3	11.0	1.1	0.0
曲 靖	119.1	19.2	0.0	0.0	0.0	0.0
玉 溪	977.6	105.2	0.1	0.4	0.0	0.0
拉 萨	154.2	10.3	0.1	0.2	0.0	0.0
西 安	3 433.0	600.0	1.6	3.0	0.0	0.0
铜 川	2.5	0.8	0.0	0.0	0.0	0.0
宝 鸡	247.9	39.7	0.4	0.4	0.0	0.0
咸 阳	40.8	3.0	0.0	0.0	0.0	0.0
渭 南	2.1	0.1	0.0	0.0	0.0	0.0
延 安	31.1	3.5	0.0	0.0	0.0	0.0
兰 州	430.3	25.0	0.2	0.2	0.0	0.0
金 昌	46.7	4.7	0.1	0.0	0.0	0.0
西 宁	814.7	75.2	0.4	0.5	0.1	0.0
银 川	201.0	19.4	0.2	0.2	0.0	0.0
石 嘴 山	13.1	1.1	0.1	0.0	0.0	0.0
乌鲁木齐	2 488.3	175.8	2.5	9.6	0.6	0.0
克拉玛依	74.7	4.7	0.1	0.2	0.0	0.0

重点城市生活垃圾处理情况（三）

（2012）　　　　　　　　　　　　　　　　　　　　　　单位：千克

城　市 名　称	渗滤液中污染物排放量					
	汞	镉	六价铬	总铬	铅	砷
总　　计	62.9	304.7	48.5	644.4	1 054.5	376.5
北　京	0.5	16.8	0.2	31.1	162.2	9.2
天　津	—	—	—	0.1	0.4	0.2
石 家 庄	0.2	0.9	0.1	1.9	3.9	1.0
唐　山	0.5	3.3	0.5	6.4	13.8	5.2
秦 皇 岛	0.0	0.0	0.0	0.0	0.0	0.0
邯　郸	1.8	14.0	0.0	27.5	69.3	17.6
保　定	0.4	3.2	0.0	5.2	12.7	4.5
太　原	0.1	0.3	0.0	0.7	1.9	0.6
大　同	0.1	0.4	0.0	0.7	1.6	0.4
阳　泉	0.2	1.2	0.0	2.4	4.5	0.9
长　治	0.0	0.0	0.0	0.0	0.0	0.0
临　汾	0.1	0.6	0.1	1.7	2.3	1.1
呼和浩特	0.2	0.0	0.0	0.0	7.5	0.0
包　头	0.0	0.0	0.0	0.0	0.0	0.0
赤　峰	0.0	0.3	0.1	0.6	1.2	0.4
沈　阳	1.7	0.1	0.3	0.3	2.1	1.3
大　连	0.1	0.0	0.5	25.2	0.4	0.6
鞍　山	2.9	2.9	0.0	1.5	7.7	0.4
抚　顺	1.0	5.0	0.0	10.0	19.0	4.0
本　溪	0.0	0.0	0.0	0.2	0.1	0.2
锦　州	0.0	0.0	0.0	0.0	0.0	0.0
长　春	0.0	0.0	0.0	4.1	15.5	0.0
吉　林	1.0	6.7	0.0	14.1	27.3	4.6
哈 尔 滨	0.2	1.0	0.4	3.3	3.4	0.0
齐齐哈尔	0.1	0.5	0.0	1.0	2.9	1.0
牡 丹 江	1.2	1.6	1.5	4.2	5.7	1.5
上　海	2.8	14.6	0.1	42.8	77.9	26.4
南　京	0.6	2.9	2.4	6.1	17.8	5.9
无　锡	0.0	0.1	0.0	0.0	0.1	0.0
徐　州	0.0	0.0	0.0	0.0	0.0	0.0
常　州	0.1	0.4	0.0	0.8	2.5	0.8
苏　州	0.0	0.2	0.0	4.5	12.1	0.5
南　通	0.1	0.4	0.0	1.2	2.4	0.7
连 云 港	10.8	3.6	0.9	11.5	2.6	1.4
扬　州	0.4	1.5	1.1	4.0	8.0	2.4
镇　江	1.0	5.0	1.2	10.0	30.0	10.0
杭　州	0.1	0.3	0.9	1.8	1.5	0.9

重点城市生活垃圾处理情况（三）（续表）

（2012）

单位：千克

城市名称	渗滤液中污染物排放量					
	汞	镉	六价铬	总铬	铅	砷
宁　波	0.1	0.0	0.3	0.3	0.3	0.3
温　州	1.2	6.1	0.0	16.6	24.4	7.6
湖　州	0.1	0.5	0.0	1.5	2.9	0.7
绍　兴	0.9	4.8	2.1	15.3	11.8	11.7
合　肥	0.3	2.2	0.0	6.2	6.9	1.9
芜　湖	0.1	0.6	0.9	1.9	1.9	0.5
马 鞍 山	0.5	1.1	0.1	20.4	20.5	3.4
福　州	0.5	2.1	0.5	13.2	9.9	5.2
厦　门	0.0	0.0	0.0	0.0	0.0	0.0
泉　州	0.3	78.5	0.0	6.0	10.8	2.7
南　昌	0.7	1.4	0.0	10.3	27.5	5.2
九　江	0.6	4.2	1.3	12.1	14.9	5.2
济　南	0.3	1.5	0.0	3.1	6.6	1.5
青　岛	0.0	0.1	0.0	0.0	0.4	0.1
淄　博	0.0	0.0	0.0	0.0	0.1	0.0
枣　庄	0.0	0.0	0.0	0.0	0.0	0.0
烟　台	0.4	1.4	0.9	2.7	2.6	3.1
潍　坊	0.2	1.3	1.4	3.2	7.8	1.7
济　宁	0.5	3.2	0.1	6.5	11.1	2.0
泰　安	0.0	0.0	0.0	0.0	0.0	0.0
日　照	0.3	1.4	1.8	2.7	5.4	1.8
郑　州	0.2	6.2	0.0	7.9	20.7	6.0
开　封	0.1	0.3	0.3	0.6	3.7	0.5
洛　阳	0.2	1.5	0.0	1.7	5.3	1.5
平 顶 山	0.1	0.1	0.0	0.0	0.1	0.0
安　阳	0.0	0.0	0.0	0.0	0.0	0.0
焦　作	1.0	0.9	0.2	0.7	2.7	0.5
三 门 峡	0.0	0.2	0.0	0.3	0.9	0.3
武　汉	0.0	0.0	0.0	0.0	0.0	0.0
宜　昌	1.1	1.0	0.0	1.6	5.0	1.5
荆　州	0.4	0.4	0.0	1.3	2.1	0.8
长　沙	0.2	0.8	0.1	1.8	1.9	1.5
株　洲	0.0	2.6	1.0	5.0	3.5	4.5
湘　潭	0.5	2.9	0.3	1.4	11.6	3.0
岳　阳	0.8	8.1	2.4	10.4	15.6	4.1
常　德	0.1	1.8	1.7	5.0	2.9	5.9
张 家 界	0.0	0.2	0.0	0.4	1.0	0.4
广　州	0.1	0.8	0.0	7.1	14.0	2.8

重点城市生活垃圾处理情况（三）（续表）

（2012）

单位：千克

城市名称	渗滤液中污染物排放量					
	汞	镉	六价铬	总铬	铅	砷
韶 关	0.2	1.5	1.3	4.0	6.2	3.7
深 圳	12.2	2.6	0.8	23.9	10.3	9.4
珠 海	0.0	1.8	0.4	0.4	0.6	0.5
汕 头	0.3	0.2	0.2	1.2	2.8	1.1
湛 江	0.1	0.1	0.0	2.7	1.3	0.7
南 宁	0.2	2.5	7.5	8.0	8.9	3.4
柳 州	0.1	0.1	0.0	2.7	1.4	0.7
桂 林	0.3	3.2	0.0	1.5	5.3	0.8
北 海	0.0	0.3	0.2	0.2	0.3	0.0
海 口	0.0	0.0	0.0	0.0	0.0	0.0
重 庆	0.6	2.3	0.7	5.2	15.1	4.2
成 都	0.2	1.2	0.0	9.2	7.7	1.7
自 贡	0.4	1.4	0.0	3.8	7.6	2.0
攀 枝 花	0.0	0.1	0.0	0.2	0.4	0.1
泸 州	0.0	0.1	0.0	0.4	0.8	0.2
德 阳	0.1	0.0	0.0	1.3	0.2	0.5
绵 阳	0.4	1.3	0.1	3.8	5.5	1.6
南 充	1.2	7.9	0.2	84.0	104.3	82.5
宜 宾	0.1	0.2	0.0	1.0	1.9	0.9
贵 阳	0.1	0.4	0.0	1.4	1.4	0.7
遵 义	0.6	3.5	0.0	9.0	13.0	3.1
昆 明	0.9	3.3	0.1	10.4	19.1	5.4
曲 靖	0.0	0.1	0.0	0.1	0.2	0.0
玉 溪	0.2	0.6	0.6	1.8	2.7	0.6
拉 萨	0.1	0.5	0.0	1.0	2.6	0.6
西 安	1.1	4.6	9.2	9.2	25.6	8.2
铜 川	0.0	0.0	0.0	0.0	0.0	0.0
宝 鸡	0.3	0.6	0.8	1.4	3.8	1.7
咸 阳	0.0	0.0	0.0	0.1	0.4	0.0
渭 南	0.0	0.0	0.0	0.0	0.0	0.1
延 安	0.0	0.1	0.0	0.1	0.3	0.0
兰 州	0.1	0.6	0.0	1.0	2.4	0.9
金 昌	0.0	0.2	0.0	0.5	1.2	0.3
西 宁	0.5	1.8	0.0	3.5	8.6	2.2
银 川	0.1	0.3	0.6	0.6	1.5	0.4
石 嘴 山	0.0	0.0	0.0	0.1	0.2	0.0
乌鲁木齐	2.8	35.1	0.0	58.0	12.4	45.9
克拉玛依	0.1	0.4	0.0	0.8	2.0	0.5

重点城市生活垃圾处理情况（四）

（2012）

单位：吨

城　市 名　称	焚烧废气中污染物排放量		
	二氧化硫	氮氧化物	烟尘
总　　计	1 056	1 572	564
北　京	34	301	41
天　津	19	150	9
石 家 庄	0	0	0
唐　山	0	0	0
秦 皇 岛	0	0	0
邯　郸	0	0	0
保　定	0	0	0
太　原	0	0	0
大　同	0	0	0
阳　泉	0	0	0
长　治	14	59	13
临　汾	0	0	0
呼 和 浩 特	0	0	0
包　头	0	0	0
赤　峰	0	0	0
沈　阳	0	0	0
大　连	0	0	0
鞍　山	0	0	0
抚　顺	0	0	0
本　溪	0	0	0
锦　州	38	8	33
长　春	0	0	0
吉　林	0	0	0
哈 尔 滨	0	0	0
齐 齐 哈 尔	0	0	0
牡 丹 江	0	0	0
上　海	—	—	—
南　京	0	0	0
无　锡	0	0	0
徐　州	0	0	0
常　州	0	0	0
苏　州	162	193	35
南　通	0	0	0
连 云 港	0	0	0
扬　州	0	0	0
镇　江	0	0	0
杭　州	31	74	114

重点城市生活垃圾处理情况（四）（续表）

（2012）　　　　　　　　　　　　　　　　　　　　单位：吨

城　市	焚烧废气中污染物排放量		
名　称	二氧化硫	氮氧化物	烟尘
宁　波	0	0	0
温　州	0	0	0
湖　州	0	0	0
绍　兴	49	43	24
合　肥	0	0	0
芜　湖	395	267	186
马鞍山	0	0	0
福　州	0	0	0
厦　门	0	0	0
泉　州	0	0	0
南　昌	0	0	0
九　江	0	0	0
济　南	27	61	13
青　岛	0	0	0
淄　博	0	0	0
枣　庄	0	0	0
烟　台	0	0	0
潍　坊	0	0	0
济　宁	31	43	2
泰　安	0	0	0
日　照	0	0	0
郑　州	0	0	0
开　封	0	0	0
洛　阳	0	0	0
平顶山	0	0	0
安　阳	0	0	0
焦　作	22	38	11
三门峡	0	0	0
武　汉	0	0	0
宜　昌	0	0	0
荆　州	0	0	0
长　沙	0	0	0
株　洲	0	0	0
湘　潭	0	0	0
岳　阳	0	0	0
常　德	0	0	0
张家界	0	0	0
广　州	0	5	0

重点城市生活垃圾处理情况（四）（续表）

（2012）

单位：吨

城 市 名 称	焚烧废气中污染物排放量		
	二氧化硫	氮氧化物	烟尘
韶 关	0	0	0
深 圳	0	0	0
珠 海	0	0	0
汕 头	0	0	0
湛 江	0	0	0
南 宁	0	0	0
柳 州	0	0	0
桂 林	0	0	0
北 海	0	0	0
海 口	0	0	0
重 庆	0	0	0
成 都	155	166	42
自 贡	0	0	0
攀 枝 花	0	0	0
泸 州	0	0	0
德 阳	74	164	37
绵 阳	0	0	0
南 充	0	0	0
宜 宾	0	0	0
贵 阳	0	0	0
遵 义	0	0	0
昆 明	6	0	3
曲 靖	0	0	0
玉 溪	0	0	0
拉 萨	0	0	0
西 安	0	0	0
铜 川	0	0	0
宝 鸡	0	0	0
咸 阳	0	0	0
渭 南	0	0	0
延 安	0	0	0
兰 州	0	0	0
金 昌	0	0	0
西 宁	0	0	0
银 川	0	0	0
石 嘴 山	0	0	0
乌鲁木齐	0	0	0
克拉玛依	0	0	0

重点城市危险（医疗）废物集中处置情况（一）

（2012）

城 市 名 称	危险废物集中 处置厂数/个	医疗废物集中 处置厂数/个	本年运行费用/ 万元	危险（医疗）废物 集中处置厂累计完 成投资/万元	新增固定资产/ 万元
总　　计	476	89	381 628.4	1 731 624.5	121 780.4
北　京	2	1	31 359.7	78 221.6	2 669.2
天　津	13	0	17 075.2	37 252.3	1 456.5
石 家 庄	4	2	1 118.4	4 650.1	520.0
唐　山	2	3	819.5	3 168.0	434.0
秦 皇 岛	5	0	1 375.0	6 100.0	141.0
邯　郸	0	2	211.3	1 905.0	0.0
保　定	2	1	1 287.0	6 532.0	139.0
太　原	2	2	792.5	3 028.0	308.9
大　同	1	1	338.0	1 620.0	99.8
阳　泉	0	1	285.0	1 500.0	0.0
长　治	0	0	0	0	0
临　汾	0	1	190.0	2 006.0	24.0
呼和浩特	0	1	160.7	381.2	0.0
包　头	1	3	468.4	21 908.5	0.0
赤　峰	1	1	331.0	1 228.1	0.0
沈　阳	3	1	4 942.8	42 316.0	427.0
大　连	2	1	2 814.0	7 631.0	2 553.0
鞍　山	0	1	231.0	1 395.0	0.0
抚　顺	3	0	1 421.0	3 358.0	0.0
本　溪	2	0	295.1	1 850.0	15.0
锦　州	5	0	0.0	10 908.0	0.0
长　春	1	1	2 377.0	4 840.0	148.0
吉　林	1	1	409.0	3 571.0	0.0
哈 尔 滨	3	1	2 130.8	8 254.0	397.0
齐齐哈尔	0	1	184.0	2 069.5	0.0
牡 丹 江	0	1	71.0	300.0	0.0
上　海	36	2	62 794.3	182 626.5	24 433.0
南　京	17	1	7 433.9	56 074.1	1 408.9
无　锡	26	0	18 266.4	71 935.8	6 111.3
徐　州	2	0	1 457.0	15 000.0	200.0
常　州	34	0	11 954.0	34 531.0	275.0
苏　州	31	0	36 584.5	124 293.0	3 863.1
南　通	17	0	3 507.2	27 916.0	172.0
连 云 港	2	0	1 250.0	9 800.0	3 800.0
扬　州	11	1	910.0	7 196.2	126.7
镇　江	5	0	3 155.0	13 950.0	1 015.0
杭　州	12	1	12 830.9	53 612.6	3 309.5

重点城市危险（医疗）废物集中处置情况（一）（续表）

（2012）

城 市名 称	危险废物集中处置厂数/个	医疗废物集中处置厂数/个	本年运行费用/万元	危险（医疗）废物集中处置厂累计完成投资/万元	新增固定资产/万元
宁　波	17	0	11 043.5	50 162.7	172.2
温　州	6	1	6 068.5	14 126.0	6 329.0
湖　州	2	0	2 100.0	7 500.0	0.0
绍　兴	7	1	5 414.0	20 535.0	3 201.0
合　肥	1	1	2 901.0	12 452.0	556.0
芜　湖	1	1	700.7	2 341.0	0.0
马 鞍 山	0	1	90.0	578.0	0.0
福　州	5	0	2 159.4	15 710.0	1 190.0
厦　门	1	0	1 529.5	3 579.5	1 548.0
泉　州	1	1	1 310.4	12 514.5	110.8
南　昌	1	1	964.0	3 880.0	0.0
九　江	4	0	400.0	124 445.0	500.0
济　南	1	0	2 142.5	4 909.6	1 566.2
青　岛	10	0	4 458.8	26 795.4	298.0
淄　博	19	1	2 554.9	12 010.0	1 004.3
枣　庄	1	0	1 200.0	2 800.0	0.0
烟　台	2	1	1 776.6	20 236.0	10 303.0
潍　坊	9	1	1 785.6	11 664.0	40.0
济　宁	1	1	660.0	1 800.0	52.0
泰　安	1	0	95.0	150.0	20.0
日　照	0	1	170.0	830.0	0.0
郑　州	0	1	2 280.0	3 500.0	260.0
开　封	0	1	480.0	1 486.0	81.0
洛　阳	0	1	1 200.0	1 850.0	0.0
平 顶 山	0	1	0.0	183.0	100.0
安　阳	0	1	181.0	1 700.0	0.0
焦　作	0	1	360.0	980.0	0.0
三 门 峡	0	1	96.0	880.0	0.0
武　汉	8	1	7 424.5	25 735.8	1 878.4
宜　昌	3	1	734.0	10 996.0	40.0
荆　州	1	1	432.8	2 031.5	184.4
长　沙	0	0	0.0	0.0	0.0
株　洲	0	1	530.0	1 931.0	35.0
湘　潭	0	0	0.0	0.0	0.0
岳　阳	16	1	400.0	16 545.0	0.0
常　德	0	1	438.0	1 960.0	232.0
张 家 界	0	1	252.0	1 191.0	0.0
广　州	20	1	7 449.1	64 353.8	9 448.0

重点城市危险（医疗）废物集中处置情况（一）（续表）

（2012）

城 市 名 称	危险废物集中处置厂数/个	医疗废物集中处置厂数/个	本年运行费用/万元	危险（医疗）废物集中处置厂累计完成投资/万元	新增固定资产/万元
韶 关	9	1	592.0	7 620.0	45.0
深 圳	7	1	17 278.0	115 300.0	1 512.0
珠 海	6	0	3 537.2	10 729.0	697.0
汕 头	1	1	1 185.6	1 780.2	78.1
湛 江	2	1	796.8	7 800.0	0.0
南 宁	4	0	20.0	3 336.0	0.0
柳 州	5	1	1 248.1	4 163.0	30.0
桂 林	0	1	286.0	2 002.0	0.0
北 海	0	1	250.0	1 349.0	0.0
海 口	1	0	823.6	3 250.2	165.0
重 庆	8	3	11 234.9	56 440.2	2 740.7
成 都	6	1	5 298.1	13 097.0	1 375.7
自 贡	0	1	0.0	3 448.0	0.0
攀 枝 花	0	0	256.4	300.0	0.0
泸 州	0	2	753.9	2 400.0	400.0
德 阳	0	1	10.8	40.0	0.0
绵 阳	7	1	4 602.6	15 204.7	31.7
南 充	2	0	0.0	80.0	0.0
宜 宾	0	1	600.2	1 468.6	11.0
贵 阳	0	1	220.0	21 859.0	0.0
遵 义	4	1	2 461.0	4 266.5	505.0
昆 明	0	1	3 618.0	12 100.0	0.0
曲 靖	0	0	0.0	0.0	0.0
玉 溪	1	1	206.0	1 692.0	115.0
拉 萨	0	0	0.0	0.0	0.0
西 安	2	0	2 229.3	5 560.5	0.0
铜 川	0	0	0.0	0.0	0.0
宝 鸡	0	1	240.0	311.2	0.0
咸 阳	0	1	544.3	1 473.3	268.8
渭 南	0	1	380.0	1 370.0	50.0
延 安	14	1	4 097.4	15 873.1	375.0
兰 州	2	0	2 283.0	23 405.3	19 503.3
金 昌	0	1	90.0	698.0	0.0
西 宁	5	1	10 129.0	16 735.0	80.0
银 川	2	0	1 182.3	17 660.6	0.0
石 嘴 山	0	1	130.0	952.0	0.0
乌鲁木齐	0	1	1 699.0	7 308.0	351.0
克拉玛依	4	1	732.8	13 282.0	250.0

重点城市危险（医疗）废物集中处置情况（二）

（2012）

城 市名 称	危险废物设计处置能力/（吨/日）	危险废物实际处置量/吨	工业危险废物处置量	医疗废物处置量	其他危险废物处置量	危险废物综合利用量/吨
总　　计	45 792	2 178 590	1 348 897	355 180	474 513	2 664 816
北　京	731	70 751	65 227	5 524	0	52 647
天　津	618	58 731	40 833	17 898	0	141 333
石 家 庄	72	9 155	8 723	432	0	704
唐　山	31	2 952	83	2 869	0	402
秦 皇 岛	345	16 835	14 960	1 743	132	1 447
邯　郸	11	1 885	0	1 885	0	0
保　定	50	7 817	5 089	2 728	0	2 843
太　原	95	9 118	598	8 520	0	515
大　同	25	1 379	0	1 379	0	1 960
阳　泉	5	427	0	427	0	0
长　治	0	0	0	0	0	0
临　汾	8	580	0	580	0	0
呼和浩特	0	2 113	0	2 113	0	0
包　头	167	779	0	779	0	0
赤　峰	5	740	0	740	0	0
沈　阳	155	22 669	17 649	4 918	102	7 400
大　连	169	10 653	6 641	4 012	0	26 839
鞍　山	8	992	0	992	0	0
抚　顺	81	0	0	0	0	12 169
本　溪	15	1 665	965	700	0	0
锦　州	434	0	0	0	0	93 927
长　春	58	15 645	8 592	5 065	1 988	373
吉　林	155	12 219	10 574	1 645	0	1 645
哈 尔 滨	129	6 473	69	6 404	0	199
齐齐哈尔	8	716	0	716	0	0
牡 丹 江	5	779	0	779	0	0
上　海	2 082	167 895	136 184	26 709	5 002	142 569
南　京	848	26 655	20 735	5 920	0	10 273
无　锡	4 243	79 559	74 956	4 602	0	536 928
徐　州	30	5 814	2 111	3 703	0	0
常　州	1 296	21 865	19 930	1 558	377	35 960
苏　州	3 040	122 851	91 422	5 472	25 957	240 596
南　通	3 298	78 806	70 518	2 221	6 067	34 714
连 云 港	49	5 152	3 797	1 356	0	0
扬　州	268	6 901	5 501	1 400	0	6 554
镇　江	382	9 762	8 785	978	0	6 884
杭　州	2 023	99 597	87 380	12 217	0	115 129

重点城市危险（医疗）废物集中处置情况（二）（续表）

（2012）

城 市 名 称	危险废物设计 处置能力/ （吨/日）	危险废物实际 处置量/吨	工业危险废物 处置量	医疗废物处置量	其他危险废物 处置量	危险废物综合 利用量/吨
宁 波	720	90 917	81 917	9 000	0	65 043
温 州	292	10 954	8 891	2 063	0	34 738
湖 州	80	9 229	3 919	5 176	134	0
绍 兴	210	14 343	10 693	3 650	0	23 628
合 肥	80	16 429	11 743	4 686	0	0
芜 湖	32	1 336	0	1 336	0	0
马 鞍 山	4	456	0	456	0	0
福 州	277	8 106	8 106	0	0	1 094
厦 门	134	9 663	6 923	2 740	0	933
泉 州	208	4 892	2 400	2 492	0	0
南 昌	19	4 335	638	3 697	0	0
九 江	380	1 564	1 564	0	0	1 200
济 南	24	5 839	1 237	4 602	0	3 232
青 岛	780	20 733	16 123	4 610	0	11 031
淄 博	647	4 745	1 289	3 457	0	99 194
枣 庄	8	2 370	590	1 780	0	0
烟 台	217	18 021	12 894	5 127	0	4 962
潍 坊	126	13 922	11 197	2 725	0	6 531
济 宁	507	6 174	3 274	2 900	0	0
泰 安	0	0	0	0	0	480
日 照	5	799	0	799	0	0
郑 州	30	6 982	0	6 982	0	0
开 封	5	1 248	0	1 248	0	0
洛 阳	24	5 544	0	5 544	0	0
平 顶 山	4	1 366	0	1 366	0	0
安 阳	5	1 220	0	1 220	0	0
焦 作	5	1 650	0	1 650	0	0
三 门 峡	5	548	0	548	0	0
武 汉	372	33 951	21 813	12 138	0	11 715
宜 昌	3 040	4 003	2 340	1 663	0	0
荆 州	16	2 896	2 131	765	0	0
长 沙	0	0	0	0	0	0
株 洲	8	2 400	600	1 800	0	0
湘 潭	0	0	0	0	0	0
岳 阳	272	1 600	0	1 600	0	35 610
常 德	41	2 403	0	2 403	0	0
张 家 界	3	698	0	698	0	0
广 州	3 479	91 938	78 052	13 886	0	176 548

重点城市危险（医疗）废物集中处置情况（二）（续表）

（2012）

城　市　名　称	危险废物设计处置能力/（吨/日）	危险废物实际处置量/吨	工业危险废物处置量	医疗废物处置量	其他危险废物处置量	危险废物综合利用量/吨
韶　关	7	2 564	0	2 564	0	0
深　圳	2 226	154 520	145 604	8 719	197	158 400
珠　海	298	18 216	17 632	0	584	17 183
汕　头	30	2 781	0	2 781	0	2 819
湛　江	60	4 172	859	3 313	0	737
南　宁	1 000	0	0	0	0	2 871
柳　州	73	9 185	6 721	2 464	0	1 277
桂　林	10	2 440	0	2 440	0	0
北　海	5	1 200	0	1 200	0	0
海　口	14	2 933	0	2 933	0	0
重　庆	2 334	42 588	25 768	9 205	7 614	317 840
成　都	171	14 804	438	14 366	0	8 260
自　贡	0	0	0	0	0	0
攀 枝 花	6	885	0	884	1	0
泸　州	17	2 823	0	2 823	0	0
德　阳	1	48	0	48	0	0
绵　阳	768	5 259	3 312	1 224	723	103 678
南　充	0	0	0	0	0	0
宜　宾	5	1 427	0	1 427	0	0
贵　阳	160	4 575	0	4 575	0	0
遵　义	124	3 785	1 751	2 019	15	3 051
昆　明	30	10 200	0	10 200	0	0
曲　靖	0	0	0	0	0	0
玉　溪	23	3 665	3 450	215	0	1 255
拉　萨	0	0	0	0	0	0
西　安	55	13 543	0	13 543	0	540
铜　川	0	0	0	0	0	0
宝　鸡	5	583	0	583	0	0
咸　阳	8	1 524	0	1 524	0	0
渭　南	10	2 241	0	2 241	0	0
延　安	2 710	33 249	32 227	1 022	0	57 279
兰　州	396	108 674	106 965	1 681	27	0
金　昌	3	438	0	438	0	438
西　宁	1 369	427 075	0	2 517	424 558	17 222
银　川	66	2 726	1 611	1 097	18	1 240
石 嘴 山	3 083	275	0	275	0	1 238
乌鲁木齐	173	16 720	12 906	2 900	914	3 850
克拉玛依	545	273	0	170	103	15 687

重点城市危险（医疗）废物集中处置情况（三）

（2012）

城 市 名 称	渗滤液中污染物排放量					
	化学需氧量/ 吨	氨氮/ 吨	油类/ 吨	总磷/ 吨	挥发酚/ 千克	氰化物/ 千克
总　　计	1 161.1	32.4	12.6	0.3	1.2	24.5
北　　京	2.7	2.5	0.0	0.0	0.0	0.0
天　　津	7.0	0.1	—	—	0.0	—
石 家 庄	0.0	0.0	0.0	0.0	0.0	0.0
唐　　山	0.0	0.0	0.0	0.0	0.0	0.0
秦 皇 岛	0.0	0.0	0.0	0.0	0.0	0.0
邯　　郸	0.0	0.0	0.0	0.0	0.0	0.0
保　　定	0.0	0.0	0.0	0.0	0.0	0.0
太　　原	0.0	0.0	0.0	0.0	0.0	0.0
大　　同	0.0	0.0	0.0	0.0	0.0	0.0
阳　　泉	0.0	0.0	0.0	0.0	0.0	0.0
长　　治	0.0	0.0	0.0	0.0	0.0	0.0
临　　汾	0.0	0.0	0.0	0.0	0.0	0.0
呼和浩特	0.0	0.0	0.0	0.0	0.0	0.0
包　　头	0.0	0.0	0.0	0.0	0.0	0.0
赤　　峰	0.0	0.0	0.0	0.0	0.0	0.0
沈　　阳	50.1	9.8	10.0	0.0	0.0	0.0
大　　连	646.1	0.1	0.0	0.0	0.0	0.0
鞍　　山	0.2	0.0	0.0	0.0	0.0	0.0
抚　　顺	0.0	0.0	0.0	0.0	0.0	0.0
本　　溪	0.0	0.0	0.0	0.0	0.0	0.0
锦　　州	0.0	0.0	0.0	0.0	0.0	0.0
长　　春	0.0	0.0	0.0	0.0	0.0	0.0
吉　　林	0.0	0.0	0.0	0.0	0.0	0.0
哈 尔 滨	0.3	0.0	0.0	0.0	0.0	0.0
齐齐哈尔	0.0	0.0	0.0	0.0	0.0	0.0
牡 丹 江	0.0	0.0	0.0	0.0	0.0	0.0
上　　海	21.8	2.0	0.1	0.1	0.0	2.8
南　　京	0.1	32.0	0.0	0.0	0.0	0.0
无　　锡	9.0	6.0	0.0	0.0	0.0	0.0
徐　　州	0.0	0.0	0.0	0.0	0.0	0.0
常　　州	3.9	0.1	0.0	0.0	0.0	0.0
苏　　州	11.3	0.9	0.5	0.0	0.0	0.0
南　　通	1.7	0.1	0.1	0.0	0.0	0.0
连 云 港	0.0	0.0	0.0	0.0	0.0	0.0
扬　　州	0.0	0.0	0.0	0.0	0.0	0.0
镇　　江	3.2	0.1	0.0	0.0	0.0	0.0
杭　　州	117.2	1.2	0.0	0.0	0.0	0.0

重点城市危险（医疗）废物集中处置情况（三）（续表）

（2012）

城 市 名 称	渗滤液中污染物排放量					
	化学需氧量/ 吨	氨氮/ 吨	油类/ 吨	总磷/ 吨	挥发酚/ 千克	氰化物/ 千克
宁 波	5.5	0.6	0.0	0.0	0.0	0.0
温 州	1.5	0.3	0.0	0.0	0.0	0.0
湖 州	0.0	0.0	0.0	0.0	0.0	0.0
绍 兴	5.0	0.9	0.0	0.0	0.0	0.0
合 肥	0.6	0.1	0.0	0.0	0.0	0.0
芜 湖	0.0	0.0	0.0	0.0	0.0	0.0
马 鞍 山	0.0	0.0	0.0	0.0	0.0	0.0
福 州	7.7	0.0	0.0	0.0	0.0	4.7
厦 门	0.0	0.0	0.0	0.0	0.0	0.0
泉 州	0.0	0.0	0.0	0.0	0.0	0.0
南 昌	0.0	0.0	0.0	0.0	0.0	0.0
九 江	0.1	0.0	0.0	0.0	0.0	0.0
济 南	0.0	0.0	0.0	0.0	0.0	0.0
青 岛	0.2	0.1	0.0	0.0	0.0	0.0
淄 博	0.5	0.0	0.0	0.0	0.0	0.0
枣 庄	0.0	0.0	0.0	0.0	0.0	0.0
烟 台	1.1	0.3	0.4	0.0	0.0	0.0
潍 坊	0.0	0.0	0.0	0.0	0.0	0.0
济 宁	0.0	0.0	0.0	0.0	0.0	0.0
泰 安	0.0	0.0	0.0	0.0	0.0	0.0
日 照	0.1	0.0	0.0	0.0	0.0	0.0
郑 州	0.0	0.0	0.0	0.0	0.0	0.0
开 封	0.0	0.0	0.0	0.0	0.0	0.0
洛 阳	0.0	0.0	0.0	0.0	0.0	0.0
平 顶 山	0.0	0.0	0.0	0.0	0.0	0.0
安 阳	0.0	0.0	0.0	0.0	0.0	0.0
焦 作	0.0	0.0	0.0	0.0	0.0	0.0
三 门 峡	0.0	0.0	0.0	0.0	0.0	0.0
武 汉	24.2	2.3	0.0	0.0	0.0	0.0
宜 昌	0.0	0.0	0.0	0.0	0.0	0.0
荆 州	0.0	0.0	0.0	0.0	0.0	0.0
长 沙	0.0	0.0	0.0	0.0	0.0	0.0
株 洲	0.0	0.0	0.0	0.0	0.0	0.0
湘 潭	0.0	0.0	0.0	0.0	0.0	0.0
岳 阳	0.0	0.0	0.0	0.0	0.0	0.0
常 德	0.9	0.0	0.5	0.0	0.0	0.0
张 家 界	0.0	0.0	0.0	0.0	0.0	0.0
广 州	8.3	0.2	0.7	0.0	0.0	1.2

重点城市危险（医疗）废物集中处置情况（三）（续表）

（2012）

城 市 名 称	渗滤液中污染物排放量					
	化学需氧量/ 吨	氨氮/ 吨	油类/ 吨	总磷/ 吨	挥发酚/ 千克	氰化物/ 千克
韶 关	0.0	0.0	0.0	0.0	0.0	0.0
深 圳	227.7	4.1	0.1	0.0	0.2	12.5
珠 海	0.1	0.0	0.0	0.0	0.0	0.0
汕 头	0.1	0.0	0.0	0.0	0.0	0.0
湛 江	0.0	0.0	0.0	0.0	0.0	0.0
南 宁	0.0	0.0	0.0	0.0	0.0	0.0
柳 州	0.0	0.0	0.0	0.0	0.0	0.0
桂 林	0.0	0.0	0.0	0.0	0.0	0.0
北 海	0.0	0.0	0.0	0.0	0.0	0.0
海 口	0.0	0.0	0.0	0.0	0.0	0.0
重 庆	1.3	—	—	—	0.1	0.0
成 都	0.0	0.0	0.1	0.0	0.0	0.0
自 贡	0.0	0.0	0.0	0.0	0.0	0.0
攀 枝 花	0.0	0.0	0.0	0.0	0.0	0.0
泸 州	0.0	0.0	0.0	0.0	0.0	0.0
德 阳	0.0	0.0	0.0	0.0	0.0	0.0
绵 阳	0.0	0.0	0.0	0.0	0.0	0.0
南 充	0.0	0.0	0.0	0.0	0.0	0.0
宜 宾	0.0	0.0	0.0	0.0	0.0	0.0
贵 阳	0.0	0.0	0.0	0.0	0.0	0.0
遵 义	0.3	0.0	0.0	0.0	0.0	0.0
昆 明	0.0	0.0	0.0	0.0	0.0	0.0
曲 靖	0.0	0.0	0.0	0.0	0.0	0.0
玉 溪	0.0	0.0	0.0	0.0	0.0	0.0
拉 萨	0.0	0.0	0.0	0.0	0.0	0.0
西 安	0.0	0.0	0.0	0.0	0.0	0.0
铜 川	0.0	0.0	0.0	0.0	0.0	0.0
宝 鸡	0.0	0.0	0.0	0.0	0.0	0.0
咸 阳	0.0	0.0	0.0	0.0	0.0	0.0
渭 南	0.0	0.0	0.0	0.0	0.0	0.0
延 安	0.0	0.0	0.0	0.0	0.0	0.0
兰 州	1.2	0.6	0.0	0.0	0.8	3.2
金 昌	0.0	0.0	0.0	0.0	0.0	0.0
西 宁	0.0	0.0	0.0	0.0	0.0	0.0
银 川	0.0	0.0	0.0	0.0	0.0	0.0
石 嘴 山	0.0	0.0	0.0	0.0	0.0	0.0
乌鲁木齐	0.0	0.0	0.0	0.0	0.0	0.0
克拉玛依	0.0	0.0	0.0	0.0	0.0	0.0

重点城市危险（医疗）废物集中处置情况（四）

（2012）

单位：千克

城 市 名 称	渗滤液中污染物排放量					
	汞	镉	六价铬	总铬	铅	砷
总 计	6.0	2.0	6.8	32.2	24.4	7.0
北 京	0.0	0.0	0.0	0.0	0.0	0.0
天 津	0.0	—	0.0	—	0.1	—
石 家 庄	0.0	0.0	0.0	0.0	0.0	0.0
唐 山	0.0	0.0	0.0	0.0	0.0	0.0
秦 皇 岛	0.0	0.0	0.0	0.0	0.0	0.0
邯 郸	0.0	0.0	0.0	0.0	0.0	0.0
保 定	0.0	0.0	0.0	0.0	0.0	0.0
太 原	0.0	0.0	0.0	0.0	0.0	0.0
大 同	0.0	0.0	0.0	0.0	0.0	0.0
阳 泉	0.0	0.0	0.0	0.0	0.0	0.0
长 治	0.0	0.0	0.0	0.0	0.0	0.0
临 汾	0.0	0.0	0.0	0.0	0.0	0.0
呼和浩特	0.0	0.0	0.0	0.0	0.0	0.0
包 头	0.0	0.0	0.0	0.0	0.0	0.0
赤 峰	0.0	0.0	0.0	0.0	0.0	0.0
沈 阳	0.0	0.8	0.7	0.7	0.0	0.1
大 连	0.0	0.0	0.0	0.0	0.0	0.0
鞍 山	0.0	0.0	0.0	0.0	0.0	0.0
抚 顺	0.0	0.0	0.0	0.0	0.0	0.0
本 溪	0.0	0.0	0.0	0.0	0.0	0.0
锦 州	0.0	0.0	0.0	0.0	0.0	0.0
长 春	0.0	0.0	0.0	0.0	0.0	0.0
吉 林	0.0	0.0	0.0	0.4	0.0	0.2
哈 尔 滨	0.0	0.0	0.0	0.0	0.0	0.0
齐齐哈尔	0.0	0.0	0.0	0.0	0.0	0.0
牡 丹 江	0.0	0.0	0.0	0.0	0.0	0.0
上 海	0.8	0.0	0.0	0.0	0.4	0.0
南 京	0.0	0.0	0.4	0.4	24.0	7.0
无 锡	0.0	0.0	0.1	0.1	0.1	0.1
徐 州	0.0	0.0	0.0	0.0	0.0	0.0
常 州	0.0	0.0	0.0	0.0	0.0	0.0
苏 州	0.0	0.0	0.0	1.8	9.3	0.3
南 通	0.0	0.0	0.0	0.1	0.1	0.1
连 云 港	0.0	0.1	0.0	0.0	0.3	0.0
扬 州	0.0	0.0	0.0	0.0	0.0	0.0
镇 江	0.0	0.0	0.0	0.0	0.0	0.0
杭 州	0.0	0.0	0.0	0.0	2.0	0.0

重点城市危险（医疗）废物集中处置情况（四）（续表）

（2012）
单位：千克

城市名称	渗滤液中污染物排放量					
	汞	镉	六价铬	总铬	铅	砷
宁　波	0.0	0.0	0.0	0.8	2.8	0.5
温　州	0.0	0.0	0.0	0.5	0.0	1.2
湖　州	0.0	0.0	0.0	0.0	0.0	0.0
绍　兴	0.0	0.0	0.0	0.0	0.0	0.5
合　肥	0.0	0.0	0.0	0.0	0.0	0.0
芜　湖	0.0	0.0	0.0	0.0	0.0	0.0
马 鞍 山	0.0	0.0	0.0	0.0	0.0	0.0
福　州	4.9	0.3	0.0	2.3	2.4	1.9
厦　门	0.0	0.0	0.0	0.0	0.0	0.0
泉　州	0.0	0.0	0.0	0.0	0.0	0.0
南　昌	0.0	0.0	0.0	0.0	0.0	0.0
九　江	0.0	0.0	0.0	0.0	0.0	0.0
济　南	0.0	0.0	0.0	0.0	0.0	0.0
青　岛	0.0	0.0	0.0	0.0	0.0	0.0
淄　博	0.0	0.0	0.0	0.0	0.0	0.0
枣　庄	0.0	0.0	0.0	0.0	0.0	0.0
烟　台	0.0	0.0	0.0	0.0	0.0	0.0
潍　坊	0.0	0.0	0.0	0.0	0.0	0.0
济　宁	0.0	0.0	0.0	0.0	0.0	0.0
泰　安	0.0	0.0	0.0	0.0	0.0	0.0
日　照	0.0	0.0	0.0	0.0	0.0	0.0
郑　州	0.0	0.0	0.0	0.0	0.0	0.0
开　封	0.0	0.0	0.0	0.0	0.0	0.0
洛　阳	0.0	0.0	0.0	0.0	0.0	0.0
平 顶 山	0.0	0.0	0.0	0.0	0.0	0.0
安　阳	0.0	0.0	0.0	0.0	0.0	0.0
焦　作	0.0	0.0	0.0	0.0	0.0	0.0
三 门 峡	0.0	0.0	0.0	0.0	0.0	0.0
武　汉	0.0	0.0	0.0	0.0	0.0	0.0
宜　昌	0.0	0.0	0.0	0.0	0.0	0.0
荆　州	0.0	0.0	0.0	0.0	0.0	0.0
长　沙	0.0	0.0	0.0	0.0	0.0	0.0
株　洲	0.0	0.0	0.0	0.0	0.0	0.0
湘　潭	0.0	0.0	0.0	0.0	0.0	0.0
岳　阳	0.0	0.0	0.0	0.0	0.0	0.0
常　德	0.0	0.0	0.0	0.0	0.0	0.0
张 家 界	0.0	0.0	0.0	0.0	0.0	0.0
广　州	0.0	0.0	0.0	0.0	0.3	0.1

重点城市危险（医疗）废物集中处置情况（四）（续表）

（2012）

单位：千克

城 市	渗滤液中污染物排放量					
名 称	汞	镉	六价铬	总铬	铅	砷
韶 关	0.0	0.0	0.0	0.0	0.0	0.0
深 圳	0.0	0.0	5.3	12.1	0.0	0.0
珠 海	0.0	0.0	0.0	1.7	0.0	0.0
汕 头	0.0	0.0	0.0	0.0	0.0	0.0
湛 江	0.0	0.0	0.0	0.0	0.0	0.0
南 宁	0.0	0.0	0.0	0.0	0.0	0.0
柳 州	0.0	0.0	0.0	0.0	0.0	0.0
桂 林	0.0	0.0	0.0	0.0	0.0	0.0
北 海	0.0	0.0	0.0	0.0	0.0	0.0
海 口	0.0	0.0	0.0	0.0	0.0	0.0
重 庆	0.0	0.2	0.3	0.8	0.2	0.1
成 都	0.0	0.0	0.0	0.0	0.0	0.0
自 贡	0.0	0.0	0.0	0.0	0.0	0.0
攀 枝 花	0.0	0.0	0.0	0.0	0.0	0.0
泸 州	0.0	0.0	0.0	0.0	0.0	0.0
德 阳	0.0	0.0	0.0	0.0	0.0	0.0
绵 阳	0.0	0.0	0.0	0.0	0.0	0.0
南 充	0.0	0.0	0.0	0.0	0.0	0.0
宜 宾	0.0	0.0	0.0	0.0	0.0	0.0
贵 阳	0.0	0.0	0.0	0.0	0.0	0.0
遵 义	0.0	0.0	0.0	0.0	0.0	0.0
昆 明	0.0	0.0	0.0	0.0	0.0	0.0
曲 靖	0.0	0.0	0.0	0.0	0.0	0.0
玉 溪	0.0	0.0	0.0	0.0	0.0	0.0
拉 萨	0.0	0.0	0.0	0.0	0.0	0.0
西 安	0.0	0.0	0.0	0.0	0.0	0.0
铜 川	0.0	0.0	0.0	0.0	0.0	0.0
宝 鸡	0.0	0.0	0.0	0.0	0.0	0.0
咸 阳	0.0	0.0	0.0	0.0	0.0	0.0
渭 南	0.0	0.0	0.0	0.0	0.0	0.0
延 安	0.0	0.0	0.0	0.0	0.0	0.0
兰 州	0.2	0.6	0.0	10.5	6.3	2.1
金 昌	0.0	0.0	0.0	0.0	0.0	0.0
西 宁	0.0	0.0	0.0	0.0	0.0	0.0
银 川	0.0	0.0	0.0	0.0	0.0	0.0
石 嘴 山	0.0	0.0	0.0	0.0	0.0	0.0
乌鲁木齐	0.0	0.0	0.0	0.0	0.0	0.0
克拉玛依	0.0	0.0	0.0	0.0	0.0	0.0

重点城市危险（医疗）废物集中处置情况（五）

（2012）

单位：吨

城 市 名 称	焚烧废气中污染物排放量		
	二氧化硫	氮氧化物	烟尘
总　　计	494.0	972.8	513.4
北　　京	0.6	20.5	2.3
天　　津	62.7	22.6	29.1
石 家 庄	9.0	5.6	6.7
唐　　山	0.9	2.1	0.6
秦 皇 岛	2.3	10.1	0.9
邯　　郸	0.7	0.1	0.0
保　　定	1.6	5.5	46.9
太　　原	0.9	10.1	0.9
大　　同	0.1	1.7	0.1
阳　　泉	0.0	0.0	0.0
长　　治	0.0	0.0	0.0
临　　汾	0.7	1.7	0.7
呼 和 浩 特	0.6	2.1	2.5
包　　头	0.0	0.6	0.1
赤　　峰	0.1	0.9	0.0
沈　　阳	1.1	3.7	3.2
大　　连	5.4	4.1	2.1
鞍　　山	0.0	0.0	0.0
抚　　顺	0.0	0.0	0.0
本　　溪	1.0	1.0	0.5
锦　　州	0.0	0.0	0.0
长　　春	0.6	11.0	2.2
吉　　林	0.5	1.5	17.8
哈 尔 滨	1.5	8.3	0.8
齐 齐 哈 尔	0.2	0.8	0.7
牡 丹 江	0.2	0.7	1.7
上　　海	59.2	235.5	52.5
南　　京	20.1	12.9	7.5
无　　锡	20.5	31.6	4.7
徐　　州	0.5	5.4	4.4
常　　州	28.9	7.2	19.2
苏　　州	23.2	146.0	27.2
南　　通	13.2	45.1	47.1
连 云 港	4.9	7.4	2.0
扬　　州	0.4	0.0	0.3
镇　　江	9.7	12.4	3.8
杭　　州	28.8	22.4	3.0

重点城市危险（医疗）废物集中处置情况（五）（续表）

（2012）   单位：吨

城 市 名 称	焚烧废气中污染物排放量		
	二氧化硫	氮氧化物	烟尘
宁　波	23.7	23.6	6.9
温　州	24.3	13.3	9.4
湖　州	19.4	5.1	5.1
绍　兴	20.4	10.2	11.9
合　肥	1.2	7.7	3.2
芜　湖	0.5	0.0	2.2
马 鞍 山	1.0	0.4	1.9
福　州	1.4	12.8	3.2
厦　门	3.0	11.0	0.9
泉　州	0.5	1.9	0.2
南　昌	1.3	4.2	0.5
九　江	5.4	10.9	0.1
济　南	0.5	8.3	0.5
青　岛	1.1	1.1	1.1
淄　博	0.6	2.1	0.7
枣　庄	0.7	2.8	0.1
烟　台	1.6	11.1	1.4
潍　坊	1.1	6.2	1.8
济　宁	0.9	2.6	1.4
泰　安	0.0	0.0	0.0
日　照	1.0	1.0	0.1
郑　州	1.4	1.2	0.4
开　封	0.2	2.4	0.4
洛　阳	5.3	2.1	2.7
平 顶 山	0.0	0.0	0.0
安　阳	0.4	1.2	0.1
焦　作	0.0	0.0	0.0
三 门 峡	0.0	0.0	0.0
武　汉	7.3	9.9	4.7
宜　昌	7.5	60.4	30.1
荆　州	0.0	0.0	0.0
长　沙	0.0	0.0	0.0
株　洲	0.0	1.2	0.1
湘　潭	0.0	0.0	0.0
岳　阳	0.5	1.5	0.8
常　德	0.9	0.9	43.0
张 家 界	0.2	0.7	4.4
广　州	14.5	18.2	6.1

重点城市危险（医疗）废物集中处置情况（六）（续表）

（2012）   单位：吨

城　市 名　称	焚烧废气中污染物排放量		
	二氧化硫	氮氧化物	烟尘
韶　关	1.3	2.0	1.3
深　圳	3.8	9.4	1.6
珠　海	1.6	0.9	0.3
汕　头	2.0	2.4	0.2
湛　江	3.0	5.3	3.1
南　宁	0.0	0.0	0.0
柳　州	1.8	9.0	4.0
桂　林	0.0	0.0	0.0
北　海	0.0	0.0	0.0
海　口	0.0	0.0	0.0
重　庆	4.4	19.2	4.7
成　都	4.0	13.1	1.2
自　贡	0.0	0.0	0.0
攀枝花	0.1	1.1	0.1
泸　州	0.9	3.1	0.7
德　阳	0.0	0.1	0.6
绵　阳	0.3	0.4	0.4
南　充	1.5	0.1	25.0
宜　宾	0.0	0.0	0.0
贵　阳	0.0	0.0	0.0
遵　义	2.0	1.0	0.9
昆　明	0.6	9.1	0.9
曲　靖	0.0	0.0	0.0
玉　溪	10.3	0.0	0.4
拉　萨	0.0	0.0	0.0
西　安	4.2	12.2	1.5
铜　川	0.0	0.0	0.0
宝　鸡	0.1	0.2	0.0
咸　阳	0.0	0.0	0.0
渭　南	0.0	0.0	0.0
延　安	0.0	0.0	0.0
兰　州	0.3	1.0	0.1
金　昌	0.0	0.0	0.0
西　宁	0.8	2.4	0.3
银　川	2.7	1.9	0.6
石嘴山	0.0	0.0	0.0
乌鲁木齐	0.7	2.6	26.9
克拉玛依	0.0	0.1	1.5

10

各工业行业环境统计

ANNUAL STATISTIC REPORT ON ENVIRONMENT IN CHINA
2012

按行业分重点调查工业废水排放及处理情况（一）

（2012）

行业名称	汇总工业企业数/个	工业废水排放量/万吨	直接排入环境的	排入污水处理厂的	工业废水处理量/万吨	废水治理设施数/套	废水治理设施处理能力/（万吨/日）	本年运行费用/万元
行业总计	147 996	2 033 626.6	1 579 913.9	453 712.8	5 274 704.9	85 673.0	26 620.0	6 677 024.6
煤炭开采和洗选业	6 516	142 220.3	134 600.4	7 619.9	171 893.6	4 970.0	1 532.8	160 053.4
石油和天然气开采业	261	9 366.6	8 392.5	974.1	90 328.8	761.0	456.5	236 602.0
黑色金属矿采选业	4 489	22 766.4	22 739.3	27.1	252 337.3	2 083.0	1 554.1	136 768.8
有色金属矿采选业	3 378	50 855.4	50 846.4	9.0	174 129.0	4 905.0	687.7	124 243.3
非金属矿采选业	992	7 367.8	6 390.3	977.5	12 432.2	477.0	92.1	20 234.4
开采辅助活动	48	72.2	34.3	38.0	221.9	26.0	2.1	758.2
其他采矿业	27	496.8	496.8	0.0	383.9	15.0	2.1	388.1
农副食品加工业	11 617	156 565.8	136 342.8	20 223.0	132 529.4	5 775.0	1 012.3	175 827.2
食品制造业	3 794	56 936.8	38 446.5	18 490.3	57 321.5	2 603.0	301.8	99 210.2
酒、饮料和精制茶制造业	4 224	74 022.1	56 406.2	17 615.9	65 817.4	2 129.0	401.4	109 259.2
烟草制品业	152	2 279.2	1 395.0	884.2	3 073.7	154.0	20.1	7 174.0
纺织业	8 887	237 252.2	113 588.4	123 663.9	232 690.1	6 438.0	1 245.3	453 206.1
纺织服装、服饰业	1 485	17 069.1	10 251.3	6 817.8	14 092.2	1 075.0	103.6	36 652.0
皮革、毛皮、羽毛及其制品和制鞋业	2 210	26 515.4	19 004.1	7 511.3	21 320.0	1 310.0	184.2	58 715.6
木材加工和木、竹、藤、棕、草制品业	2 634	4 775.8	4 369.7	406.2	3 452.4	612.0	19.9	7 360.5
家具制造业	465	644.6	363.3	281.3	455.1	190.0	2.5	1 534.0
造纸和纸制品业	5 235	342 717.3	301 484.0	41 233.3	438 658.3	4 574.0	2 821.2	603 670.7
印刷和记录媒介复制业	727	1 420.3	692.1	728.1	1 047.6	266.0	9.4	4 493.5
文教、工美、体育和娱乐用品制造业	694	2 204.9	1 271.2	933.6	1 817.3	441.0	9.7	5 631.5
石油加工、炼焦和核燃料加工业	1 307	87 474.1	73 899.9	13 574.2	109 640.6	1 483.0	441.0	418 801.6
化学原料和化学制品制造业	12 043	274 344.4	218 269.9	56 074.5	451 616.8	9 674.0	2 612.9	963 343.0
医药制造业	3 264	57 217.9	33 786.5	23 431.4	50 518.6	2 683.0	224.0	330 265.9
化学纤维制造业	483	35 307.5	28 073.0	7 234.5	33 565.3	403.0	151.8	77 485.4
橡胶和塑料制品业	2 539	12 797.5	7 273.3	5 524.1	16 983.0	994.0	78.8	28 510.6
非金属矿物制品业	33 164	29 439.8	25 159.0	4 280.8	84 802.2	5 323.0	550.5	135 224.8
黑色金属冶炼和压延加工业	4 599	106 148.1	96 315.2	9 832.9	2 315 068.5	4 466.0	9 242.1	1 239 531.3
有色金属冶炼和压延加工业	4 190	28 835.4	25 955.2	2 880.2	172 013.3	2 979.0	645.1	143 716.6
金属制品业	8 617	33 589.3	22 620.2	10 969.0	52 386.4	5 917.0	461.5	191 711.7
通用设备制造业	3 092	10 158.8	6 216.8	3 941.9	9 538.8	1 312.0	47.6	27 949.4
专用设备制造业	1 583	7 920.6	4 200.4	3 720.1	6 101.1	749.0	34.3	12 848.8
汽车制造业	1 853	16 320.2	7 003.9	9 316.3	14 943.4	1 369.0	90.4	63 675.9
铁路、船舶、航空航天和其他运输设备制造业	1 012	12 484.2	8 478.0	4 006.2	9 833.0	876.0	59.1	23 469.3
电气机械和器材制造业	1 881	9 366.9	4 270.0	5 096.9	8 556.1	1 316.0	88.1	39 614.5
计算机、通信和其他电子设备制造业	2 586	48 173.4	19 820.2	28 353.2	49 205.2	2 671.0	249.4	469 961.6
仪器仪表制造业	386	2 450.6	1 024.2	1 426.4	2 171.6	324.0	10.9	6 739.6
其他制造业	1 159	5 199.0	3 210.6	1 988.4	4 588.5	518.0	42.0	12 559.2
废弃资源综合利用业	494	2 291.5	1 826.8	464.7	2 042.6	193.0	19.1	6 193.1
金属制品、机械和设备修理业	180	1 111.1	644.3	466.8	1 018.3	137.0	6.6	3 088.8
电力、热力生产和供应业	5 525	95 575.0	82 971.1	12 603.9	204 208.2	3 350.0	1 092.9	232 901.0
燃气生产和供应业	59	952.9	872.5	80.4	1 087.0	33.0	4.9	6 907.6
水的生产和供应业	4	12.4	12.4	0.0	2.4	6.0	0.3	6.0
其他行业	141	906.8	895.5	11.2	812.3	93.0	8.0	736.2

按行业分重点调查工业废水排放及处理情况（二）

（2012）

单位：吨

行业名称	工业废水中污染物产生量				
	化学需氧量	氨氮	石油类	挥发酚	氰化物
行业总计	23 032 919.1	1 374 834.7	296 813.7	65 801.3	5 180.2
煤炭开采和洗选业	380 711.0	9 410.7	7 586.5	273.2	11.4
石油和天然气开采业	280 603.6	6 093.2	48 490.6	93.8	—
黑色金属矿采选业	176 128.8	519.1	444.7	5.0	10.0
有色金属矿采选业	194 177.8	4 729.2	69.4	0.1	17.2
非金属矿采选业	70 802.3	886.3	255.5	0.0	0.0
开采辅助活动	492.9	30.8	9.6	—	—
其他采矿业	495.5	7.8	0.8	—	—
农副食品加工业	2 461 481.9	64 559.7	732.8	3.4	1.6
食品制造业	1 214 490.3	81 978.2	236.5	2.1	4.2
酒、饮料和精制茶制造业	2 895 267.4	33 888.8	143.5	0.1	0.0
烟草制品业	9 525.1	516.8	22.3	0.0	—
纺织业	2 089 853.6	63 699.7	380.8	12.4	11.5
纺织服装、服饰业	51 699.4	2 881.3	30.7	0.1	0.0
皮革、毛皮、羽毛及其制品和制鞋业	415 302.7	21 990.2	2 252.7	2.7	0.0
木材加工和木、竹、藤、棕、草制品业	64 360.0	1 013.2	7.2	0.9	—
家具制造业	1 703.0	122.0	51.8	0.0	4.5
造纸和纸制品业	5 700 614.9	56 081.8	361.4	239.2	0.1
印刷和记录媒介复制业	5 495.9	363.2	64.4	0.0	0.1
文教、工美、体育和娱乐用品制造业	7 471.9	460.2	78.4	5.0	36.0
石油加工、炼焦和核燃料加工业	763 809.5	188 820.0	115 275.8	40 162.0	1 006.5
化学原料和化学制品制造业	2 553 555.6	616 520.0	17 806.7	5 201.7	377.3
医药制造业	790 873.9	30 816.3	2 214.4	25.0	0.4
化学纤维制造业	475 024.7	7 650.1	347.4	2.9	0.6
橡胶和塑料制品业	50 757.1	2 748.7	13 119.0	1.3	5.5
非金属矿物制品业	252 733.9	6 356.6	1 284.4	26.0	0.3
黑色金属冶炼和压延加工业	1 106 555.4	49 302.9	60 908.8	18 183.5	1 718.4
有色金属冶炼和压延加工业	143 421.8	77 667.3	2 865.8	1 139.0	36.2
金属制品业	110 110.9	6 848.1	7 255.6	4.1	1 639.8
通用设备制造业	31 355.8	1 607.1	1 578.5	0.2	25.3
专用设备制造业	23 915.1	1 395.4	937.5	0.0	10.2
汽车制造业	111 267.8	3 642.3	6 150.7	3.6	14.3
铁路、船舶、航空航天和其他运输设备制造业	41 931.8	2 477.7	2 849.7	0.7	7.1
电气机械和器材制造业	31 393.4	1 777.0	463.1	0.3	10.9
计算机、通信和其他电子设备制造业	218 057.7	12 739.6	727.4	2.1	97.7
仪器仪表制造业	8 655.7	666.5	46.9	0.0	20.4
其他制造业	37 630.0	959.7	260.3	0.2	96.1
废弃资源综合利用业	10 861.0	2 970.0	109.3	0.1	3.4
金属制品、机械和设备修理业	4 503.9	192.9	339.9	0.2	0.8
电力、热力生产和供应业	199 595.6	8 792.3	765.9	1.0	0.3
燃气生产和供应业	12 989.3	936.5	273.0	409.4	12.1
水的生产和供应业	9.5	0.0	0.0	—	0.0
其他行业	33 231.8	715.5	14.0	—	—

按行业分重点调查工业废水排放及处理情况（三）

（2012）　　　　　　　　　　　　　　　　　　　　　　　　　　　　　　单位：吨

行业名称	工业废水中污染物产生量					
	汞	镉	六价铬	总铬	铅	砷
行业总计	19.961	2 318.538	3 273.441	7 322.903	3 260.838	11 300.867
煤炭开采和洗选业	—	0.002	0.003	0.003	0.059	0.025
石油和天然气开采业	—	—	—	—	—	—
黑色金属矿采选业	0.007	0.815	0.207	7.797	3.897	11.795
有色金属矿采选业	1.382	23.610	0.928	1.727	191.275	154.391
非金属矿采选业	0.097	0.376	0.393	0.920	6.959	3.848
开采辅助活动	—	—	—	0.000	—	—
其他采矿业	0.000	0.000	0.001	0.001	0.010	0.018
农副食品加工业	—	0.001	0.055	0.553	—	0.001
食品制造业	—	0.009	0.001	0.136	0.000	—
酒、饮料和精制茶制造业	—	—	—	0.003	—	—
烟草制品业	—	—	—	—	—	—
纺织业	0.023	0.266	3.538	41.192	0.555	0.055
纺织服装、服饰业	—	0.000	0.055	0.354	0.000	0.000
皮革、毛皮、羽毛及其制品和制鞋业	0.000	0.750	136.398	820.726	0.014	0.001
木材加工和木、竹、藤、棕、草制品业	0.000	0.000	0.000	0.000	0.001	0.000
家具制造业	—	—	1.347	3.110	—	—
造纸和纸制品业	—	0.004	0.035	0.211	0.000	0.666
印刷和记录媒介复制业	0.000	0.000	2.290	4.519	0.007	0.001
文教、工美、体育和娱乐用品制造业	0.000	0.001	9.056	78.755	0.004	—
石油加工、炼焦和核燃料加工业	0.005	0.010	0.251	0.261	0.006	0.066
化学原料和化学制品制造业	13.305	81.292	157.993	316.405	271.548	1 599.479
医药制造业	0.000	0.000	0.244	5.868	0.000	14.798
化学纤维制造业	0.000	—	—	0.001	—	0.000
橡胶和塑料制品业	—	0.000	2.965	60.073	0.002	—
非金属矿物制品业	0.017	0.259	0.143	0.231	1.180	0.639
黑色金属冶炼和压延加工业	0.085	16.349	67.553	88.089	30.128	39.789
有色金属冶炼和压延加工业	4.527	2 092.778	21.533	40.957	2 533.294	9 291.746
金属制品业	0.161	97.060	2 536.882	5 370.586	141.857	167.432
通用设备制造业	—	0.006	48.119	69.079	0.456	0.004
专用设备制造业	0.004	0.016	45.451	49.201	1.796	0.044
汽车制造业	0.000	0.094	101.104	158.897	0.309	0.011
铁路、船舶、航空航天和其他运输设备制造业	0.000	0.127	55.416	61.445	0.009	0.000
电气机械和器材制造业	0.015	4.388	13.745	26.321	63.828	1.485
计算机、通信和其他电子设备制造业	0.093	0.257	30.668	51.657	10.430	1.829
仪器仪表制造业	0.000	0.008	21.770	31.157	0.136	0.341
其他制造业	0.112	0.001	6.866	16.816	2.121	0.004
废弃资源综合利用业	0.001	0.029	0.029	0.204	0.854	0.051
金属制品、机械和设备修理业	—	0.014	8.066	14.275	0.002	—
电力、热力生产和供应业	0.127	0.018	0.036	1.068	0.100	12.350
燃气生产和供应业	—	—	—	—	—	—
水的生产和供应业	—	—	—	0.300	0.300	—
其他行业	—	—	—	—	—	—

按行业分重点调查工业废水排放及处理情况（四）

（2012）

单位：吨

行业名称	工业废水中污染物排放量				
	化学需氧量	氨氮	石油类	挥发酚	氰化物
行业总计	3 038 760.6	242 235.9	17 327.2	1 481.4	171.8
煤炭开采和洗选业	122 355.9	3 677.7	2 539.2	0.2	1.5
石油和天然气开采业	13 032.7	851.5	842.5	4.0	—
黑色金属矿采选业	9 485.8	160.6	126.9	—	—
有色金属矿采选业	44 737.8	1 207.9	37.7	0.1	3.0
非金属矿采选业	7 360.9	393.5	11.8	0.0	0.0
开采辅助活动	128.3	9.6	0.5	—	—
其他采矿业	270.2	3.5	0.4	—	—
农副食品加工业	510 284.7	19 395.0	166.6	3.1	0.8
食品制造业	112 292.7	9 196.3	47.1	0.2	0.2
酒、饮料和精制茶制造业	233 397.1	10 601.3	34.4	0.1	0.0
烟草制品业	2 472.7	174.8	7.8	0.0	—
纺织业	277 437.0	19 337.3	86.6	1.9	0.1
纺织服装、服饰业	15 931.0	1 459.4	8.6	0.1	0.0
皮革、毛皮、羽毛及其制品和制鞋业	62 114.8	6 051.3	332.2	1.0	0.0
木材加工和木、竹、藤、棕、草制品业	15 863.7	495.9	2.5	0.2	—
家具制造业	555.9	45.8	2.7	—	0.0
造纸和纸制品业	623 220.5	20 698.9	71.7	59.9	0.0
印刷和记录媒介复制业	2 202.9	114.7	35.0	0.0	0.0
文教、工美、体育和娱乐用品制造业	1 802.8	148.7	27.7	0.2	0.4
石油加工、炼焦和核燃料加工业	80 418.1	14 821.5	1 897.5	1 234.9	37.7
化学原料和化学制品制造业	325 062.6	84 095.1	2 322.1	91.1	54.1
医药制造业	96 452.1	7 365.0	363.3	3.3	0.1
化学纤维制造业	146 941.3	3 740.7	119.3	0.5	0.2
橡胶和塑料制品业	13 392.1	1 089.8	571.3	0.2	0.2
非金属矿物制品业	32 351.9	1 835.8	196.2	14.7	0.1
黑色金属冶炼和压延加工业	75 473.1	6 491.9	2 642.2	48.0	31.0
有色金属冶炼和压延加工业	27 791.7	15 457.4	386.0	6.3	2.3
金属制品业	32 421.7	2 734.7	1 203.3	3.2	33.0
通用设备制造业	10 620.8	668.3	570.8	0.2	0.5
专用设备制造业	6 624.6	604.4	278.4	0.0	0.2
汽车制造业	15 940.7	1 031.7	715.5	3.6	0.3
铁路、船舶、航空航天和其他运输设备制造业	16 174.7	1 123.8	1 046.8	0.3	0.3
电气机械和器材制造业	7 700.3	607.9	114.3	0.2	0.4
计算机、通信和其他电子设备制造业	33 498.2	2 942.3	165.3	2.0	3.9
仪器仪表制造业	1 814.7	140.5	20.8	0.0	0.2
其他制造业	5 808.1	430.7	54.8	0.0	0.2
废弃资源综合利用业	3 310.8	181.6	8.8	0.1	0.1
金属制品、机械和设备修理业	1 012.7	56.4	60.0	0.2	0.0
电力、热力生产和供应业	30 906.0	2 023.4	195.1	0.7	0.2
燃气生产和供应业	1 002.5	156.7	12.5	0.9	0.7
水的生产和供应业	9.5	0.0	0.0	—	0.0
其他行业	19 084.9	612.6	1.0	—	—

按行业分重点调查工业废水排放及处理情况（五）

单位：吨

行业名称	工业废水中污染物排放量					
	汞	镉	六价铬	总铬	铅	砷
行业总计	1.082	26.662	70.375	188.636	97.137	127.692
煤炭开采和洗选业	—	0.002	0.003	0.003	0.046	0.024
石油和天然气开采业	—	—	—	—	—	—
黑色金属矿采选业	0.000	0.343	0.126	0.930	1.413	0.172
有色金属矿采选业	0.298	2.602	0.400	0.594	26.037	30.623
非金属矿采选业	0.065	0.232	0.038	0.042	0.416	1.606
开采辅助活动	—	—	—	0.000	—	—
其他采矿业	—	0.000	0.001	0.001	0.002	0.004
农副食品加工业	—	0.000	0.001	0.009	—	0.001
食品制造业	—	0.002	0.001	0.007	0.000	—
酒、饮料和精制茶制造业	—	—	—	0.003	—	—
烟草制品业	—	—	—	—	—	—
纺织业	0.004	0.003	0.604	1.181	0.006	0.002
纺织服装、服饰业	—	0.000	0.021	0.125	0.000	0.000
皮革、毛皮、羽毛及其制品和制鞋业	0.000	0.750	2.666	74.046	0.003	0.001
木材加工和木、竹、藤、棕、草制品业	0.000	0.000	0.000	0.000	0.001	0.000
家具制造业	—	—	0.021	0.036	—	—
造纸和纸制品业	—	0.002	0.000	0.093	0.000	0.666
印刷和记录媒介复制业	0.000	0.000	0.055	0.073	0.006	0.001
文教、工美、体育和娱乐用品制造业	0.000	0.001	0.206	0.248	0.003	—
石油加工、炼焦和核燃料加工业	0.005	0.010	0.070	0.079	0.006	0.066
化学原料和化学制品制造业	0.460	1.633	0.790	6.301	8.528	44.009
医药制造业	0.000	0.000	0.004	0.036	0.000	0.007
化学纤维制造业	0.000	—	—	0.001	—	0.000
橡胶和塑料制品业	—	0.000	0.124	0.399	0.002	—
非金属矿物制品业	0.016	0.003	0.122	0.188	0.153	0.328
黑色金属冶炼和压延加工业	0.007	0.165	3.906	4.927	1.200	4.373
有色金属冶炼和压延加工业	0.178	20.435	2.366	3.152	51.446	35.971
金属制品业	0.020	0.347	44.746	77.293	1.630	0.294
通用设备制造业	—	0.005	1.095	1.724	0.082	0.004
专用设备制造业	0.001	0.013	0.667	1.040	0.166	0.041
汽车制造业	—	0.012	10.212	10.796	0.054	0.011
铁路、船舶、航空航天和其他运输设备制造业	0.000	0.010	0.640	0.965	0.009	0.000
电气机械和器材制造业	0.006	0.045	0.264	0.462	2.944	0.067
计算机、通信和其他电子设备制造业	0.010	0.032	0.550	2.542	2.673	0.060
仪器仪表制造业	0.000	0.005	0.276	0.540	0.036	0.341
其他制造业	0.008	0.001	0.063	0.329	0.137	0.004
废弃资源综合利用业	0.001	0.007	0.001	0.022	0.116	0.003
金属制品、机械和设备修理业	—	0.001	0.017	0.102	0.001	—
电力、热力生产和供应业	0.002	0.001	0.021	0.046	0.023	9.013
燃气生产和供应业	—	—	—	—	—	—
水的生产和供应业	—	—	—	0.300	0.300	—
其他行业						

按行业分重点调查工业废气排放及处理情况（一）

（2012）

行业名称	工业废气排放总量（标态）/亿米³	废气治理设施数/套	废气治理设施处理能力（标态）/（万米³/时）	废气治理设施运行费用/万元
行业总计	635 519.1	225 913	1 649 353.3	14 522 519.8
煤炭开采和洗选业	3 248.6	5 500	7 458.1	41 797.3
石油和天然气开采业	1 896.0	202	302.6	14 398.6
黑色金属矿采选业	2 946.0	1 257	1 963.6	18 208.4
有色金属矿采选业	1 120.8	708	1 481.3	18 406.7
非金属矿采选业	849.0	569	1 444.5	13 124.9
开采辅助活动	47.5	26	30.7	386.0
其他采矿业	583.9	24	348.6	317.3
农副食品加工业	4 500.2	6 960	16 450.1	63 856.4
食品制造业	2 119.0	3 462	10 155.9	38 294.5
酒、饮料和精制茶制造业	2 060.5	2 900	65 961.5	39 198.7
烟草制品业	557.7	1 081	2 642.4	12 556.7
纺织业	3 164.1	8 881	12 292.0	92 153.0
纺织服装、服饰业	204.1	1 153	1 059.6	7 732.7
皮革、毛皮、羽毛及其制品和制鞋业	328.6	1 566	1 260.0	6 357.7
木材加工和木、竹、藤、棕、草制品业	2 713.2	2 966	7 135.4	26 753.8
家具制造业	283.9	723	2 302.4	2 812.7
造纸和纸制品业	6 146.0	5 489	17 510.1	162 604.7
印刷和记录媒介复制业	246.2	351	787.5	3 813.6
文教、工美、体育和娱乐用品制造业	190.5	609	694.6	2 706.8
石油加工、炼焦和核燃料加工业	20 375.8	3 261	24 610.3	522 758.5
化学原料和化学制品制造业	30 613.8	20 076	65 142.3	665 946.3
医药制造业	3 988.8	3 733	4 776.9	48 382.6
化学纤维制造业	2 192.0	1 026	3 217.2	34 270.1
橡胶和塑料制品业	2 936.0	4 096	9 276.7	52 483.6
非金属矿物制品业	123 285.2	65 810	265 266.2	1 214 941.7
黑色金属冶炼和压延加工业	160 874.9	17 009	371 693.5	3 029 739.6
有色金属冶炼和压延加工业	31 799.3	9 366	54 172.6	786 778.8
金属制品业	5 078.8	8 417	12 579.2	75 329.1
通用设备制造业	1 128.8	2 865	5 379.9	17 155.4
专用设备制造业	1 249.7	2 302	6 246.0	17 682.3
汽车制造业	3 972.6	2 767	8 352.2	59 852.0
铁路、船舶、航空航天和其他运输设备制造业	1 691.9	4 724	7 710.4	28 907.4
电气机械和器材制造业	2 007.8	4 500	5 776.7	42 134.4
计算机、通信和其他电子设备制造业	5 744.6	6 419	14 839.7	99 794.2
仪器仪表制造业	296.2	605	952.1	3 643.9
其他制造业	488.6	1 014	97 575.4	11 628.2
废弃资源综合利用业	336.3	398	721.1	6 055.5
金属制品、机械和设备修理业	424.3	212	753.8	1 999.8
电力、热力生产和供应业	203 436.4	22 642	547 619.5	7 231 127.3
燃气生产和供应业	341.3	210	748.9	6 041.1
水的生产和供应业	0.0	2	0.0	2.1
其他行业	50.2	34	46.7	385.5

按行业分重点调查工业废气排放及处理情况（二）

（2012）

单位：万吨

行业名称	二氧化硫产生量	二氧化硫排放量	氮氧化物产生量	氮氧化物排放量	烟粉尘产生量	烟粉尘排放量
行业总计	5 712.3	1 775.8	1 709.3	1 580.8	75 268.2	957.4
煤炭开采和洗选业	16.6	12.5	4.7	4.5	136.2	33.3
石油和天然气开采业	4.5	2.2	3.0	3.0	4.2	0.7
黑色金属矿采选业	2.8	2.4	0.7	0.7	34.0	10.3
有色金属矿采选业	4.0	2.4	0.6	0.5	45.3	2.2
非金属矿采选业	4.4	3.9	1.1	1.1	57.4	3.7
开采辅助活动	0.3	0.3	0.2	0.2	0.4	0.1
其他采矿业	0.1	0.0	0.0	0.0	3.0	0.2
农副食品加工业	31.1	23.8	9.4	9.2	191.9	18.2
食品制造业	18.9	14.7	4.7	4.7	120.3	5.8
酒、饮料和精制茶制造业	16.7	12.9	4.1	4.1	155.3	6.6
烟草制品业	1.6	1.1	0.4	0.4	4.6	0.7
纺织业	31.8	27.0	8.8	7.7	116.2	9.2
纺织服装、服饰业	1.9	1.7	0.5	0.5	5.2	0.8
皮革、毛皮、羽毛及其制品和制鞋业	2.8	2.7	0.6	0.6	5.6	1.0
木材加工和木、竹、藤、棕、草制品业	4.4	4.3	1.5	1.4	150.2	15.7
家具制造业	0.3	0.3	0.1	0.1	1.4	0.3
造纸和纸制品业	75.0	49.7	21.0	20.7	468.8	16.7
印刷和记录媒介复制业	0.5	0.5	0.2	0.2	3.9	0.2
文教、工美、体育和娱乐用品制造业	0.2	0.2	0.1	0.1	0.6	0.2
石油加工、炼焦和核燃料加工业	350.9	80.2	38.3	37.6	1 122.5	44.2
化学原料和化学制品制造业	242.8	126.2	53.6	50.2	1 502.0	58.3
医药制造业	13.9	10.8	3.2	3.1	98.4	4.4
化学纤维制造业	15.4	10.1	4.9	4.9	127.5	2.2
橡胶和塑料制品业	12.0	8.8	2.8	2.8	45.6	3.3
非金属矿物制品业	228.7	199.8	279.2	274.2	22 381.8	255.2
黑色金属冶炼和压延加工业	325.5	240.6	98.3	97.2	7 756.3	181.3
有色金属冶炼和压延加工业	1 025.5	114.4	23.7	23.0	1 501.8	31.9
金属制品业	21.2	7.6	2.4	2.4	36.5	8.2
通用设备制造业	2.8	2.3	0.9	0.9	12.9	3.2
专用设备制造业	2.8	1.9	1.0	1.0	24.1	2.2
汽车制造业	1.8	1.4	0.7	0.6	15.5	2.1
铁路、船舶、航空航天和其他运输设备制造业	2.4	1.7	1.0	1.0	30.0	5.2
电气机械和器材制造业	1.8	1.1	0.5	0.5	4.4	0.7
计算机、通信和其他电子设备制造业	0.8	0.8	0.6	0.5	3.3	1.3
仪器仪表制造业	0.1	0.1	0.0	0.0	0.3	0.1
其他制造业	6.5	6.2	1.4	1.3	15.9	2.5
废弃资源综合利用业	0.6	0.4	0.1	0.1	13.3	0.4
金属制品、机械和设备修理业	0.1	0.1	0.0	0.0	7.3	0.9
电力、热力生产和供应业	3 236.4	797.0	1 133.9	1 018.7	39 041.0	222.8
燃气生产和供应业	2.2	1.7	1.2	1.2	22.1	0.7
水的生产和供应业	0.0	0.0	0.0	0.0	—	—
其他行业	0.1	0.1	0.0	0.0	0.9	0.5

按行业分重点调查一般工业固体废物产生及处置利用情况

（2012）

单位：万吨

行业名称	一般工业固体废物产生量	一般工业固体废物综合利用量	一般工业固体废物处置量	一般工业固体废物贮存量	一般工业固体废物倾倒丢弃量	一般工业固体废弃物综合利用率/%
行业总计	314 073	191 787	67 992	58 175	131	60.5
煤炭开采和洗选业	38 537	30 440	6 420	2 750	18	77.7
石油和天然气开采业	127	100	24	3	0	79.1
黑色金属矿采选业	70 602	15 622	27 148	27 936	27	22.1
有色金属矿采选业	40 290	14 489	14 062	12 439	17	35.4
非金属矿采选业	3 598	1 969	794	842	22	54.6
开采辅助活动	102	10	92	0	0	10.1
其他采矿业	74	39	152	5	0	53.0
农副食品加工业	2 209	2 087	116	26	4	93.5
食品制造业	563	541	22	0	0	95.9
酒、饮料和精制茶制造业	941	894	51	7	1	93.8
烟草制品业	92	61	14	36	0	55.7
纺织业	691	591	108	1	0	84.4
纺织服装、服饰业	28	24	4	0	0	83.3
皮革、毛皮、羽毛及其制品和制鞋业	63	53	11	0	0	82.2
木材加工和木、竹、藤、棕、草制品业	229	224	5	0	0	97.8
家具制造业	12	10	2	0	0	84.2
造纸和纸制品业	2 168	1 926	231	13	0	88.7
印刷和记录媒介复制业	21	15	6	0	0	71.2
文教、工美、体育和娱乐用品制造业	6	5	1	0	0	84.9
石油加工、炼焦和核燃料加工业	3 671	2 358	1 234	97	1	63.9
化学原料和化学制品制造业	26 644	17 375	3 992	5 755	4	64.4
医药制造业	312	278	32	3	1	88.9
化学纤维制造业	323	286	36	2	0	88.2
橡胶和塑料制品业	238	226	12	0	0	94.9
非金属矿物制品业	6 780	6 388	419	92	4	92.6
黑色金属冶炼和压延加工业	42 047	37 528	3 099	1 733	13	88.6
有色金属冶炼和压延加工业	9 978	5 122	3 289	1 645	1	51.1
金属制品业	523	436	54	33	0	83.5
通用设备制造业	186	142	35	9	0	76.4
专用设备制造业	205	162	38	5	0	79.2
汽车制造业	290	255	34	1	0	87.9
铁路、船舶、航空航天和其他运输设备制造业	244	209	35	0	0	85.4
电气机械和器材制造业	69	57	12	0	0	82.3
计算机、通信和其他电子设备制造业	343	297	46	0	0	86.5
仪器仪表制造业	8	6	2	0	0	78.4
其他制造业	73	50	22	0	0	69.5
废弃资源综合利用业	246	223	15	9	0	90.5
金属制品、机械和设备修理业	16	14	2	0	0	89.2
电力、热力生产和供应业	61 456	51 209	6 320	4 733	16	82.2
燃气生产和供应业	59	58	1	0	0	97.8
水的生产和供应业	—	—	0	0	0	100.0
其他行业	8	8	0	0	0	100.0

按行业分重点调查危险废物产生及处置利用情况

（2012）

行业名称	危险废物产生量/万吨	危险废物综合利用量/万吨	危险废物处置量/万吨	危险废物贮存量/万吨	危险废物倾倒丢弃量/吨	危险废物处置利用率/%
行业总计	3 465	2 005	698	847	16	76.1
煤炭开采和洗选业	0	0	0	0	0	92.2
石油和天然气开采业	39	10	29	0	0	99.4
黑色金属矿采选业	0	0	0	0	0	92.4
有色金属矿采选业	73	57	5	19	0	76.3
非金属矿采选业	712	46	3	663	0	6.9
开采辅助活动	1	1	0	0	0	100.0
其他采矿业	0	0	0	0	0	0.0
农副食品加工业	9	8	0	0	0	99.8
食品制造业	1	0	0	0	0	99.9
酒、饮料和精制茶制造业	0	0	0	0	0	99.1
烟草制品业	0	0	0	0	0	94.7
纺织业	2	1	2	0	0	97.7
纺织服装、服饰业	0	0	0	0	0	99.3
皮革、毛皮、羽毛及其制品和制鞋业	3	1	2	0	0	98.2
木材加工和木、竹、藤、棕、草制品业	0	0	0	0	0	99.6
家具制造业	0	0	0	0	0	99.1
造纸和纸制品业	715	667	48	0	0	100.0
印刷和记录媒介复制业	1	0	1	0	0	99.9
文教、工美、体育和娱乐用品制造业	1	0	1	0	0	99.6
石油加工、炼焦和核燃料加工业	139	98	40	2	0	98.8
化学原料和化学制品制造业	662	447	229	36	0	95.0
医药制造业	51	33	18	1	0	98.7
化学纤维制造业	40	37	3	0	0	99.9
橡胶和塑料制品业	5	1	4	0	0	99.5
非金属矿物制品业	34	12	22	0	0	99.6
黑色金属冶炼和压延加工业	161	141	26	2	0	98.6
有色金属冶炼和压延加工业	454	286	68	113	8	75.8
金属制品业	47	21	26	1	2	98.6
通用设备制造业	10	2	8	0	0	98.5
专用设备制造业	4	1	3	0	0	96.8
汽车制造业	27	7	20	0	2	99.5
铁路、船舶、航空航天和其他运输设备制造业	12	7	5	0	0	99.6
电气机械和器材制造业	21	6	14	0	0	97.8
计算机、通信和其他电子设备制造业	147	80	67	1	4	99.6
仪器仪表制造业	10	0	2	7	0	26.9
其他制造业	3	0	3	0	0	99.2
废弃资源综合利用业	5	3	2	0	0	91.4
金属制品、机械和设备修理业	1	0	0	0	0	97.8
电力、热力生产和供应业	73	30	43	0	0	99.4
燃气生产和供应业	2	—	1	—	0	100.0
水的生产和供应业	—	0	—	0	0	100.0
其他行业	—	0	—	0	0	100.0

按行业分重点调查工业汇总情况（一）

（2012）

行业名称	汇总工业企业数/个	工业总产值（现价）/万元	工业煤炭消耗量/万吨	燃料煤	工业用水总量/万吨	取水量	重复用水量
行业总计	147 996	368 365.3	361 405.9	274 930.0	36 758 273.1	4 721 197.7	32 037 075.4
煤炭开采和洗选业	6 516	13 405.3	21 078.7	1 612.6	233 919.8	97 750.7	136 169.1
石油和天然气开采业	261	5 675.8	77.8	75.5	130 471.9	31 704.6	98 767.3
黑色金属矿采选业	4 489	2 357.8	151.1	144.3	416 506.1	80 195.7	336 310.4
有色金属矿采选业	3 378	2 121.2	152.7	150.4	257 807.9	69 889.9	187 918.0
非金属矿采选业	992	1 516.4	274.2	260.9	53 712.5	11 907.0	41 805.5
开采辅助活动	48	399.9	13.0	11.6	1 867.4	612.2	1 255.2
其他采矿业	27	15.6	21.5	21.5	683.1	593.9	89.2
农副食品加工业	11 617	13 170.7	2 158.5	2 146.3	436 472.2	188 456.0	248 016.1
食品制造业	3 794	5 733.6	1 206.4	1 202.5	157 507.2	71 608.0	85 899.1
酒、饮料和精制茶制造业	4 224	7 268.7	1 126.5	1 122.8	236 375.1	101 336.8	135 038.3
烟草制品业	152	7 239.2	67.9	67.2	16 666.1	3 921.1	12 745.0
纺织业	8 887	11 949.2	2 342.9	2 314.5	376 729.7	290 148.4	86 581.3
纺织服装、服饰业	1 485	1 147.4	137.3	136.5	23 881.9	19 552.7	4 329.2
皮革、毛皮、羽毛及其制品和制鞋业	2 210	1 647.3	183.5	182.7	36 193.9	32 252.5	3 941.4
木材加工和木、竹、藤、棕、草制品业	2 634	1 548.4	266.8	233.6	10 808.2	6 864.5	3 943.7
家具制造业	465	518.2	14.4	14.4	1 201.2	1 080.3	120.9
造纸和纸制品业	5 235	7 125.8	4 944.1	4 932.5	1 212 955.4	407 844.2	805 111.2
印刷和记录媒介复制业	727	889.0	29.3	28.9	3 743.5	2 109.6	1 633.8
文教、工美、体育和娱乐用品制造业	694	1 232.0	12.2	12.2	3 799.1	2 879.9	919.2
石油加工、炼焦和核燃料加工业	1 307	36 038.2	34 585.1	7 505.0	2 415 566.0	174 747.5	2 240 818.5
化学原料和化学制品制造业	12 043	30 249.4	19 045.2	10 782.9	5 759 611.9	469 397.6	5 290 214.3
医药制造业	3 264	7 549.3	898.8	893.5	297 852.4	72 585.3	225 267.1
化学纤维制造业	483	3 590.0	1 060.4	1 031.6	205 471.0	44 608.3	160 862.8
橡胶和塑料制品业	2 539	6 715.3	725.5	721.0	83 829.3	18 096.2	65 733.0
非金属矿物制品业	33 164	16 741.1	30 652.2	21 846.6	398 605.5	79 242.9	319 362.7
黑色金属冶炼和压延加工业	4 599	45 575.5	30 602.5	9 970.7	7 764 125.6	314 679.8	7 449 445.8
有色金属冶炼和压延加工业	4 190	17 040.5	5 269.8	4 587.6	758 824.7	69 724.4	689 100.4
金属制品业	8 617	10 614.0	556.2	439.8	340 944.9	59 489.7	281 455.2
通用设备制造业	3 092	8 616.0	186.6	178.5	33 813.5	13 485.6	20 327.9
专用设备制造业	1 584	6 458.3	188.2	163.7	53 448.6	11 129.7	42 319.0
汽车制造业	1 853	22 897.1	122.7	119.7	121 542.2	21 324.9	100 217.3
铁路、船舶、航空航天和其他运输设备制造业	1 012	8 465.7	211.7	206.2	38 973.1	16 661.1	22 312.0
电气机械和器材制造业	1 881	10 997.4	80.0	68.7	44 307.5	13 592.8	30 714.7
计算机、通信和其他电子设备制造业	2 586	27 433.6	48.4	47.9	250 819.5	60 972.5	189 847.0
仪器仪表制造业	386	2 697.7	8.2	7.7	8 058.4	3 682.9	4 375.5
其他制造业	1 159	2 430.1	391.1	217.7	19 436.8	7 012.4	12 424.5
废弃资源综合利用业	494	489.1	27.7	27.2	5 627.4	3 501.5	2 125.9
金属制品、机械和设备修理业	180	592.3	6.9	6.9	1 911.3	1 583.6	327.6
电力、热力生产和供应业	5 525	17 757.6	201 571.2	201 265.8	14 468 677.4	1 840 180.6	12 628 496.8
燃气生产和供应业	59	290.9	901.9	163.4	74 344.1	3 829.0	70 515.1
水的生产和供应业	3	0.2	0.0	0.0	2.4	2.4	0.0
其他行业	141	164.0	6.9	6.8	1 177.8	959.3	218.4

按行业分重点调查工业汇总情况（二）

（2012）

行业名称	工业炉窑数/台	工业锅炉数/台	35 蒸吨及以上的	20（含）～35 蒸吨之间的	10（含）～20 蒸吨之间的	10 蒸吨以下的
行业总计	99 205	103 469	13 683	7 400	13 726	68 660
煤炭开采和洗选业	127	7 285	76	208	841	6 160
石油和天然气开采业	2 759	2 510	33	305	130	2 042
黑色金属矿采选业	146	564	14	14	79	457
有色金属矿采选业	233	516	9	14	48	445
非金属矿采选业	404	199	15	18	20	146
开采辅助活动	26	206	6	14	15	171
其他采矿业	10	11	0	0	5	6
农副食品加工业	454	8 841	578	488	628	7 147
食品制造业	116	4 234	167	217	555	3 295
酒、饮料和精制茶制造业	248	4 340	130	324	753	3 133
烟草制品业	16	360	29	70	150	111
纺织业	242	8 812	165	258	1 531	6 858
纺织服装、服饰业	7	1 442	0	12	123	1 307
皮革、毛皮、羽毛及其制品和制鞋业	27	1 804	4	9	81	1 710
木材加工和木、竹、藤、棕、草制品业	184	2 753	28	66	241	2 418
家具制造业	52	261	0	5	13	243
造纸和纸制品业	398	5 690	536	217	793	4 144
印刷和记录媒介复制业	9	393	0	2	24	367
文教、工美、体育和娱乐用品制造业	30	251	0	0	9	242
石油加工、炼焦和核燃料加工业	3 643	1 838	539	283	420	596
化学原料和化学制品制造业	6 459	10 585	1 289	591	1 311	7 394
医药制造业	69	3 395	93	110	407	2 785
化学纤维制造业	102	814	113	109	204	388
橡胶和塑料制品业	61	2 583	72	157	343	2011
非金属矿物制品业	42 103	3 625	92	175	412	2 946
黑色金属冶炼和压延加工业	12 353	1 702	536	197	179	790
有色金属冶炼和压延加工业	16 335	1 575	177	116	227	1 055
金属制品业	4 066	2 081	11	32	97	1 941
通用设备制造业	2 349	917	5	31	81	800
专用设备制造业	1 611	729	21	43	147	518
汽车制造业	1 326	1 058	9	76	116	857
铁路、船舶、航空航天和其他运输设备制造业	910	854	59	160	137	498
电气机械和器材制造业	432	689	2	17	71	599
计算机、通信和其他电子设备制造业	444	1 037	16	18	108	895
仪器仪表制造业	252	119	0	0	14	105
其他制造业	745	497	3	16	62	416
废弃资源综合利用业	156	110	0	2	6	102
金属制品、机械和设备修理业	23	121	1	1	13	106
电力、热力生产和供应业	147	18 487	8 803	3 001	3 309	3 374
燃气生产和供应业	47	140	52	24	22	42
水的生产和供应业	0	0	0	0	0	0
其他行业	84	41	0	0	1	40

各地区独立火电厂污染排放及处理情况（一）

（2012）

地 区 名 称	汇总工业企业数/个	机组数/台	废气治理设施数/套	脱硫设施数	脱硝设施数	除尘设施数
全 国	1 824	4 625	9 243	3 465	438	5 171
北 京	9	29	58	12	26	20
天 津	17	42	117	50	7	60
河 北	104	233	501	216	15	267
山 西	114	279	505	217	24	263
内蒙古	93	250	461	170	8	283
辽 宁	74	225	435	143	12	257
吉 林	38	101	200	49	6	138
黑龙江	76	205	383	69	5	309
上 海	22	60	90	36	8	46
江 苏	187	453	1 082	455	57	549
浙 江	148	401	948	398	46	467
安 徽	53	118	243	87	18	138
福 建	32	80	168	64	26	76
江 西	19	40	90	29	8	53
山 东	270	736	1 529	613	22	884
河 南	93	201	443	187	22	231
湖 北	47	107	181	70	9	101
湖 南	30	63	118	41	14	61
广 东	84	211	402	145	59	180
广 西	17	31	71	30	3	35
海 南	8	23	27	11	4	12
重 庆	33	57	130	49	0	78
四 川	44	77	130	47	5	77
贵 州	23	70	133	55	5	71
云 南	15	40	90	38	5	47
西 藏	2	21	0	0	0	0
陕 西	54	142	224	53	9	139
甘 肃	22	64	121	39	9	70
青 海	7	22	12	3	0	9
宁 夏	24	58	120	46	6	68
新 疆	65	186	231	43	0	182

各地区独立火电厂污染排放及处理情况（二）

（2012）

地 区 名 称	燃料煤消耗量/ 万吨			燃料煤平均 含硫量/%	燃料油消耗量/ 万吨	燃料油平均 含硫量/%
		发电消耗量	供热消耗量			
全　国	190 599.4	174 352.8	16 246.7	0.96	679.1	0.13
北　京	888.8	639.2	249.7	0.57	0.0	0.00
天　津	2 802.1	2 550.3	251.8	0.73	0.1	0.11
河　北	9 946.1	8 944.9	1 001.2	1.01	0.7	0.44
山　西	12 689.2	11 964.2	725.0	1.18	1.7	0.37
内蒙古	18 579.5	17 132.4	1 447.1	0.78	1.5	0.75
辽　宁	8 630.6	7 069.9	1 560.7	0.64	4.7	0.30
吉　林	4 541.5	3 518.3	1 023.2	0.41	0.5	1.39
黑龙江	6 035.2	4 759.9	1 275.3	0.35	0.6	0.72
上　海	3 169.9	3 051.9	118.0	0.64	7.3	0.92
江　苏	16 850.7	14 993.1	1 857.6	0.75	555.4	0.01
浙　江	9 847.2	8 267.9	1 579.3	0.71	83.5	0.36
安　徽	7 846.2	7 587.2	259.0	0.51	0.2	0.68
福　建	4 373.3	4 199.5	173.8	0.67	0.8	0.57
江　西	2 678.2	2 678.2	0.0	1.11	0.3	0.49
山　东	15 898.2	13 473.6	2 424.6	1.13	0.6	0.42
河　南	12 340.1	11 632.3	707.8	0.93	2.1	0.33
湖　北	3 689.4	3 545.7	143.7	1.38	0.6	13.50
湖　南	3 252.7	3 166.1	86.6	1.10	0.6	0.52
广　东	11 392.9	11 133.9	259.0	0.71	1.8	0.82
广　西	2 241.7	2 221.8	19.9	1.76	0.1	0.14
海　南	584.7	584.7	0.0	0.90	0.0	0.92
重　庆	1 934.4	1 873.7	60.6	2.70	0.5	0.50
四　川	3 178.1	3 146.6	31.5	2.02	0.9	0.33
贵　州	5 518.6	5 515.2	3.4	2.40	0.9	28.83
云　南	2 836.7	2 828.2	8.6	1.58	1.1	1.30
西　藏	0.0	0.0	0.0	0.00	11.1	0.04
陕　西	6 109.1	5 959.8	149.3	1.28	0.4	0.32
甘　肃	3 257.1	2 969.2	288.0	0.74	0.2	0.38
青　海	360.2	360.2	0.0	0.87	0.0	0.21
宁　夏	4 950.5	4 819.8	130.6	1.12	0.4	0.45
新　疆	4 176.7	3 765.4	411.3	1.41	0.5	1.16

各地区独立火电厂污染排放及处理情况（三）

（2012）

地 区名 称	工业废气排放量/亿米³	工业二氧化硫产生量/吨	工业二氧化硫排放量/吨	工业氮氧化物产生量/吨	工业氮氧化物排放量/吨	工业烟粉尘产生量/吨	工业烟粉尘排放量/吨
全 国	165 059	31 026 365	7 063 467	10 931 787	9 815 720	366 164 211	1 441 955
北 京	1 285	83 520	9 993	53 493	41 578	1 903 863	3 793
天 津	2 210	391 869	60 854	190 057	170 520	5 465 083	6 660
河 北	10 352	1 813 442	303 122	707 429	637 495	21 263 787	49 572
山 西	9 780	2 518 631	590 170	677 455	612 152	34 625 839	163 162
内蒙古	13 568	2 461 754	631 894	889 412	836 518	32 385 758	205 332
辽 宁	7 996	993 915	305 938	427 735	410 948	13 783 972	91 072
吉 林	2 735	317 245	131 126	250 477	241 805	7 682 247	45 665
黑龙江	4 184	340 630	184 478	329 336	323 127	9 937 573	158 695
上 海	3 056	344 495	54 687	222 872	183 037	4 380 812	10 009
江 苏	13 831	2 264 422	405 692	918 282	742 489	25 208 878	69 131
浙 江	8 198	1 178 121	241 402	504 522	419 763	11 096 447	36 713
安 徽	8 674	664 783	129 696	413 472	366 093	20 148 040	38 158
福 建	4 316	498 419	75 040	216 267	169 742	4 409 186	11 795
江 西	2 068	522 347	117 916	180 530	170 319	4 889 005	17 531
山 东	14 898	3 050 060	691 036	772 947	719 044	33 156 641	95 224
河 南	9 437	1 905 305	399 580	763 218	715 799	30 654 718	53 495
湖 北	4 336	747 009	201 628	241 841	231 170	10 572 446	31 634
湖 南	2 742	607 533	121 786	348 429	245 161	10 475 103	16 132
广 东	10 154	1 372 414	265 466	516 652	422 439	10 028 182	36 077
广 西	1 646	659 715	121 532	127 259	119 782	5 397 546	7 748
海 南	446	87 741	15 526	49 274	35 500	419 072	1 339
重 庆	1 690	854 609	207 980	147 770	147 770	6 967 887	56 162
四 川	2 926	989 275	251 934	189 415	171 861	9 081 903	33 848
贵 州	4 334	2 189 471	568 090	384 134	375 473	11 733 361	27 836
云 南	1 823	756 418	168 255	156 893	154 064	7 887 763	14 821
西 藏	12	477	477	743	743	28	28
陕 西	5 066	1 395 947	339 789	473 967	414 805	9 313 083	52 649
甘 肃	4 415	473 526	105 478	222 345	207 889	5 526 635	20 990
青 海	348	53 514	24 330	18 312	18 312	364 648	6 697
宁 夏	4 496	974 530	151 137	281 530	254 602	11 862 869	34 664
新 疆	4 037	515 227	187 436	255 720	255 720	5 541 835	45 326

各地区水泥行业污染排放及处理情况（一）

（2012）

地 区 名 称	汇总工业企业数/个	水泥窑数/台	新型干法窑	废气治理设施数/套	脱硝设施数	除尘设施数	煤炭消耗量/万吨
全 国	3 288	3 615	1 667	49 848	84	48 339	22 403.1
北 京	8	10	10	539	1	538	120.9
天 津	7	3	3	215	0	214	27.1
河 北	129	125	80	2 311	11	2 296	906.3
山 西	173	146	60	2 490	0	2 414	497.9
内蒙古	58	84	57	1 542	0	1 537	563.6
辽 宁	72	70	45	1 444	1	1 438	557.2
吉 林	28	39	31	747	6	511	884.4
黑龙江	58	61	27	1 068	1	1 067	288.9
上 海	7	11	1	104	0	104	8.3
江 苏	216	159	71	1 755	5	1 630	818.2
浙 江	108	123	80	3 216	3	3 212	863.0
安 徽	72	116	90	1 418	1	1 415	1 922.3
福 建	98	126	41	2 150	12	2 138	668.8
江 西	121	156	62	1 876	1	1 820	716.6
山 东	251	250	92	2 729	0	2 713	1 232.2
河 南	231	181	104	2 897	0	2 791	1 140.0
湖 北	134	127	65	2 161	1	2 156	853.5
湖 南	211	217	62	2 049	5	2 022	1 254.8
广 东	197	287	61	2 054	6	1 928	1 329.5
广 西	162	200	56	2 597	3	2 568	1 177.7
海 南	5	10	10	198	0	198	191.9
重 庆	117	155	47	1 790	0	1 789	691.9
四 川	187	222	99	2 879	4	2 795	1 387.2
贵 州	141	142	58	977	8	947	723.6
云 南	173	194	114	2 475	1	2 190	1 461.5
西 藏	9	18	4	226	0	142	35.9
陕 西	102	104	63	1 845	12	1 681	587.5
甘 肃	79	106	46	1 147	0	1 144	437.1
青 海	18	22	15	494	2	492	197.8
宁 夏	18	28	25	289	0	283	196.3
新 疆	98	123	88	2 166	0	2 166	661.3

各地区水泥行业污染排放及处理情况（二）

（2012）

地 区 名 称	工业废气排放量/ 万米³	工业二氧化硫 产生量/吨	工业二氧化硫 排放量/吨	工业氮氧化物 产生量/吨	工业氮氧化物 排放量/吨	工业烟粉尘 产生量/吨	工业烟粉尘 排放量/吨
全 国	623 384 230.2	347 110.9	331 590.7	2 017 298.7	1 979 397.0	182 496 778.2	670 685.2
北 京	3 876 526.3	948.9	948.9	11 929.6	11 929.6	1 645 530.7	2 780.1
天 津	1 038 289.7	247.3	247.3	3 092.2	3 092.2	380 979.4	1 880.3
河 北	41 532 244.7	10 012.5	7 559.1	95 881.3	92 869.0	8 964 853.0	23 370.0
山 西	14 792 284.1	10 235.3	6 795.7	50 719.4	50 719.4	5 865 781.4	31 988.1
内蒙古	17 499 400.6	8 357.6	8 357.6	59 148.2	59 148.2	5 281 769.5	29 658.2
辽 宁	15 097 624.6	6 010.1	6 000.1	56 126.4	55 780.2	6 748 580.4	63 816.3
吉 林	10 638 706.1	2 736.7	2 736.7	42 758.4	42 690.4	2 581 897.4	13 999.1
黑龙江	6 461 457.6	3 250.1	3 054.8	27 839.0	27 741.0	2 508 222.1	17 787.2
上 海	553 815.0	210.6	210.6	741.4	741.4	59 298.0	847.7
江 苏	18 770 289.8	9 030.7	6 503.5	76 649.5	76 649.5	4 883 141.6	11 433.2
浙 江	24 360 542.6	12 180.2	12 180.2	93 467.0	93 359.0	10 619 935.7	19 993.7
安 徽	53 004 363.2	18 556.0	17 746.6	181 973.1	179 413.0	12 480 201.5	51 670.1
福 建	22 423 897.9	9 541.5	9 541.5	72 713.4	71 584.3	9 287 423.2	30 800.6
江 西	25 464 773.4	13 862.5	13 862.5	67 376.5	67 376.5	6 726 011.1	25 347.4
山 东	31 481 272.9	14 814.6	14 800.3	130 644.8	130 644.8	12 874 146.1	34 260.5
河 南	31 824 886.0	15 177.5	15 001.5	115 637.8	115 637.8	11 184 500.1	25 362.8
湖 北	20 963 801.7	7 812.0	7 807.7	88 448.2	88 277.1	9 999 225.9	18 556.0
湖 南	24 361 148.3	26 919.6	25 656.6	79 564.7	76 894.1	7 224 735.9	51 907.6
广 东	23 283 770.8	21 954.7	21 249.5	117 753.2	99 009.8	5 843 134.6	20 993.7
广 西	84 102 408.3	13 285.0	11 188.8	104 291.7	104 289.5	5 483 457.5	19 057.0
海 南	4 615 578.8	3 701.6	3 701.6	20 709.5	20 709.5	971 630.1	1 929.3
重 庆	23 376 995.0	56 359.1	56 180.7	61 487.7	61 487.7	7 894 254.9	23 555.8
四 川	31 672 844.7	18 678.6	18 119.9	119 005.3	114 312.0	12 251 540.4	24 585.9
贵 州	18 193 849.1	19 258.0	18 775.8	64 162.5	60 965.7	4 915 021.5	29 640.7
云 南	22 241 695.0	14 681.2	14 387.1	80 188.0	80 152.0	8 209 872.4	25 980.0
西 藏	383 135.0	171.7	162.2	2 400.0	2 400.0	188 508.0	339.4
陕 西	14 027 648.3	8 484.8	8 484.8	59 827.9	58 761.4	3 221 196.8	22 359.9
甘 肃	11 158 431.0	9 254.0	8 996.0	39 283.7	39 283.7	3 882 991.2	13 430.3
青 海	3 282 276.1	2 185.0	2 185.3	14 537.1	14 537.1	1 657 549.0	5 462.5
宁 夏	4 781 821.1	2 501.3	2 501.3	19 436.9	19 436.9	1 883 240.6	6 183.0
新 疆	18 118 452.7	6 691.3	6 646.7	59 504.3	59 504.3	6 778 148.2	21 708.7

各地区钢铁行业污染排放及处理情况（一）

（2012）

地 区 名 称	汇总工业 企业数/ 个	烧结机数/ 台	有脱硫 设施的	有脱硝 设施的	有除尘 设施的	球团设备数/ 套	有脱硫 设施的	有脱硝 设施的	有除尘 设施的
全 国	1 180	1 127	333	21	749	657	45	14	386
北 京	0	0	0	0	0	0	0	0	0
天 津	10	12	0	0	12	6	0	0	6
河 北	216	317	142	14	244	211	5	4	163
山 西	116	138	22	2	59	38	1	0	16
内 蒙 古	18	36	11	0	22	19	3	0	12
辽 宁	75	59	11	0	43	41	0	0	24
吉 林	8	12	1	0	13	23	0	0	9
黑 龙 江	9	13	0	0	6	6	0	0	3
上 海	4	6	6	0	6	0	0	0	0
江 苏	71	89	19	0	41	24	3	2	10
浙 江	29	8	8	2	10	4	2	0	4
安 徽	44	39	6	0	12	50	9	6	25
福 建	27	22	13	0	15	4	1	0	3
江 西	20	16	3	0	16	13	2	2	10
山 东	56	80	36	0	70	20	1	0	12
河 南	34	44	0	0	30	23	4	0	16
湖 北	44	22	7	0	21	17	1	0	10
湖 南	107	21	6	0	13	7	1	0	3
广 东	14	10	6	0	6	0	0	0	0
广 西	26	12	8	2	10	3	0	0	2
海 南	1	0	0	0	1	0	0	0	0
重 庆	9	3	3	0	3	1	0	0	1
四 川	54	33	10	0	17	50	12	0	19
贵 州	20	6	2	0	7	25	0	0	20
云 南	53	55	11	1	22	16	0	0	6
西 藏									
陕 西	7	15	2	0	12	9	0	0	0
甘 肃	26	10	0	0	6	17	0	0	5
青 海	4	1	0	0	1	3	0	0	1
宁 夏	5	9	0	0	6	11	0	0	0
新 疆	73	39	0	0	25	16	0	0	6

各地区钢铁行业污染排放及处理情况（二）

（2012）

地 区 名 称	铁精矿 消耗量/ 亿吨	铁精矿平均 含硫率/ %	炼焦煤 消耗量/ 万吨	炼焦煤平均 含硫率/ %	高炉 喷煤量/ 万吨	高炉喷煤 平均含硫率/ %	焦炭 消耗量/ 万吨	焦炭平均 含硫率/ %
全 国	181.4	0.10	15 753	0.77	10 191	0.58	22 393	0.65
北 京	0.0	0.00	0	0.00	0	0.00	0	0.00
天 津	0.2	0.05	0	0.00	176	0.47	517	0.73
河 北	89.8	0.10	1 771	0.81	2 855	0.56	6 076	0.68
山 西	0.6	0.11	756	0.94	595	0.64	1 330	0.75
内蒙古	0.2	0.33	632	0.55	170	1.44	601	0.58
辽 宁	39.1	0.06	1 588	0.79	805	0.44	2 126	0.67
吉 林	0.2	0.16	365	0.56	146	0.53	471	0.48
黑龙江	0.1	0.12	237	0.36	80	0.31	261	0.29
上 海	0.2	0.02	861	0.75	271	0.43	109	0.66
江 苏	0.9	0.19	1 552	0.70	935	0.55	1 347	0.60
浙 江	0.2	0.09	400	0.58	162	0.55	306	0.61
安 徽	0.3	0.09	714	0.67	291	0.44	776	0.62
福 建	0.2	0.25	106	0.64	175	0.62	284	0.60
江 西	0.3	0.20	541	0.83	372	0.59	819	0.73
山 东	0.8	0.06	1 374	0.93	941	0.47	2 384	0.70
河 南	0.3	0.17	276	0.50	308	0.51	685	0.59
湖 北	0.4	0.11	1 359	0.83	369	0.48	310	0.70
湖 南	0.3	0.11	357	0.89	281	0.78	299	0.44
广 东	0.2	0.11	251	0.73	156	0.56	370	0.70
广 西	0.2	0.09	584	0.74	209	0.49	468	0.58
海 南	0.0	0.00	0	0.00	0	0.00	0	0.00
重 庆	0.1	0.11	332	1.07	64	0.51	229	0.79
四 川	6.8	0.29	271	0.67	198	0.61	675	0.64
贵 州	0.2	0.09	195	0.67	70	0.71	212	0.82
云 南	32.2	0.21	301	0.68	164	0.69	573	0.44
西 藏								
陕 西	0.1	0.15	54	0.70	121	0.46	139	0.66
甘 肃	4.2	0.21	425	0.70	102	0.38	332	0.61
青 海	0.0	0.22	0	0.00	15	0.31	61	0.76
宁 夏	0.0	0.21	23	1.20	13	0.64	26	0.76
新 疆	3.5	0.24	430	0.66	147	3.17	608	0.62

各地区钢铁行业污染排放及处理情况（三）

（2012）

地区名称	工业废气排放量/万米³	工业二氧化硫产生量/吨	工业二氧化硫排放量/吨	工业氮氧化物产生量/吨	工业氮氧化物排放量/吨	工业烟粉尘产生量/吨	工业烟粉尘排放量/吨
全 国	509 397 084	2 513 024	2 000 238	549 274	542 945	22 664 666	601 087
北 京	0	0	0	0	0	0	0
天 津	10 522 424	23 668	23 668	11 542	11 542	453 328	7 950
河 北	146 527 948	574 055	433 499	157 420	157 420	6 872 420	210 139
山 西	37 440 042	165 073	124 577	30 943	29 983	1 739 255	78 677
内蒙古	10 377 788	147 875	112 670	12 586	12 586	530 213	16 198
辽 宁	40 173 707	105 243	89 339	42 626	39 158	1 791 967	42 921
吉 林	10 039 787	46 567	42 895	9 607	9 607	484 302	7 844
黑龙江	4 349 366	21 320	21 320	4 463	4 463	196 281	3 778
上 海	11 616 965	19 072	16 232	10 625	10 625	509 942	7 765
江 苏	30 685 215	144 110	92 634	44 453	44 445	1 092 468	18 112
浙 江	7 720 358	30 949	27 758	4 800	4 800	245 821	5 413
安 徽	22 472 241	62 142	51 948	18 307	18 307	760 640	10 546
福 建	7 491 875	71 206	49 469	10 019	10 019	420 018	18 751
江 西	11 957 416	114 313	105 688	15 928	15 928	515 407	9 371
山 东	46 995 177	184 115	119 547	50 043	50 043	2 010 900	41 390
河 南	12 215 772	105 832	105 832	14 747	14 736	466 004	9 387
湖 北	11 480 817	82 559	70 585	17 296	17 296	267 509	8 504
湖 南	8 188 691	71 670	55 519	16 158	16 158	420 220	8 643
广 东	5 121 087	37 938	30 023	6 746	6 746	183 622	3 104
广 西	9 311 188	33 247	24 439	8 986	8 986	423 246	12 893
海 南	78 605	21	21	20	20	0	0
重 庆	7 717 638	27 566	25 200	6 360	6 077	530 736	6 650
四 川	21 636 332	180 616	126 732	18 796	17 197	964 297	19 385
贵 州	3 050 723	17 764	17 764	1 868	1 868	159 533	3 128
云 南	7 286 735	89 905	76 680	10 771	10 771	373 866	10 734
西 藏							
陕 西	5 600 071	35 080	35 080	5 880	5 880	392 785	6 101
甘 肃	6 980 085	39 212	39 212	4 895	4 895	286 715	7 899
青 海	1 203 044	7 849	7 849	365	365	51 733	608
宁 夏	1 633 749	6 945	6 945	1 332	1 332	55 964	7 180
新 疆	9 522 238	67 112	67 112	11 692	11 692	465 475	18 017

各地区制浆及造纸行业污染排放及处理情况（一）

（2012）

地 区 名 称	汇总工业 企业数/ 个	造纸 生产线// 条	化学浆 生产线/ 条	化机浆 生产线/ 条	废纸浆 生产线/ 条	取水量/ 万吨	造纸 生产线	化学浆 生产线	化机浆 生产线	废纸浆 生产线
全 国	5 235	4 623	629	420	2 030	407 844	219 071	53 282	6 896	52 023
北 京	18	8	0	0	1	121	67	0	0	28
天 津	46	28	1	1	14	2 032	425	0	20	844
河 北	394	375	26	15	177	34 347	18 358	592	87	8 970
山 西	56	39	7	4	14	6 215	3 069	538	0	326
内 蒙 古	31	26	5	2	10	1 350	543	552	0	46
辽 宁	76	69	15	9	27	6 988	4 205	0	285	1 278
吉 林	47	47	7	4	13	7 736	2 997	3 377	0	367
黑 龙 江	47	46	18	12	22	3 783	1 534	906	40	88
上 海	59	46	0	0	3	1 251	1 014	0	0	191
江 苏	207	176	44	44	84	22 999	11 870	737	1 377	2 816
浙 江	726	636	156	148	198	44 316	35 912	10	0	2 473
安 徽	157	138	4	8	79	7 870	4 592	0	45	2 056
福 建	396	372	25	10	211	18 184	6 324	2 306	174	5 456
江 西	182	174	13	5	95	17 563	11 221	1 432	59	2 545
山 东	270	254	30	7	72	36 592	16 762	9 985	939	2 947
河 南	268	224	29	16	108	31 434	17 661	4 308	656	3 363
湖 北	95	81	15	13	46	13 165	5 731	1 378	153	2 747
湖 南	391	375	60	41	202	33 609	20 393	4 769	1 068	3 363
广 东	754	560	56	27	212	36 198	23 394	2 075	18	2 733
广 西	264	242	27	7	82	27 285	10 843	10 838	787	2 015
海 南	5	4	1	1	2	3 987	881	0	0	2
重 庆	79	71	7	5	44	5 987	3 084	1 125	0	1 507
四 川	304	291	33	15	137	15 753	8 249	2 489	37	2 973
贵 州	41	38	11	10	22	441	130	50	0	65
云 南	109	109	7	1	48	4 881	1 074	1 389	0	363
西 藏	1	1	0	0	1	32	32	0	0	0
陕 西	58	54	13	11	32	12 609	4 347	849	476	1 530
甘 肃	17	17	5	2	10	1 964	845	702	0	153
青 海	0	0	0	0	0	0	0	0	0	0
宁 夏	19	15	9	1	4	6 281	2 796	2 325	676	109
新 疆	118	107	5	1	60	2 872	714	548	0	669

各地区制浆及造纸行业污染排放及处理情况（二）

（2012）

地 区 名 称	废水 产生量/ 万吨	造纸 生产线	化学浆 生产线	化机浆 生产线	废纸浆 生产线	化学需氧量 产生量/ 吨	造纸 生产线	化学浆 生产线	化机浆 生产线	废纸浆 生产线
全 国	409 861	265 627	56 600	5 733	81 901	5 343 364	3 502 055	966 873	129 242	745 193
北 京	79	60	0	0	20	385	338	0	0	47
天 津	1 147	438	0	66	643	21 116	4 165	0	778	16 173
河 北	33 085	22 801	1 190	85	9 009	329 920	220 358	14 956	1 106	93 501
山 西	3 991	2 899	735	0	358	23 832	11 890	7 623	0	4 320
内 蒙 古	1 294	535	733	0	26	40 670	6 359	33 168	28	1 114
辽 宁	5 741	3 507	407	390	1 437	63 369	44 039	4 883	5 736	8 711
吉 林	6 844	3 030	3 419	0	395	97 814	35 588	57 479	0	4 747
黑 龙 江	2 658	1 741	744	40	132	41 770	37 002	3 838	270	659
上 海	1 418	1 113	0	0	305	11 304	8 609	0	0	2 694
江 苏	34 485	10 899	722	0	22 863	290 311	213 330	20 885	1 011	55 084
浙 江	78 415	75 337	10	0	3 068	1 038 602	1 010 614	734	0	27 254
安 徽	7 239	5 149	128	37	1 925	103 785	81 209	307	686	21 583
福 建	20 467	12 519	1 778	159	6 010	320 241	208 650	25 355	5 240	80 996
江 西	11 328	7 001	1 434	59	2 834	152 621	117 461	1 001	240	33 919
山 东	37 696	20 744	11 665	1 107	4 180	688 131	362 401	221 771	14 399	89 559
河 南	26 213	16 288	5 596	818	3 511	455 999	253 221	136 462	26 170	40 146
湖 北	10 375	6 020	1 101	254	3 000	131 014	61 174	23 235	11 685	34 920
湖 南	23 863	15 028	3 778	1 090	3 967	233 154	120 867	55 208	7 966	49 112
广 东	34 557	26 280	1 629	17	6 631	429 743	350 914	17 970	487	60 373
广 西	23 112	10 543	9 935	499	2 135	248 829	112 532	110 600	7 669	18 028
海 南	2 777	657	2 119	0	1	47 915	6 265	41 637	0	14
重 庆	5 626	2 125	1 115	0	2 386	89 775	21 697	33 822	0	34 255
四 川	16 378	9 728	2 455	69	4 126	129 548	74 661	30 912	488	23 486
贵 州	424	103	300	0	21	6 139	3 741	1 843	0	555
云 南	3 773	1 944	1 296	0	532	58 185	32 140	20 925	0	5 120
西 藏	10	10	0	0	0	210	210	0	0	0
陕 西	8 023	5 279	812	434	1 498	101 597	49 451	15 093	10 235	26 818
甘 肃	1 633	847	628	0	158	27 498	17 427	5 888	0	4 184
青 海										
宁 夏	5 245	2 041	2 461	609	134	142 881	29 631	76 611	35 047	1 592
新 疆	1 966	960	411	0	595	17 007	6 112	4 668	0	6 227

11

流域及入海陆源废水
排放统计

ANNUAL STATISTIC REPORT ON ENVIRONMENT IN CHINA
2012

十大水系接纳工业废水及处理情况（一）

（2012）

流域	地区名称	工业废水排放量/万吨	直接排入环境的	排入污水处理厂的	工业废水处理量/万吨	废水治理设施数/套	废水治理设施治理能力/（万吨/日）	废水治理设施运行费用/万元
	总 计	2 215 857	1 732 817	483 039	5 274 705	85 673	26 620	6 677 024.6
辽河	内蒙古	7 030	4 037	2 993	8 462	207	46	8 162.9
	辽 宁	87 168	71 969	15 198	235 649	2 387	1 212	259 558.7
	吉 林	16 589	16 005	584	69 815	255	338	25 790.0
	合 计	110 786	92011	18 775	313 926	2 849	1 596	293 511.6
海河	北 京	9 190	2 232	6 958	11 081	521	65	51 338.7
	天 津	19 117	8 027	11 090	40 464	967	171	127 294.3
	河 北	121 136	85 695	35 441	775 778	4 770	3 723	549 859.2
	山 西	18 947	17 475	1 472	35 475	1 436	223	48 797.4
	内蒙古	564	508	55	258	16	98	386.4
	山 东	27 213	19 277	7 936	32 055	592	337	73 137.1
	河 南	32 941	30 894	2 046	60 817	856	241	67 437.2
	合 计	229 107	164 109	64 998	955 928	9 158	4 857	918 250.3
淮河	江 苏	78 357	65 771	12 587	122 710	2 305	693	213 017.1
	安 徽	31 914	28 073	3 841	38 540	892	316	65 663.5
	山 东	146 364	91 980	54 384	265 020	4 366	1 407	399 629.2
	河 南	68 008	56 564	11 444	61 745	1 226	431	79 374.8
	合 计	324 643	242 388	82 256	488 014	8 789	2 848	757 684.6
松花江	内蒙古	6 422	3 532	2 890	8 271	151	47	10 345.8
	吉 林	28 254	22 459	5 795	29 490	422	147	66 640.7
	黑龙江	58 355	52 926	5 429	92 346	1 199	1 214	233 221.7
	合 计	93 031	78 917	14 114	130 107	1 772	1 408	310 208.2
珠江	江 西	349	349	0	228	18	1	507.3
	湖 南	2 757	2 757	0	2012	96	13	1 486.4
	广 东	186 126	149 391	36 735	287 598	10 608	1 655	518 526.4
	广 西	109 731	106 324	3 408	344 119	2 309	1 270	180 894.6
	贵 州	6 914	6 914	0	10 390	451	108	13 693.2
	云 南	7 448	7 270	178	73 336	948	434	52 045.8
	海 南	7 465	6 530	935	9 144	364	40	33 048.1
	福 建	2 708	2 537	171	6 157	234	53	16 540.2
	合 计	323 497	282 071	41 427	732 984	15 028	3 574	816 742.0

十大水系接纳工业废水及处理情况（一）（续表）

（2012）

流域	地 区	工业废水排放量/万吨	直接排入环境的	排入污水处理厂的	工业废水处理量/万吨	废水治理设施数/套	废水治理设施治理能力/（万吨/日）	废水治理设施运行费用/万元
长江	上 海	46 359	19 728	26 631	78 777	1 802	321	328 406.1
	江 苏	157 737	91 988	65 749	282 782	5 190	1 260	753 706.3
	浙 江	39 572	10 167	29 405	83 135	2 002	448	122 815.2
	安 徽	34 273	30 316	3 957	167 045	1 427	699	142 000.1
	江 西	67 523	66 084	1 439	175 101	2 470	810	157 954.5
	河 南	7 785	7 785	0	13 427	241	59	16 976.9
	湖 北	91 609	80 109	11 499	281 273	2 110	1 023	176 058.4
	湖 南	94 376	91 767	2 609	267 483	3 035	1 109	177 130.3
	广 西	940	940	0	1 187	83	5	1 445.6
	重 庆	30 611	29 466	1 145	44 824	1 578	395	79 899.4
	四 川	69 949	63 634	6 315	207 431	4 245	999	284 349.8
	贵 州	16 485	16 485	0	82 333	1 212	335	44 632.6
	云 南	7 310	6 890	419	29 322	763	137	35 546.0
	西 藏	0	0	0	0	0	0	0.0
	陕 西	4 661	4 537	124	14 558	436	93	13 861.7
	甘 肃	1 784	1 784	0	2 079	90	12	2 313.9
	青 海	—	0	0	—	0	0	0.0
	合 计	670 972	521 681	149 292	1 730 757	26 684	7 705	2 337 096.8
黄河	山 西	29 161	27 257	1 904	168 901	2 131	790	209 201.3
	内蒙古	15 015	6 358	8 657	85 953	658	425	79 219.4
	山 东	10 057	7 486	2 571	68 354	359	263	59 523.9
	河 南	28 622	25 736	2 886	40 647	1 287	239	66 224.3
	四 川	35	35	0	24	10	0	27.0
	陕 西	33 376	27 317	6 059	49 146	1 770	240	79 135.8
	甘 肃	9 613	8 217	1 396	20 509	335	129	21 771.2
	青 海	3 935	3 935	0	17 708	101	57	11 074.4
	宁 夏	16 548	15 202	1 346	25 385	377	159	40 555.0
	合 计	146 363	121 544	24 819	476 627	7 028	2 303	566 732.3

十大水系接纳工业废水及处理情况（一）（续表）

（2012）

流域	地 区名 称	工业废水排放量/万吨	直接排入环境的	排入污水处理厂的	工业废水处理量/万吨	废水治理设施数/套	废水治理设施治理能力/（万吨/日）	废水治理设施运行费用/万元
东南诸河	浙 江	135 843	67 374	68 469	173 419	6 570	914	381 340.5
	安 徽	989	778	211	845	93	5	1 290.6
	福 建	103 611	91 525	12 086	167 631	3 227	683	145 184.3
	合 计	240 444	159 678	80 766	341 895	9 890	1 601	527 815.4
西北诸河	河 北	1 509	1 279	230	342	10	2	390.0
	内蒙古	4 588	3 635	953	1 823	70	43	2 667.7
	西 藏	0	0	0	0	0	0	0.0
	甘 肃	7 791	7 687	104	10 413	206	65	12 636.6
	青 海	4 982	4 982	0	2 077	84	10	3 437.9
	新 疆	29 738	25 150	4 589	48 221	908	238	95 713.6
	合 计	48 607	42 732	5 875	62 876	1 278	358	114 845.8
西南诸河	云 南	28 054	27 336	718	41 035	3 167	363	33 475.2
	西 藏	353	353	0	556	30	6	662.4
	青 海	0	0	0	0	0	0	0.0
	合 计	28 407	27 688	718	41 591	3 197	369	34 137.6

十大水系接纳工业废水及处理情况（二）

（2012）

单位：吨

流域	地区名称	工业废水中污染物产生量				
		化学需氧量	氨氮	石油类	挥发酚	氰化物
	总 计	24 305 917.4	1 460 908.1	296 813.7	65 801.3	5 180.2
辽河	内蒙古	116 667.7	3 624.6	41.0	7.2	0.0
	辽 宁	508 531.7	54 472.1	16 175.9	1 479.3	152.8
	吉 林	94 870.3	2 408.9	412.4	662.9	8.4
	合 计	720 069.8	60 505.5	16 629.4	2 149.4	161.2
海河	北 京	61 769.1	2 725.3	3 298.4	876.9	5.9
	天 津	130 525.3	7 368.7	1 135.9	2.6	0.5
	河 北	1 787 917.3	98 718.5	21 828.8	11 702.5	558.8
	山 西	183 011.9	17 558.2	2 499.3	2 853.3	97.2
	内蒙古	3 367.1	43.7	0.9	0.0	0.0
	山 东	551 067.3	22 687.2	944.0	1.7	9.1
	河 南	579 823.4	13 492.9	2 318.0	1 130.8	51.9
	合 计	3 297 481.3	162 594.5	32 025.2	16 567.8	723.4
淮河	江 苏	931 514.4	24 158.1	3 569.2	698.9	62.2
	安 徽	357 177.9	21 730.9	1 155.7	45.1	14.8
	山 东	2 121 132.8	93 286.2	35 286.0	4 143.0	122.2
	河 南	629 150.3	24 001.3	3 005.7	870.3	44.4
	合 计	4 038 975.3	163 176.5	43 016.6	5 757.2	243.6
松花江	内蒙古	187 145.3	19 561.1	252.0	6.7	8.0
	吉 林	501 032.0	15 130.6	1 604.5	211.3	11.3
	黑龙江	1 161 474.6	34 380.8	39 539.5	5 271.8	84.6
	合 计	1 849 651.9	69 072.4	41 396.0	5 489.8	103.9
珠江	江 西	2 092.8	8 435.6	16.2	0.0	0.0
	湖 南	28 223.7	1 309.4	137.9	0.0	1.8
	广 东	1 472 798.8	50 465.1	4 968.2	144.8	628.3
	广 西	1 169 103.0	65 910.1	3 761.1	2 080.9	19.9
	贵 州	36 611.9	4 367.0	1 772.0	875.0	25.7
	云 南	150 747.3	48 929.9	1 998.6	1 569.6	81.0
	海 南	92 870.1	2 202.2	162.6	0.1	0.0
	福 建	14 616.0	842.3	144.7	12.2	5.4
	合 计	2 967 063.7	182 461.6	12 961.4	4 682.6	762.1

十大水系接纳工业废水及处理情况（二）（续表）

（2012）

单位：吨

流域	地 区名 称	工业废水中污染物产生量				
		化学需氧量	氨氮	石油类	挥发酚	氰化物
长 江	上 海	312 601.6	11 889.5	7 690.6	1 066.4	217.7
	江 苏	1 140 021.7	51 914.3	9 555.2	3 195.8	316.3
	浙 江	441 513.8	10 455.0	1 005.3	195.3	2.9
	安 徽	345 283.8	11 827.8	6 600.5	1 750.3	273.8
	福 建	0	0	0	0	0
	江 西	479 823.6	33 235.0	5 900.7	2 514.3	66.8
	河 南	275 514.5	3 536.2	750.2	0.5	1.1
	湖 北	589 765.1	35 872.9	4 316.0	794.5	58.8
	湖 南	598 012.7	65 253.2	4 264.3	1 180.6	356.7
	广 东	0	0	0	0	0
	广 西	6 043.1	52.2	0.0	0.0	0.0
	重 庆	288 043.8	28 232.9	2 727.8	849.6	84.5
	四 川	669 304.7	38 425.7	3 981.4	962.2	77.6
	贵 州	146 670.4	47 524.9	1 164.6	271.6	43.1
	云 南	80 600.0	25 145.9	747.0	603.8	27.1
	西 藏	0.4	0.0	0.0	0.0	0.0
	陕 西	50 491.8	6 128.8	231.7	1.6	0.8
	甘 肃	6 542.0	24.7	5.4	0.0	0.0
	青 海	0.0	0.0	0.0	0.0	0.0
	合 计	5 430 232.9	369 518.9	48 940.7	13 386.4	1 527.3
黄 河	山 西	259 376.8	80 270.2	5 681.2	4 932.3	266.2
	内蒙古	193 489.5	24 046.4	2 977.2	989.3	20.1
	山 东	129 001.0	2 630.1	2 969.2	1 343.0	138.6
	河 南	279 627.4	11 264.4	35 829.8	538.0	22.0
	四 川	239.0	12.7	0.0	0.0	0.0
	陕 西	457 284.3	33 350.9	6 521.9	5 006.6	163.7
	甘 肃	115 928.2	12 382.2	1 968.2	114.0	1.8
	青 海	51 092.8	1 290.5	501.3	2.3	0.0
	宁 夏	497 532.7	72 394.9	3 189.4	538.6	22.5
	合 计	1 983 571.7	237 642.3	59 638.3	13 464.0	635.0

313

十大水系接纳工业废水及处理情况（二）（续表）

（2012）

单位：吨

流域	地区名称	工业废水中污染物产生量				
		化学需氧量	氨氮	石油类	挥发酚	氰化物
东南诸河	浙 江	1 996 116.2	74 666.8	20 588.1	91.5	759.1
	安 徽	4 968.4	208.1	27.9	0.0	0.0
	福 建	934 489.4	40 716.6	11 190.4	276.7	125.5
	合 计	2 935 574.1	115 591.4	31 806.4	368.2	884.7
西北诸河	河 北	11 037.1	205.1	5.0	0.0	0.0
	内蒙古	37 333.7	7 639.4	421.9	658.5	9.5
	西 藏	0.0	0.0	0.0	0.0	0.0
	甘 肃	85 298.3	6 517.8	398.8	83.7	1.2
	青 海	32 286.1	1 706.6	760.9	2.0	31.2
	新 疆	472 437.1	81 048.4	8 463.6	3 186.6	87.0
	合 计	638 392.3	97 117.2	10 050.2	3 930.8	128.9
西南诸河	云 南	438 367.2	2 958.2	345.3	5.1	10.0
	西 藏	6 537.0	269.4	4.2	0.0	0.0
	青 海	0.0	0.0	0.0	0.0	0.0
	合 计	444 904.3	3 227.6	349.6	5.1	10.0

十大水系接纳工业废水及处理情况（三）

（2012）

单位：吨

流域	地区名称	工业废水中污染物产生量					
		汞	镉	六价铬	总铬	铅	砷
总 计		19.961	2 318.538	3 273.441	7 322.903	3 260.838	11 300.867
辽河	内蒙古	0.092	62.261	0.020	0.025	52.366	159.962
	辽 宁	0.134	41.802	18.415	22.901	41.636	39.446
	吉 林	0.210	0.004	0.014	0.017	0.234	0.036
	合 计	0.436	104.067	18.449	22.943	94.236	199.444
海河	北 京	0.000	0.001	10.270	11.038	0.269	2.344
	天 津	0.039	0.024	8.561	13.436	15.271	0.949
	河 北	0.042	1.269	312.680	381.079	11.112	42.040
	山 西	0.357	0.761	2.699	3.007	0.676	0.977
	内蒙古	0.000	0.000	0.000	0.000	0.000	0.000
	山 东	0.122	247.212	0.248	30.130	242.331	640.121
	河 南	0.359	81.591	4.332	15.260	45.080	23.543
	合 计	0.919	330.858	338.790	453.950	314.739	709.974
淮河	江 苏	0.375	0.038	71.597	114.885	15.013	0.961
	安 徽	0.001	0.006	6.649	11.336	2.198	2.121
	山 东	0.749	24.066	43.513	95.804	23.931	211.577
	河 南	0.167	1.347	9.547	123.619	3.087	7.695
	合 计	1.293	25.457	131.306	345.643	44.228	222.354
松花江	内蒙古	0.061	0.023	0.000	0.001	1.059	0.150
	吉 林	0.005	0.014	11.555	11.562	0.058	2.821
	黑龙江	0.160	0.051	6.597	7.280	6.592	1.515
	合 计	0.225	0.088	18.153	18.843	7.709	4.486
珠江	江 西	0.004	0.036	0.588	0.066	0.602	0.305
	湖 南	0.003	0.452	0.066	9.605	8.852	5.713
	广 东	0.135	1.811	0.000	1 599.586	16.689	11.337
	广 西	0.517	93.655	125.204	55.723	96.934	149.215
	贵 州	0.099	0.042	26.111	0.018	0.339	0.612
	云 南	0.064	26.043	0.000	3.921	29.538	114.837
	海 南	0.000	0.014	0.018	16.896	0.003	0.014
	福 建	0.012	9.160	3.891	0.662	20.163	22.553
	合 计	0.836	131.212	155.879	1 686.478	173.121	304.585

十大水系接纳工业废水及处理情况（三）（续表）

（2012） 单位：吨

流域	地区名称	工业废水中污染物产生量					
		汞	镉	六价铬	总铬	铅	砷
	上　海	0.003	0.027	101.576	130.581	1.380	0.818
	江　苏	0.025	2.457	216.203	347.794	6.111	5.383
	浙　江	0.000	1.307	44.255	228.991	4.010	0.536
	安　徽	2.244	13.534	12.821	15.190	55.036	554.568
	福　建	0.000	0.000	0.000	0.000	0.000	0.000
	江　西	0.308	398.938	47.503	51.552	95.215	5 826.509
	河　南	0.001	0.004	0.682	9.902	0.132	0.128
	湖　北	0.426	8.433	50.332	92.570	20.301	97.804
	湖　南	4.474	202.683	14.048	28.793	485.390	468.422
长	广　东	0.000	0.000	0.000	0.000	0.000	0.000
江	广　西	0.000	0.001	0.000	0.000	0.019	0.015
	重　庆	0.000	0.001	85.731	163.953	6.990	2.998
	四　川	0.968	45.370	52.045	135.526	71.586	488.677
	贵　州	0.050	4.964	0.618	3.333	9.348	21.359
	云　南	0.743	148.848	0.034	0.781	252.511	575.755
	西　藏	0.000	0.000	0.000	0.000	0.000	0.000
	陕　西	0.343	74.935	3.672	3.829	111.081	36.806
	甘　肃	0.020	5.056	0.000	0.000	6.514	8.047
	青　海	0.000	0.000	0.000	0.000	0.000	0.000
	合　计	9.605	906.559	629.519	1 212.796	1 125.625	8 087.824
	山　西	0.006	14.754	12.512	12.784	24.473	58.423
	内蒙古	0.796	14.489	1.123	1.237	33.289	79.338
	山　东	0.082	0.114	5.026	6.694	0.756	0.126
	河　南	0.984	184.112	4.992	48.449	296.605	84.204
黄	四　川	0.000	0.000	0.000	0.000	0.000	0.000
河	陕　西	0.042	0.051	6.071	8.275	0.925	26.577
	甘　肃	0.053	98.832	0.520	33.896	149.000	154.852
	青　海	0.367	11.246	1.192	1.302	10.114	70.492
	宁　夏	0.710	3.757	1.077	1.434	4.873	86.819
	合　计	3.041	327.356	32.513	114.070	520.035	560.831

十大水系接纳工业废水及处理情况（三）（续表）

（2012）

单位：吨

流域	地区名称	工业废水中污染物产生量					
		汞	镉	六价铬	总铬	铅	砷
东南诸河	浙 江	0.030	2.946	1 530.949	2 676.773	3.239	30.800
	安 徽	0.008	0.033	0.087	0.457	0.043	0.012
	福 建	0.071	3.246	410.958	762.532	9.979	27.522
	合 计	0.109	6.225	1 941.995	3 439.763	13.261	58.334
西北诸河	河 北	0.000	0.000	0.000	0.000	0.000	0.000
	内蒙古	0.005	8.730	0.010	0.010	6.821	8.484
	西 藏	0.000	0.000	0.000	0.000	0.000	0.000
	甘 肃	0.178	290.969	2.778	7.981	663.500	839.579
	青 海	0.072	16.884	0.000	0.000	11.961	6.940
	新 疆	2.953	2.396	3.984	14.216	4.351	30.364
	合 计	3.207	318.978	6.772	22.208	686.633	885.367
西南诸河	云 南	0.232	167.709	0.064	0.071	280.548	252.774
	西 藏	0.059	0.029	0.000	6.138	0.703	14.894
	青 海	0.000	0.000	0.000	0.000	0.000	0.000
	合 计	0.291	167.738	0.064	6.209	281.251	267.667

十大水系接纳工业废水及处理情况（四）

（2012）

单位：吨

流域	地区名称	工业废水中污染物排放量				
		化学需氧量	氨氮	石油类	挥发酚	氰化物
总　计		3 384 539.3	264 142.3	17 327.2	1 481.4	171.8
辽河	内蒙古	29 393.2	2 968.4	12.9	0.0	0.0
	辽　宁	102 213.2	8 366.8	698.4	29.2	4.0
	吉　林	22 347.7	936.0	118.9	2.1	2.1
	合　计	153 954.1	12 271.2	830.2	31.3	6.1
海河	北　京	6 265.8	352.0	49.1	0.2	0.1
	天　津	26 533.3	3 294.5	138.2	1.2	0.0
	河　北	189 482.6	15 763.4	972.7	119.0	12.8
	山　西	33 527.1	3 762.8	577.3	356.0	17.1
	内蒙古	1 644.1	18.8	0.9	0.0	0.0
	山　东	23 816.2	1 872.6	29.1	0.3	0.6
	河　南	46 977.1	3 068.5	131.0	0.7	3.9
	合　计	328 246.2	28 132.6	1 898.3	477.4	34.6
淮河	江　苏	116 667.1	8 801.8	668.8	25.6	6.8
	安　徽	38 393.8	5 542.5	367.5	2.7	2.1
	山　东	107 617.3	8 824.4	963.6	36.4	5.3
	河　南	84 811.7	6 156.3	686.0	124.5	9.8
	合　计	347 489.8	29 325.0	2 685.9	189.1	23.9
松花江	内蒙古	18 919.4	1 040.7	31.6	6.7	0.0
	吉　林	53 971.7	3 452.0	171.8	1.4	0.1
	黑龙江	98 187.3	5 846.8	292.7	6.3	1.6
	合　计	171 078.4	10 339.5	496.2	14.3	1.7
珠江	江　西	552.7	962.2	4.7	0.0	0.0
	湖　南	14 109.2	1 072.4	122.9	0.0	0.7
	广　东	238 503.3	14 937.3	684.3	11.3	10.9
	广　西	197 201.7	8 534.0	283.5	14.8	4.9
	贵　州	15 569.5	486.8	131.3	0.1	0.3
	云　南	28 929.3	1 801.7	74.8	1.4	0.9
	海　南	12 542.6	883.2	3.4	0.0	0.0
	福　建	3 791.0	524.0	18.0	0.0	0.1
	合　计	511 199.5	29 201.8	1 322.9	27.6	17.7

十大水系接纳工业废水及处理情况（四）（续表）

（2012）

单位：吨

流域	地区名称	工业废水中污染物排放量				
		化学需氧量	氨氮	石油类	挥发酚	氰化物
	上 海	26 159.5	2 281.0	646.2	2.7	2.2
	江 苏	114 780.0	7 457.5	531.1	26.2	8.7
	浙 江	34 437.9	3 343.0	42.4	0.5	0.3
	安 徽	49 480.2	2 694.0	349.5	2.6	4.2
	福 建	0.0	0.0	0.0	0.0	0.0
	江 西	100 146.8	9 245.7	564.4	8.9	7.8
	河 南	11 947.1	1 246.1	42.0	0.0	0.0
	湖 北	135 278.1	14 769.9	965.4	15.4	7.8
	湖 南	135 512.9	24 729.6	648.9	18.2	10.6
长	广 东	0.0	0.0	0.0	0.0	0.0
江	广 西	2 416.6	40.1	0.0	0.0	0.0
	重 庆	49 157.2	3 077.2	353.1	9.5	1.5
	四 川	118 298.6	5 585.3	418.8	1.3	1.5
	贵 州	49 149.1	3 119.1	328.8	0.9	1.2
	云 南	14 588.8	611.5	120.2	0.5	0.4
	西 藏	0.4	0.0	0.0	0.0	0.0
	陕 西	14 028.4	2 400.8	17.0	0.0	0.1
	甘 肃	3 118.7	4.3	5.4	0.0	0.0
	青 海	0.0	0.0	0.0	0.0	0.0
	合 计	858 500.3	80 605.0	5 033.2	86.8	46.2
	山 西	53 142.4	4 572.6	620.5	371.0	13.1
	内蒙古	25 593.1	5 329.3	592.8	1.0	1.3
	山 东	7 970.6	284.5	88.1	1.4	0.1
	河 南	35 251.3	2 556.0	281.6	9.8	2.4
黄	四 川	79.3	5.9	0.0	0.0	0.0
河	陕 西	84 427.8	6 252.3	719.2	2.5	5.4
	甘 肃	43 197.0	8 011.3	187.0	0.6	0.0
	青 海	25 064.0	714.0	40.4	0.6	0.0
	宁 夏	105 438.1	8 609.1	178.5	10.6	2.0
	合 计	380 163.6	36 334.9	2 708.2	397.5	24.4

十大水系接纳工业废水及处理情况（四）（续表）

（2012）

单位：吨

流域	地区名称	工业废水中污染物排放量				
		化学需氧量	氨氮	石油类	挥发酚	氰化物
东南诸河	浙 江	151 233.4	8 746.5	634.8	22.7	7.3
	安 徽	1 549.8	117.5	13.6	0.0	0.0
	福 建	86 761.5	6 233.9	403.2	9.1	5.9
	合 计	239 544.7	15 097.9	1 051.6	31.8	13.3
西北诸河	河 北	2 296.5	62.3	5.0	0.0	0.0
	内蒙古	15 252.3	1 789.1	155.0	208.6	0.0
	西 藏	0.0	0.0	0.0	0.0	0.0
	甘 肃	46 691.0	5 903.6	71.1	1.2	0.2
	青 海	16 782.1	1 318.2	274.1	0.7	0.0
	新 疆	186 302.6	11 792.7	588.0	14.7	3.7
	合 计	267 324.5	20 865.9	1 093.3	225.3	3.9
西南诸河	云 南	126 114.8	1 886.0	207.0	0.1	0.0
	西 藏	923.5	82.6	0.5	0.0	0.0
	青 海	0.0	0.0	0.0	0.0	0.0
	合 计	127 038.3	1 968.5	207.4	0.1	0.0

十大水系接纳工业废水及处理情况（五）

（2012）

流域	地 区 名 称	工业废水中污染物排放量					
		汞	镉	六价铬	总铬	铅	砷
	总 计	1.082	26.662	70.375	188.636	97.137	127.692
辽 河	内蒙古	0.034	0.102	0.000	0.005	1.216	1.900
	辽 宁	0.001	0.036	0.510	0.679	0.479	0.331
	吉 林	0.001	0.001	0.014	0.016	0.083	0.031
	合 计	0.037	0.140	0.524	0.701	1.777	2.263
海 河	北 京	0.000	0.001	0.326	0.429	0.054	0.012
	天 津	0.004	0.010	0.169	0.454	1.004	0.019
	河 北	0.001	0.001	2.870	5.914	0.260	0.030
	山 西	0.001	0.754	0.308	0.313	0.275	0.364
	内蒙古	0.000	0.000	0.000	0.000	0.000	0.000
	山 东	0.008	0.978	0.025	5.240	0.445	1.835
	河 南	0.000	0.256	0.152	3.594	0.749	0.068
	合 计	0.015	1.999	3.850	15.943	2.786	2.329
淮 河	江 苏	0.096	0.006	1.056	5.036	1.303	0.027
	安 徽	0.001	0.003	0.080	0.440	0.162	2.116
	山 东	0.004	0.008	0.498	1.725	0.194	0.224
	河 南	0.002	0.538	0.751	8.152	1.815	0.631
	合 计	0.103	0.555	2.385	15.352	3.474	2.998
松 花 江	内蒙古	0.037	0.006	0.000	0.001	0.313	0.059
	吉 林	0.001	0.002	0.094	0.099	0.006	0.964
	黑龙江	0.000	0.001	0.368	0.368	0.019	0.002
	合 计	0.038	0.010	0.462	0.468	0.339	1.024
珠 江	江 西	0.000	0.025	0.006	0.004	0.129	0.019
	湖 南	0.001	0.030	0.004	9.605	0.803	1.309
	广 东	0.016	0.778	0.000	28.352	4.784	0.730
	广 西	0.045	1.397	8.959	1.747	5.390	6.630
	贵 州	0.015	0.021	0.719	0.003	0.152	0.438
	云 南	0.000	0.121	0.000	0.010	0.184	0.771
	海 南	0.000	0.001	0.003	0.131	0.002	0.013
	福 建	0.009	0.060	0.010	0.020	0.637	0.156
	合 计	0.087	2.433	9.701	39.873	12.080	10.067

十大水系接纳工业废水及处理情况（五）（续表）

（2012）

单位：吨

流域	地 区名 称	工业废水中污染物排放量					
		汞	镉	六价铬	总铬	铅	砷
	上 海	0.000	0.001	1.011	2.773	0.243	0.073
	江 苏	0.004	0.016	3.478	6.256	0.920	0.530
	浙 江	0.000	0.013	0.476	4.330	0.189	0.007
	安 徽	0.001	0.094	2.197	2.995	1.440	2.844
	福 建	0.000	0.000	0.000	0.000	0.000	0.000
	江 西	0.088	2.162	17.144	17.321	6.442	8.632
	河 南	0.000	0.001	0.034	0.358	0.033	0.027
	湖 北	0.209	0.623	11.557	12.402	3.187	9.553
	湖 南	0.225	13.437	2.077	8.446	37.672	52.157
长	广 东	0.000	0.000	0.000	0.000	0.000	0.000
江	广 西	0.000	0.000	0.000	0.000	0.000	0.000
	重 庆	0.000	0.000	0.204	0.507	0.073	1.358
	四 川	0.068	0.127	0.840	3.478	1.482	2.541
	贵 州	0.007	0.094	0.077	0.143	0.101	0.111
	云 南	0.001	1.077	0.000	0.007	5.126	4.380
	西 藏	0.000	0.000	0.000	0.000	0.000	0.000
	陕 西	0.028	0.609	0.012	0.018	1.573	0.421
	甘 肃	0.010	0.158	0.000	0.000	1.574	1.092
	青 海	0.000	0.000	0.000	0.000	0.000	0.000
	合 计	0.640	18.411	39.107	59.036	60.055	83.726
	山 西	0.003	0.040	0.157	0.178	0.158	0.205
	内蒙古	0.003	0.194	0.002	0.005	1.175	2.314
	山 东	0.000	0.001	0.006	0.101	0.014	0.111
	河 南	0.011	0.498	0.065	20.450	1.959	0.614
黄	四 川	0.000	0.000	0.000	0.000	0.000	0.000
河	陕 西	0.000	0.007	0.282	1.670	0.055	0.217
	甘 肃	0.035	0.319	0.271	0.570	2.737	0.631
	青 海	0.003	0.080	0.007	0.009	0.314	0.795
	宁 夏	0.004	0.023	0.109	0.338	0.087	0.167
	合 计	0.061	1.164	0.901	23.322	6.499	5.054

十大水系接纳工业废水及处理情况（五）（续表）

单位：吨

流域	地区名称	工业废水中污染物排放量					
		汞	镉	六价铬	总铬	铅	砷
东南诸河	浙 江	0.005	0.179	8.879	15.115	0.207	0.155
	安 徽	0.001	0.007	0.002	0.016	0.010	0.012
	福 建	0.011	0.201	2.759	11.703	2.395	1.056
	合 计	0.017	0.387	11.640	26.834	2.612	1.223
西北诸河	河 北	0.000	0.000	0.000	0.000	0.000	0.000
	内蒙古	0.000	0.101	0.001	0.001	0.287	0.792
	西 藏	0.000	0.000	0.000	0.000	0.000	0.000
	甘 肃	0.043	0.822	0.110	4.438	2.462	2.022
	青 海	0.006	0.037	0.000	0.000	0.397	0.654
	新 疆	0.028	0.166	1.690	2.661	0.856	1.280
	合 计	0.077	1.126	1.801	7.100	4.002	4.749
西南诸河	云 南	0.007	0.438	0.004	0.009	3.512	5.318
	西 藏	0.000	0.000	0.000	0.000	0.000	8.942
	青 海	0.000	0.000	0.000	0.000	0.000	0.000
	合 计	0.007	0.438	0.004	0.009	3.512	14.260

十大水系工业废水污染防治投资情况

（2012）

单位：个

流域	地区名称	工业废水治理施工项目数/个	工业废水治理竣工项目数/个	工业废水治理项目完成投资/万元	废水治理竣工项目新增处理能力/（万吨/日）
总 计		1 790	2 017	1 384 333.3	478.7
辽河	内蒙古	1	9	19 020.6	4.8
	辽 宁	16	33	27 776.9	3.9
	吉 林	2	0	11 095.6	5.9
	合 计	19	42	57 893.0	14.7
海河	北 京	10	13	3 011.9	0.9
	天 津	50	20	11 305.9	4.0
	河 北	35	45	52 159.3	12.5
	山 西	14	14	8 450.0	5.1
	内蒙古	0	1	1 608.0	0.0
	山 东	15	28	30 418.2	9.9
	河 南	24	14	8 146.6	6.3
	合 计	148	135	115 099.9	38.8
淮河	江 苏	28	42	32 455.4	10.4
	安 徽	12	11	11 458.7	3.4
	山 东	151	180	221 117.3	42.0
	河 南	22	17	11 788.2	1.9
	合 计	213	250	276 819.6	57.8
松花江	内蒙古	2	2	760.0	0.0
	吉 林	5	5	3 606.2	0.4
	黑龙江	11	9	7 349.7	2.8
	合 计	18	16	11 715.9	3.2
珠江	江 西	0	0	0.0	0.0
	湖 南	7	8	582.0	0.1
	广 东	169	185	57 889.4	24.6
	广 西	95	92	43 969.2	28.7
	贵 州	5	15	4 265.2	2.5
	云 南	14	16	11 738.1	3.9
	海 南	20	21	25 036.9	7.1
	福 建	11	18	3 641.5	3.1
	合 计	321	355	147 122.3	70.1

十大水系工业废水污染防治投资情况（续表）

（2012）

流域	地区名称	工业废水治理施工项目数/个	工业废水治理竣工项目数/个	工业废水治理项目完成投资/万元	废水治理竣工项目新增处理能力/（万吨/日）
	上 海	39	41	5 336.1	1.1
	江 苏	102	116	36 416.4	12.9
	浙 江	40	43	16 936.7	2.7
	安 徽	28	35	10 017.2	0.8
	福 建	0	0	0.0	0.0
	江 西	39	45	16 574.6	4.1
	河 南	13	13	4 932.7	1.0
	湖 北	36	38	32 031.8	6.5
	湖 南	40	47	47 210.1	17.3
长江	广 东	0	0	0.0	0.0
	广 西	0	0	0.0	0.0
	重 庆	49	78	17 510.5	3.8
	四 川	93	114	53 196.0	10.5
	贵 州	30	107	16 775.5	6.2
	云 南	24	20	53 685.2	1.3
	西 藏	0	0	0.0	0.0
	陕 西	10	11	8 565.7	0.3
	甘 肃	6	10	11 076.6	0.7
	青 海	0	0	0.0	0.0
	合 计	549	718	330 265.1	69.3
	山 西	58	56	22 073.1	13.7
	内蒙古	40	38	17 205.4	3.0
	山 东	11	12	10 251.0	3.6
	河 南	8	9	5 584.5	2.6
黄河	四 川	1	1	450.4	0.0
	陕 西	59	53	98 400.0	14.8
	甘 肃	5	8	11 128.6	0.9
	青 海	5	4	3 030.1	0.4
	宁 夏	9	13	14 298.3	2.4
	合 计	196	194	182 421.3	41.5

十大水系工业废水污染防治投资情况（续表）

（2012）

流域	地 区 名 称	工业废水治理 施工项目数/ 个	工业废水治理 竣工项目数/ 个	工业废水治理 项目完成投资/ 万元	废水治理竣工项目 新增处理能力/ （万吨/日）
东 南 诸 河	浙 江	156	164	81 775.3	19.2
	安 徽	0	0	0.0	0.0
	福 建	98	84	99 241.4	31.3
	合 计	254	248	181 016.6	50.6
西 北 诸 河	河 北	0	0	0.0	0.0
	内蒙古	1	0	105.0	0.0
	西 藏	0	0	0.0	0.0
	甘 肃	18	12	7 304.4	2.7
	青 海	2	0	232.5	0.8
	新 疆	22	8	37 784.8	96.3
	合 计	43	20	45 426.7	99.8
西 南 诸 河	云 南	27	36	35 631.0	32.8
	西 藏	2	3	922.0	0.3
	青 海	0	0	0.0	0.0
	合 计	29	39	36 553.0	33.1

十大水系农业污染排放情况

（2012）

单位：万吨

流域	地区名称	农业污染物排放（流失）总量			
		化学需氧量	氨氮	总氮	总磷
	总 计	1 153.8	80.6	451.4	48.9
辽河	内蒙古	16.5	0.4	4.2	0.4
	辽 宁	86.3	3.4	20.5	2.8
	吉 林	10.9	0.5	2.9	0.4
	合 计	113.7	4.3	27.6	3.5
海河	北 京	7.8	0.5	3.3	0.4
	天 津	11.4	0.6	3.3	0.4
	河 北	90.9	4.4	35.4	3.8
	山 西	6.8	0.4	2.8	0.2
	内蒙古	2.8	0.0	0.6	0.0
	山 东	35.4	1.8	21.6	1.9
	河 南	16.1	1.2	7.3	1.0
	合 计	171.3	8.9	74.2	7.8
淮河	江 苏	28.5	2.7	12.4	1.3
	安 徽	27.0	2.6	13.0	1.4
	山 东	89.3	5.2	31.1	3.7
	河 南	47.1	3.9	25.2	2.9
	合 计	192.0	14.4	81.7	9.3
松花江	内蒙古	15.9	0.3	15.2	1.2
	吉 林	39.5	1.3	10.0	1.2
	黑龙江	105.2	3.4	24.8	2.4
	合 计	160.6	5.1	50.0	4.7
珠江	江 西	0.5	0.1	0.3	0.0
	湖 南	0.4	0.1	0.2	0.0
	广 东	60.0	5.8	19.5	2.5
	广 西	21.2	2.6	11.3	1.3
	贵 州	1.1	0.2	1.2	0.1
	云 南	2.6	0.4	1.9	0.2
	海 南	10.3	0.9	4.1	0.5
	福 建	3.5	0.4	1.2	0.1
	合 计	99.7	10.5	39.6	4.8

十大水系农业污染排放情况（续表）

（2012）

单位：万吨

| 流域 | 地区名称 | 农业污染物排放（流失）总量 | | | |
		化学需氧量	氨氮	总氮	总磷
长江	上 海	3.3	0.3	1.5	0.2
	江 苏	10.2	1.2	5.1	0.6
	浙 江	6.1	0.8	2.5	0.3
	安 徽	11.0	1.1	4.9	0.6
	福 建	0	0	0.0	0.0
	江 西	23.4	2.9	11.1	1.3
	河 南	7.4	0.6	4.6	0.5
	湖 北	47.6	4.7	19.5	2.3
	湖 南	56.3	6.2	21.8	2.3
	广 东	0	0	0.0	0.0
	广 西	0.3	0.1	0.3	0.0
	重 庆	12.4	1.3	5.4	0.6
	四 川	54.0	5.7	22.1	2.5
	贵 州	5.2	0.6	3.5	0.3
	云 南	2.2	0.4	2.6	0.2
	西 藏	0.0	0.0	0.0	0.0
	陕 西	4.3	0.6	2.7	0.3
	甘 肃	2.3	0.1	0.8	0.1
	青 海	0.0	0.0	0.0	0.0
	合 计	245.9	26.7	108.6	12.2
黄河	山 西	11.2	0.9	5.7	0.6
	内蒙古	15.4	0.3	4.2	0.3
	山 东	9.7	0.4	3.6	0.4
	河 南	9.6	0.7	4.7	0.5
	四 川	0.0	0.0	0.0	0.0
	陕 西	15.4	0.9	5.8	0.5
	甘 肃	6.5	0.3	2.3	0.2
	青 海	1.9	0.1	0.6	0.0
	宁 夏	10.2	0.2	2.7	0.2
	合 计	79.9	3.8	29.7	2.8

十大水系农业污染排放情况（续表）

（2012）

单位：万吨

流域	地区名称	农业污染物排放（流失）总量			
		化学需氧量	氨氮	总氮	总磷
东南诸河	浙 江	14.3	1.9	6.7	0.8
	安 徽	0.4	0.1	0.2	0.0
	福 建	17.9	2.9	8.4	1.1
	合 计	32.6	4.9	219.3	1.9
西北诸河	河 北	0.9	0.0	0.6	0.0
	内蒙古	11.3	0.1	2.5	0.2
	西 藏	0.0	0.0	0.0	0.0
	甘 肃	5.8	0.2	1.6	0.1
	青 海	0.3	0.0	0.1	0.0
	新 疆	36.7	1.3	16.2	1.2
	合 计	55.1	1.6	21.0	1.5
西南诸河	云 南	2.7	0.4	3.2	0.3
	西 藏	0.4	0.0	0.6	0.0
	青 海	0.0	0.0	0.0	0.0
	合 计	3.1	0.5	3.7	0.4

十大水系城镇生活污染排放及处理情况（一）

（2012）

流域	地区名称	城镇人口/万人	生活用水总量/万吨	污水处理厂数/座	污水处理厂设计处理能力/（万吨/日）	生活污水实际处理量/万吨
	总　计	71 704	5 455 535	4 628	15 314	3 576 313
辽河	内蒙古	313	17 849	24	78	14 444
	辽　宁	2 882	187 510	127	605	124 419
	吉　林	306	18 048	12	48	11 237
	合　计	3 500	223 407	163	731	150 099
海河	北　京	1 784	156 113	130	431	120 340
	天　津	1 152	68 329	57	269	57 431
	河　北	3 360	206 962	244	820	159 125
	山　西	627	38 506	59	114	25 606
	内蒙古	27	1 293	5	5	358
	山　东	641	37 718	46	160	33 493
	河　南	696	52 492	31	148	38 964
	合　计	8 288	561 415	572	1 947	435 317
淮河	江　苏	2 189	163 986	206	396	88 856
	安　徽	1 431	97 937	59	184	55 858
	山　东	4 059	271 952	266	1 009	210 820
	河　南	2 271	179 378	96	427	130 863
	合　计	9 951	713 253	627	2 016	486 397
松花江	内蒙古	256	14 929	19	42	5 732
	吉　林	1 171	69 233	39	221	47 086
	黑龙江	2 157	130 219	64	295	57 408
	合　计	3 584	214 381	122	558	110 226
珠江	江　西	17	1 207	2	2	7 140
	湖　南	32	2 206	2	3	590
	广　东	7 140	772 947	410	1 949	887
	广　西	2 013	156 211	107	379	527 573
	贵　州	180	10 697	28	22	92 893
	云　南	489	29 688	25	64	20 412
	海　南	458	36 398	51	133	4 812
	福　建	105	10 010	5	22	13 698
	合　计	10 435	1 019 365	630	2 573	668 005

十大水系城镇生活污染排放及处理情况（一）（续表）

（2012）

流域	地区	城镇人口/万人	生活用水总量/万吨	污水处理厂数/座	污水处理厂设计处理能力/（万吨/日）	生活污水实际处理量/万吨
	上 海	2 125	191 700	55	682	147 117
	江 苏	2 803	262 692	405	1 038	193 948
	浙 江	698	68 371	72	381	40 582
	安 徽	1 293	113 172	62	285	80 709
	福 建	0	0	0		
	江 西	2 123	155 828	119	300	90 485
	河 南	375	26 093	14	50	11 729
	湖 北	3 092	227 073	157	601	158 305
	湖 南	3 066	240 886	132	512	145 175
长	广 东	0	0	0		
江	广 西	25	1 932	4	4	481
	重 庆	1 678	120 134	131	274	82 468
	四 川	3 641	249 772	285	507	145 748
	贵 州	1 089	76 224	83	145	46 523
	云 南	717	70 377	36	156	49 425
	西 藏	1	48	0	0	0
	陕 西	318	16 413	18	34	8 365
	甘 肃	64	2 894	2	3	360
	青 海	9	512	0	0	0
	合 计	23 118	1 824 120	1 575	4 970	1 201 419
	山 西	1 224	72 474	111	189	43 594
	内蒙古	690	40 867	54	149	26 620
	山 东	379	25 949	21	62	16 140
	河 南	654	49 725	56	137	37 480
黄	四 川	3	168	0	0	0
河	陕 西	1 558	94 473	96	275	56 056
	甘 肃	725	38 441	29	91	14 660
	青 海	214	11 470	15	31	6 868
	宁 夏	328	25 306	19	70	15 965
	合 计	5 773	358 873	401	1 004	217 383

十大水系城镇生活污染排放及处理情况（一）（续表）

（2012）

流域	地 区 名 称	城镇人口/ 万人	生活用水总量/ 万吨	污水处理厂数/ 座	污水处理厂设计 处理能力/ （万吨/日）	生活污水实际 处理量/ 万吨
东 南 诸 河	浙 江	2 763	229 715	252	744	143 961
	安 徽	59	4 911	8	15	3 712
	福 建	2 129	163 974	142	391	97 396
	合 计	4 951	398 600	402	1 150	245 069
西 北 诸 河	河 北	24	1 288	3	6	1 079
	内蒙古	151	7 916	25	28	2 541
	西 藏	2	93	0	0	0
	甘 肃	206	11 536	13	53	6 893
	青 海	45	4 012	5	8	2 045
	新 疆	992	77 274	58	219	38 578
	合 计	1 421	102 119	104	314	51 136
西 南 诸 河	云 南	588	34 976	30	46	9 223
	西 藏	91	4 960	2	6	2 038
	青 海	4	66	0	0	0
	合 计	683	40 002	32	52	11 261

十大水系城镇生活污染排放及处理情况（二）

（2012）

流域	地 区 名 称	城镇生活污水排放量/万吨	城镇生活化学需氧量产生量/吨	城镇生活氨氮产生量/吨	城镇生活化学需氧量排放量/吨	城镇生活氨氮排放量/吨
	总　计	4 626 445	17 326 955	2 220 636	9 126 894	1 446 196
辽河	内蒙古	16 399	73 077	9 135	25 188	7 466
	辽　宁	151 495	672 167	92 414	334 062	64 018
	吉　林	15 360	65 820	8 652	43 894	7 499
	合　计	183 254	811 063	110 201	403 145	78 982
海河	北　京	130 985	514 451	63 167	95 661	14 762
	天　津	63 651	289 314	39 472	88 544	16 138
	河　北	181 948	730 142	98 086	228 574	49 015
	山　西	31 216	137 093	19 499	65 951	11 809
	内蒙古	1 098	5 848	733	3 500	512
	山　东	35 680	152 854	19 677	62 265	10 852
	河　南	46 455	153 800	19 733	61 552	11 954
	合　计	491 033	1 983 503	260 366	606 046	115 041
淮河	江　苏	144 253	551 623	71 932	366 816	53 194
	安　徽	87 698	331 788	39 921	239 886	30 472
	山　东	239 702	954 357	128 924	328 010	64 605
	河　南	154 581	519 701	68 146	249 774	42 890
	合　计	626 234	2 357 469	308 923	1 184 486	191 160
松花江	内蒙古	13 627	53 626	7 341	34 957	5 535
	吉　林	59 247	274 618	34 917	150 930	25 419
	黑龙江	104 213	476 277	63 145	346 894	52 222
	合　计	177 086	804 521	105 403	532 781	83 176
珠江	江　西	1 026	4 254	500	3 581	419
	湖　南	1 765	7 767	941	6 330	857
	广　东	651 883	1 824 867	234 656	950 426	149 475
	广　西	133 135	486 086	59 186	356 093	46 312
	贵　州	8 606	39 571	5 046	29 676	3 953
	云　南	23 855	126 054	15 018	96 994	12 729
	海　南	29 587	125 061	14 891	80 233	12 276
	福　建	8 563	25 661	3 332	13 883	2 398
	合　计	858 420	2 639 321	333 570	1 537 217	228 418

十大水系城镇生活污染排放及处理情况（二）（续表）

（2012）

流域	地区名称	城镇生活污水排放量/万吨	城镇生活化学需氧量产生量/吨	城镇生活氨氮产生量/吨	城镇生活化学需氧量排放量/吨	城镇生活氨氮排放量/吨
	上 海	172 530	589 475	75 236	177 504	41 370
	江 苏	217 583	708 796	92 141	205 686	43 854
	浙 江	51 083	181 587	23 147	54 898	13 412
	安 徽	94 671	309 969	36 466	190 674	27 313
	福 建	0	0	0	0	0
	江 西	132 033	526 272	60 390	394 631	49 888
	河 南	21 830	73 738	9 172	46 016	7 342
	湖 北	198 337	701 501	87 938	461 247	65 919
	湖 南	204 915	729 496	93 684	524 068	70 574
长	广 东	0	0	0	0	0
江	广 西	1 651	5 624	742	5 010	625
	重 庆	101 677	416 507	53 288	229 448	37 215
	四 川	213 295	908 683	110 658	604 053	76 896
	贵 州	59 398	249 992	32 096	170 061	22 683
	云 南	59 413	183 740	21 165	57 477	11 351
	西 藏	40	253	30	253	30
	陕 西	15 328	70 453	9 403	54 278	8 235
	甘 肃	2 465	13 557	1 737	13 384	1 704
	青 海	341	1 927	262	1 927	262
	合 计	1 546 589	5 671 571	707 553	3 190 615	478 674
	山 西	54 946	283 909	38 496	140 952	23 960
	内蒙古	31 685	162 206	19 879	86 087	12 394
	山 东	19 753	89 874	11 826	42 754	7 511
	河 南	42 990	143 306	18 808	45 640	9 460
黄	四 川	148	740	90	740	90
河	陕 西	75 297	350 771	45 047	180 261	28 667
	甘 肃	31 930	150 609	20 159	115 595	15 977
	青 海	9 612	46 738	6 077	28 654	5 159
	宁 夏	22 389	74 216	9 337	20 062	6 617
	合 计	288 750	1 302 370	169 719	660 743	109 834

十大水系城镇生活污染排放及处理情况（二）（续表）

（2012）

流域	地名	区称	城镇生活污水排放量/万吨	城镇生活化学需氧量产生量/吨	城镇生活氨氮产生量/吨	城镇生活化学需氧量排放量/吨	城镇生活氨氮排放量/吨
东	浙	江	193 966	731 236	94 889	334 277	59 090
南	安	徽	4 611	13 915	1 588	9 655	1 233
诸	福	建	141 117	520 924	67 642	337 380	50 360
河	合	计	339 695	1 266 075	164 119	681 312	110 683
	河	北	1 095	5 243	629	2 537	424
西	内蒙古		5 977	30 840	4 023	21 827	3 210
北	西	藏	79	424	55	424	55
诸	甘	肃	9 055	43 100	5 708	20 319	3 511
	青	海	2 965	9 383	1 541	5 956	1 204
河	新	疆	64 032	225 027	27 899	121 162	22 352
	合	计	83 203	314 017	39 854	172 226	30 756
西	云	南	27 820	154 642	18 195	137 258	16 811
南	西	藏	4 210	21 462	2 628	20 126	2 556
诸	青	海	152	941	104	941	104
河	合	计	32 181	177 045	20 928	158 324	19 472

重点流域接纳工业废水及处理情况（一）

（2012）

流域	地 区名 称	工业废水排放量/万吨	直接排入环境的	排入污水处理厂的	工业废水处理量/万吨	废水治理设施数/套	废水治理设施治理能力/（万吨/日）	废水治理设施运行费用/万元
	总计	1 142 256	909 041	233 214	3 133 234	38 787	15 887	3 893 889.0
辽河	内蒙古	6 451	3 688	2 763	7 481	178	32	7 497.0
	辽 宁	44 589	33 882	10 707	197 055	1 467	1 034	217 171.6
	吉 林	2 411	2 175	236	2 509	113	17	5 428.4
	合 计	53 451	39 745	13 706	207 044	1 758	1 084	230 097.0
海河	北 京	7 582	2 232	5 350	9 714	449	57	39 700.8
	天 津	19 117	8 027	11 090	40 464	967	171	127 294.3
	河 北	111 918	79 758	32 160	715 282	4 528	3 491	515 700.9
	山 西	18 931	18 028	903	35 488	1 443	222	48 255.6
	内蒙古	576	521	55	361	26	99	698.1
	山 东	68 071	45 915	22 156	121 885	1 765	929	224 745.7
	河 南	31 412	30 130	1 282	59 532	813	235	62 929.2
	合 计	257 607	184 611	72 995	982 726	9 991	5 203	1 019 324.6
淮河	江 苏	69 712	60 902	8 809	116 250	1 976	652	173 559.1
	安 徽	31 112	27 276	3 837	37 345	868	312	65 218.3
	山 东	48 323	39 835	8 488	60 348	1 214	302	96 687.2
	河 南	66 806	56 384	10 423	61 036	1 171	430	76 923.2
	合 计	215 953	184 396	31 557	274 979	5 229	1 696	412 387.8
松花江	内蒙古	3 528	1 291	2 237	2 668	76	17	3 842.7
	吉 林	25 325	19 702	5 623	25 857	414	120	63 006.2
	黑龙江	51 489	46 345	5 144	89 501	1 002	1 183	229 112.2
	合 计	80 342	67 337	13 005	118 026	1 492	1 319	295 961.1

重点流域接纳工业废水及处理情况（一）（续表）

（2012）

流域	地区	工业废水排放量/万吨	直接排入环境的	排入污水处理厂的	工业废水处理量/万吨	废水治理设施数/套	废水治理设施治理能力/（万吨/日）	废水治理设施运行费用/万元
长江中下游	上　海	92 716	39 456	53 260	157 553	3 602	642	656 806.6
	江　苏	41 874	25 749	16 124	137 570	1 587	459	169 889.9
	安　徽	27 505	27 020	485	139 824	1 078	536	114 934.3
	江　西	67 358	65 919	1 439	175 096	2 468	810	157 941.5
	河　南	7 749	7 749		8 929	239	61	18 066.3
	湖　北	70 371	62 768	7 604	261 477	1 626	933	145 574.6
	湖　南	93 480	91 041	2 439	267 481	3 054	1 112	175 081.6
	广　西	1 481	1 481		1 977	112	9	1 767.6
	合　计	402 534	321 183	81 351	1 149 907	13 766	4 562	1 440 062.4
黄河中上游	山　西	29 065	27 466	1 599	168 528	2 086	790	207 820.1
	内蒙古	15 668	6 590	9 079	85 834	673	424	79 099.2
	河　南	32 823	30 734	2 089	44 036	1 376	272	69 472.7
	陕　西	26 492	21 401	5 092	41 967	1 608	205	69 175.8
	甘　肃	9 613	8 217	1 396	20 509	335	129	21 771.2
	青　海	4 071	4 071		17 928	130	58	12 050.4
	宁　夏	14 636	13 290	1 346	21 750	343	147	36 666.8
	合　计	132 369	111 768	20 601	400 552	6 551	2 024	496 056.2

重点流域接纳工业废水及处理情况（二）

（2012）

单位：吨

流域	地区名称	工业废水中污染物产生量				
		化学需氧量	氨氮	石油类	挥发酚	氰化物
	总　计	13 376 999.9	787 551.7	206 278.7	50 625.3	3 040.1
辽河	内蒙古	110 187.1	3 293.2	34.5	7.2	0.0
	辽　宁	302 831.4	17 979.5	13 554.5	1 223.5	150.6
	吉　林	41 874.5	378.1	144.4	—	—
	合　计	454 892.9	21 650.7	13 733.5	1 230.7	150.6
海河	北　京	57 397.6	2 084.3	3 292.0	876.9	5.9
	天　津	130 525.3	7 368.7	1 135.9	2.6	0.5
	河　北	1 667 592.3	93 337.0	19 839.6	9 307.7	505.8
	山　西	173 009.1	16 636.8	2 509.0	2 853.3	97.2
	内蒙古	3 496.0	68.8	1.1	0.0	—
	山　东	1 186 689.5	65 686.4	13 603.7	2 665.8	70.2
	河　南	569 964.9	12 722.5	2 256.7	1 128.7	51.9
	合　计	3 788 674.7	197 904.6	42 637.9	16 835.0	731.5
淮河	江　苏	824 800.9	21 596.8	2 997.6	697.1	60.4
	安　徽	354 532.8	21 611.4	1 129.8	45.1	14.8
	山　东	806 461.0	24 634.7	4 194.3	1 355.3	24.1
	河　南	568 866.1	23 556.6	3 276.0	870.3	44.4
	合　计	2 554 660.8	91 399.6	11 597.6	2 967.8	143.7
松花江	内蒙古	175 734.1	19 164.0	143.6	6.7	7.9
	吉　林	470 109.6	15 155.4	1 567.3	208.2	11.3
	黑龙江	1 061 003.3	28 164.8	39 500.4	5 271.8	84.6
	合　计	1 706 847.0	62 484.2	41 211.3	5 486.7	103.8

338

重点流域接纳工业废水及处理情况（二）（续表）

（2012）

单位：吨

流域	地 区名 称	工业废水中污染物产生量				
		化学需氧量	氨氮	石油类	挥发酚	氰化物
长江中下游	上 海	625 202.8	23 778.8	15 381.1	2 132.8	435.3
	江 苏	369 221.2	20 091.8	5 922.6	2 962.5	193.1
	安 徽	251 694.3	6 924.9	5 384.9	1 750.2	270.8
	江 西	479 508.1	33 214.4	5 900.7	2 514.3	66.5
	河 南	280 102.7	3 931.5	696.7	0.2	0.0
	湖 北	450 339.1	28 777.7	3 374.6	773.4	53.4
	湖 南	569 029.2	65 901.2	4 388.6	1 173.6	358.5
	广 西	8 758.3	403.2	30.7	——	——
	合 计	3 033 855.7	183 023.5	41 080.0	11 307.0	1 377.7
黄河中上游	山 西	257 045.7	80 134.9	5 679.7	4 932.3	266.2
	内蒙古	197 997.5	24 677.9	3 267.4	1 647.9	29.6
	河 南	354 222.4	13 555.5	35 912.5	562.0	27.4
	陕 西	398 617.5	27 866.4	5 979.1	5 001.0	163.4
	甘 肃	115 928.2	12 382.2	1 968.2	114.0	1.8
	青 海	51 674.7	1 315.2	503.9	2.9	22.4
	宁 夏	462 582.8	71 157.1	2 707.7	538.0	21.8
	合 计	1 838 068.7	231 089.1	56 018.5	12 798.1	532.6

重点流域接纳工业废水及处理情况（三）

（2012）

单位：吨

流域	地区名称	工业废水中污染物产生量					
		汞	镉	六价铬	总铬	铅	砷
总 计		11.856	1 343.300	831.415	1 354.286	1 580.268	8 392.058
辽河	内蒙古	0.091	62.260	0.020	0.025	52.317	159.953
	辽 宁	0.052	0.300	14.171	18.468	0.812	1.574
	吉 林	0.208	0.003	0.000	0.000	0.214	0.000
	合 计	0.351	62.563	14.191	18.493	53.343	161.526
海河	北 京	0.000	0.001	9.985	10.750	0.269	0.854
	天 津	0.039	0.024	8.561	13.436	15.271	0.949
	河 北	0.042	1.269	258.005	305.383	10.764	42.040
	山 西	0.357	0.761	2.699	3.007	0.676	0.977
	内蒙古	0.000	0.000	0.000	0.000	0.000	0.000
	山 东	0.122	247.212	6.858	51.969	243.326	641.204
	河 南	0.359	81.591	4.332	15.260	45.080	23.543
	合 计	0.919	330.858	290.440	399.805	315.386	709.567
淮河	江 苏	0.007	0.020	70.056	95.590	14.412	0.941
	安 徽	0.001	0.006	6.649	11.336	2.198	2.121
	山 东	0.017	0.178	4.895	20.399	2.255	0.647
	河 南	0.167	1.347	9.545	123.608	3.087	7.695
	合 计	0.193	1.550	91.145	250.932	21.951	11.405
松花江	内蒙古	0.000	0.003	0.000	0.001	0.052	0.041
	吉 林	0.002	0.002	11.560	11.566	0.021	0.024
	黑龙江	0.159	0.049	6.589	7.272	6.496	1.491
	合 计	0.162	0.054	18.149	18.839	6.569	1.556

重点流域接纳工业废水及处理情况（三）（续表）

（2012）

单位：吨

流域	地 区名 称	工业废水中污染物产生量					
		汞	镉	六价铬	总铬	铅	砷
长江中下游	上 海	0.006	0.053	203.153	261.163	2.760	1.636
	江 苏	0.025	0.195	84.453	110.010	3.843	1.687
	安 徽	2.244	13.476	12.764	15.133	54.500	554.300
	江 西	0.308	398.938	47.191	51.216	95.215	5 826.509
	河 南	0.001	0.006	0.189	9.401	0.126	5.254
	湖 北	0.209	8.412	36.402	71.942	17.101	95.689
	湖 南	4.476	203.121	13.996	38.327	494.240	472.109
	广 西	0.000	0.017	0.092	0.092	0.025	0.023
	合 计	7.268	624.220	398.240	557.283	667.810	6 957.207
黄河中上游	山 西	0.006	14.754	2.389	2.661	24.473	58.423
	内蒙古	0.798	10.851	1.123	1.237	27.923	68.110
	河 南	0.985	184.553	6.983	60.397	297.889	85.394
	陕 西	0.044	0.061	5.967	8.006	0.935	26.586
	甘 肃	0.053	98.832	0.520	33.896	149.000	154.852
	青 海	0.367	11.246	1.192	1.302	10.115	70.612
	宁 夏	0.710	3.757	1.077	1.434	4.873	86.819
	合 计	2.962	324.055	19.251	108.933	515.209	550.797

重点流域接纳工业废水及处理情况（四）

（2012）

单位：吨

流域	地区名称	工业废水中污染物排放量				
		化学需氧量	氨氮	石油类	挥发酚	氰化物
总计		1 699 553.2	160 224.8	11 289.4	1 241.6	124.7
辽河	内蒙古	25 819.4	2 740.4	12.9	0.0	0.0
	辽宁	56 662.5	3 025.8	402.3	17.7	3.8
	吉林	5 424.4	198.2	8.9	—	—
	合计	87 906.2	5 964.4	424.1	17.7	3.8
海河	北京	5 527.3	313.0	43.9	0.2	0.1
	天津	26 533.3	3 294.5	138.2	1.2	0.0
	河北	178 785.1	14 768.6	902.6	16.3	10.7
	山西	32 932.5	3 713.3	578.8	356.0	17.1
	内蒙古	1 677.0	19.9	0.9	0.0	—
	山东	57 471.1	4 736.9	221.8	13.5	4.9
	河南	44 887.6	2 947.4	110.6	0.7	3.9
	合计	347 814.0	29 793.5	1 996.6	387.9	36.8
淮河	江苏	103 696.6	7 883.0	548.5	23.8	6.6
	安徽	37 430.1	5 489.5	353.6	2.7	2.1
	山东	33 750.3	2 010.9	169.4	2.6	0.4
	河南	83 870.2	6 075.6	681.2	124.5	9.8
	合计	258 747.2	21 459.0	1 752.8	153.6	19.0
松花江	内蒙古	17 416.8	1 020.2	25.0	6.6	0.0
	吉林	42 480.5	3 325.3	157.4	1.2	0.1
	黑龙江	79 559.4	5 299.7	282.4	6.3	1.6
	合计	139 456.7	9 645.2	464.9	14.1	1.7

重点流域接纳工业废水及处理情况（四）（续表）

（2012）

单位：吨

流域	地区名称	工业废水中污染物排放量				
		化学需氧量	氨氮	石油类	挥发酚	氰化物
长江中下游	上 海	52 318.7	4 561.9	1 292.5	5.5	4.4
	江 苏	45 602.6	2 835.1	326.7	16.8	2.9
	安 徽	38 915.1	2 217.9	311.4	2.6	4.2
	江 西	99 844.5	9 227.1	564.4	8.9	7.7
	河 南	11 997.1	1 282.9	41.4	0.0	0.0
	湖 北	104 283.4	12 015.2	789.1	13.5	7.2
	湖 南	142 748.8	25 533.1	760.2	15.6	11.2
	广 西	3 203.6	244.4	15.7	—	—
	合 计	498 913.6	57 917.6	4 101.4	62.9	37.7
黄河中上游	山 西	53 354.4	4 514.9	619.2	371.0	13.1
	内蒙古	29 515.4	5 556.8	685.7	209.6	1.3
	河 南	41 149.1	3 079.9	303.7	10.4	4.3
	陕 西	71 210.7	5 207.3	611.2	2.5	5.4
	甘 肃	43 197.0	8 011.3	187.0	0.6	0.0
	青 海	25 254.5	731.2	41.9	0.6	—
	宁 夏	103 034.3	8 343.7	101.1	10.6	1.6
	合 计	366 715.4	35 445.1	2 549.7	605.3	25.7

重点流域接纳工业废水及处理情况（五）

（2012）

单位：吨

流域	地区名称	工业废水中污染物排放量					
		汞	镉	六价铬	总铬	铅	砷
	总　计	0.637	20.300	36.286	106.355	64.752	86.384
辽河	内蒙古	0.034	0.102	0.000	0.005	1.216	1.900
	辽　宁	0.000	0.021	0.406	0.552	0.181	0.112
	吉　林	0.001	0.001	0.000	0.000	0.064	0.000
	合　计	0.036	0.125	0.406	0.557	1.461	2.012
海河	北　京	0.000	0.001	0.168	0.269	0.054	0.010
	天　津	0.004	0.010	0.169	0.454	1.004	0.019
	河　北	0.001	0.001	2.397	5.272	0.245	0.030
	山　西	0.001	0.754	0.308	0.313	0.275	0.364
	内蒙古	0.000	0.000	0.000	0.000	0.000	0.000
	山　东	0.008	0.978	0.117	5.637	0.536	1.901
	河　南	0.000	0.256	0.152	3.594	0.749	0.068
	合　计	0.015	1.999	3.312	15.538	2.863	2.391
淮河	江　苏	0.006	0.004	1.027	2.882	0.997	0.025
	安　徽	0.001	0.003	0.080	0.440	0.162	2.116
	山　东	0.003	0.006	0.075	0.350	0.041	0.114
	河　南	0.002	0.538	0.751	8.151	1.815	0.631
	合　计	0.012	0.551	1.932	11.823	3.015	2.886
松花江	内蒙古	0.000	0.002	0.000	0.001	0.003	0.033
	吉　林	0.000	0.000	0.108	0.113	0.001	0.000
	黑龙江	0.000	0.001	0.368	0.368	0.019	0.001
	合　计	0.000	0.003	0.476	0.482	0.023	0.034

重点流域接纳工业废水及处理情况（五）（续表）

（2012）

单位：吨

流域	地区名称	工业废水中污染物排放量					
		汞	镉	六价铬	总铬	铅	砷
长江中下游	上 海	0.001	0.002	2.022	5.545	0.486	0.147
	江 苏	0.004	0.011	2.275	3.234	0.797	0.510
	安 徽	0.001	0.092	2.191	2.990	1.412	2.831
	江 西	0.088	2.162	17.136	17.301	6.442	8.632
	河 南	0.000	0.001	0.014	0.333	0.028	0.037
	湖 北	0.195	0.603	3.583	4.282	2.578	8.054
	湖 南	0.225	13.454	2.076	18.050	38.475	53.378
	广 西	0.000	0.003	0.002	0.002	0.002	0.002
	合 计	0.514	16.327	29.299	51.738	50.220	73.590
黄河中上游	山 西	0.003	0.001	0.168	0.269	0.054	0.010
	内蒙古	0.002	0.010	0.169	0.454	1.004	0.019
	河 南	0.012	0.001	2.397	5.272	0.245	0.030
	陕 西	0.001	0.754	0.308	0.313	0.275	0.364
	甘 肃	0.035	0.000	0.000	0.000	0.000	0.000
	青 海	0.004	0.978	0.117	5.637	0.536	1.901
	宁 夏	0.004	0.256	0.152	3.594	0.749	0.068
	合 计	0.060	1.999	3.312	15.538	2.863	2.391

重点流域工业废水污染防治投资情况

（2012）

流域	地 区名 称	工业废水治理施工项目数/个	工业废水治理竣工项目数/个	工业废水治理项目完成投资/万元	废水治理竣工项目新增处理能力/（万吨/日）
	总 计	777	825	659 577.1	161.4
辽河	内蒙古	1	9	19 020.6	4.8
	辽 宁	8	23	15 147.5	2.9
	吉 林	0	0	700.0	0.7
	合 计	9	32	34 868.1	8.5
海河	北 京	10	13	3 011.9	0.9
	天 津	50	20	11 305.9	4.0
	河 北	31	38	48 465.8	10.4
	山 西	12	13	7 430.0	5.0
	内蒙古	0	1	1 608.0	0.0
	山 东	60	77	111 043.5	15.9
	河 南	23	11	7 157.2	2.6
	合 计	186	173	190 022.3	38.8
淮河	江 苏	23	37	28 438.4	10.2
	安 徽	12	11	11 458.7	3.4
	山 东	50	67	85 045.8	25.6
	河 南	17	12	9 732.2	1.3
	合 计	102	127	134 675.1	40.6
松花江	内蒙古	2	2	760.0	0.0
	吉 林	5	5	3 606.2	0.4
	黑龙江	10	8	7 111.9	2.7
	合 计	17	15	11 478.1	3.1

重点流域工业废水污染防治投资情况（续表）

（2012）

流域	地区名称	工业废水治理施工项目数/个	工业废水治理竣工项目数/个	工业废水治理项目完成投资/万元	废水治理竣工项目新增处理能力/（万吨/日）
长江中下游	上 海	78	82	10 672.2	2.1
	江 苏	61	64	11 780.3	5.5
	安 徽	12	16	6 530.8	0.4
	江 西	39	45	16 574.6	4.1
	河 南	15	12	5 728.7	0.7
	湖 北	20	24	18 142.0	3.3
	湖 南	46	55	46 442.1	16.8
	广 西	0	0	0	0
	合 计	271	298	115 870.7	33.0
黄河中上游	山 西	58	56	22 073.1	13.7
	内蒙古	40	38	17 205.4	3.0
	河 南	19	10	9 234.8	4.4
	陕 西	54	51	95 460.0	11.8
	甘 肃	5	8	11 128.6	0.9
	青 海	7	4	3 262.6	1.2
	宁 夏	9	13	14 298.3	2.4
	合 计	192	180	172 662.8	37.4

重点流域农业污染排放情况

（2012）

单位：万吨

流域	地区名称	农业污染物排放（流失）总量			
		化学需氧量	氨氮	总氮	总磷
总　计		767.7	47.7	289.4	31.2
辽河	内蒙古	14.9	0.4	3.9	0.4
	辽　宁	69.4	2.7	17.0	2.2
	吉　林	6.0	0.3	1.7	0.2
	合　计	90.3	3.4	22.6	2.8
海河	北　京	7.8	0.5	3.3	0.4
	天　津	11.4	0.6	3.3	0.4
	河　北	87.4	4.3	34.4	3.7
	山　西	6.9	0.4	2.8	0.2
	内蒙古	3.3	0.0	0.7	0.0
	山　东	50.4	2.5	27.6	2.7
	河　南	15.8	1.2	7.2	0.9
	合　计	183.0	9.5	79.2	8.4
淮河	江　苏	27.2	2.5	11.9	1.3
	安　徽	26.9	2.6	12.9	1.4
	山　东	34.2	2.2	13.5	1.5
	河　南	47.8	3.9	25.6	2.9
	合　计	136.1	11.3	63.9	7.0
松花江	内蒙古	11.3	0.3	2.8	0.2
	吉　林	40.5	1.4	10.3	1.3
	黑龙江	87.1	2.6	20.1	1.9
	合　计	138.9	4.3	33.2	3.4

重点流域农业污染排放情况（续表）

单位：万吨

流域	地 区 名 称	农业污染物排放（流失）总量			
		化学需氧量	氨氮	总氮	总磷
长江中下游	上 海	3.3	0.3	1.5	0.2
	江 苏	5.2	0.7	2.8	0.3
	安 徽	6.8	0.8	3.5	0.4
	江 西	23.3	2.9	11.1	1.3
	河 南	7.2	0.6	4.5	0.4
	湖 北	43.4	4.2	17.4	2.1
	湖 南	55.6	6.1	21.6	2.3
	广 西	0.5	0.1	0.4	0.0
	合 计	145.3	15.7	62.8	7.1
黄河中上游	山 西	11.3	0.9	5.7	0.6
	内蒙古	16.2	0.3	4.5	0.3
	河 南	13.6	0.9	6.3	0.7
	陕 西	14.6	0.9	5.6	0.5
	甘 肃	6.5	0.3	2.3	0.2
	青 海	2.0	0.1	0.6	0.0
	宁 夏	9.9	0.2	2.7	0.2
	合 计	74.0	3.6	27.7	2.5

重点流域城镇生活污染排放及处理情况（一）

（2012）

流域	地 区 名 称	城镇人口/ 万人	生活用水总量/ 万吨	污水处理厂数/ 座	污水处理厂设计 处理能力/ （万吨/日）	生活污水实际 处理量/ 万吨
	总　计	39 264	2 773 580	2 288	8 319	1 983 199
辽河	内蒙古	297	17 013	22	72	13 436
	辽　宁	2 028	133 300	80	410	95 888
	吉　林	138	8 400	5	25	6 087
	合　计	2 463	158 714	107	506	115 411
海河	北　京	1 775	155 385	127	419	118 663
	天　津	1 152	68 329	57	269	57 431
	河　北	3 249	199 381	227	777	153 495
	山　西	621	38 544	60	111	25 576
	内蒙古	32	1 511	6	5	358
	山　东	1 480	94 979	101	369	82 612
	河　南	668	50 437	31	148	40 838
	合　计	8 978	608 567	609	2 098	478 974
淮河	江　苏	2 038	152 716	180	333	78 752
	安　徽	1 419	97 113	57	173	49 162
	山　东	1 410	90 889	79	245	61 781
	河　南	2 154	166 844	89	325	94 777
	合　计	7 021	507 562	405	1 075	284 472
松花江	内蒙古	174	9 749	12	24	3 478
	吉　林	1 122	66 602	38	214	44 005
	黑龙江	1 917	113 670	56	285	55 093
	合　计	3 213	190 021	106	524	102 575

重点流域城镇生活污染排放及处理情况（一）（续表）

（2012）

流域	地 区	城镇人口/万人	生活用水总量/万吨	污水处理厂数/座	污水处理厂设计处理能力/（万吨/日）	生活污水实际处理量/万吨
长江中下游	上 海	2 055	185 415	110	1 363	294 235
	江 苏	1 154	99 579	142	339	86 174
	安 徽	873	75 556	44	168	48 195
	江 西	2 117	155 420	118	299	89 690
	河 南	352	24 583	13	46	10 954
	湖 北	2 638	196 689	110	470	125 407
	湖 南	3 030	238 307	131	506	144 809
	广 西	34	2 677	5	7	1 093
	合 计	12 255	978 227	673	3 199	800 557
黄河中上游	山 西	1 204	71 218	112	188	43 210
	内蒙古	705	41 696	59	151	27 102
	河 南	719	55 829	58	132	37 234
	陕 西	1 439	86 491	93	253	55 921
	甘 肃	725	38 441	29	91	14 666
	青 海	225	11 996	18	33	7 112
	宁 夏	318	24 819	19	70	15 964
	合 计	5 335	330 490	388	918	201 211

重点流域城镇生活污染排放及处理情况（二）

（2012）

流域	地 区名 称	城镇生活污水排放量/万吨	城镇生活化学需氧量产生量/吨	城镇生活氨氮产生量/吨	城镇生活化学需氧量排放量/吨	城镇生活氨氮排放量/吨
	总　计	2 366 926	9 301 262	1 198 626	4 801 683	775 999
辽河	内蒙古	15 731	69 489	8 686	23 830	7 165
	辽　宁	104 675	472 460	64 074	222 708	47 123
	吉　林	7 233	30 119	3 884	15 723	3 314
	合　计	127 639	572 068	76 644	262 261	57 602
海河	北　京	130 374	512 051	62 872	95 215	14 693
	天　津	63 651	289 314	39 472	88 544	16 138
	河　北	175 333	707 539	94 761	221 502	47 630
	山　西	31 227	137 962	19 612	67 615	12011
	内蒙古	1 283	7 080	873	4 732	653
	山　东	86 877	344 991	46 463	117 140	20 589
	河　南	44 655	147 815	18 965	58 021	11 455
	合　计	533 399	2 146 752	283 020	652 768	123 169
淮河	江　苏	134 438	513 391	66 900	348 232	49 696
	安　徽	86 952	328 863	39 575	237 474	30 170
	山　东	78 641	334 731	44 781	160 838	29 690
	河　南	143 704	493 948	64 422	245 010	41 359
	合　计	443 735	1 670 934	215 678	991 554	150 915
松花江	内蒙古	8 956	36 545	4 910	26 477	4 025
	吉　林	56 840	262 684	33 426	146 532	24 352
	黑龙江	92 563	423 590	56 270	300 999	45 663
	合　计	158 359	722 819	94 606	474 007	74 040

重点流域城镇生活污染排放及处理情况（二）（续表）

（2012）

流域	地区名称		城镇生活污水排放量/万吨	城镇生活化学需氧量产生量/吨	城镇生活氨氮产生量/吨	城镇生活化学需氧量排放量/吨	城镇生活氨氮排放量/吨
长江中下游	上	海	166 873	570 148	72 769	171 684	40 013
	江	苏	82 918	291 448	37 991	126 955	24 886
	安	徽	60 317	207 028	24 285	140 899	19 788
	江	西	131 695	524 869	60 239	393 574	49 751
	河	南	20 571	69 522	8 650	43 807	6 985
	湖	北	170 114	596 600	74 936	402 437	56 002
	湖	南	202 930	721 289	92 350	517 876	69 863
	广	西	2 287	7 791	1 028	6 470	839
	合 计		837 704	2 988 695	372 247	1 803 702	268 128
黄河中上游	山	西	53 998	279 246	37 913	139 521	23 597
	内蒙古		32 142	165 849	20 283	87 133	12 592
	河	南	46 786	159 876	21 069	60 954	11 646
	陕	西	69 180	323 544	41 535	165 551	26 564
	甘	肃	31 930	150 609	20 159	115 595	15 977
	青	海	10 079	48 823	6 410	30 644	5 426
	宁	夏	21 973	72 048	9 064	17 993	6 344
	合 计		266 088	1 199 995	156 433	617 391	102 145

湖泊水库接纳工业废水及处理情况（一）

（2012）

流域	地区名称	工业废水排放量/万吨	直接排入环境的	排入污水处理厂的	工业废水处理量/万吨	废水治理设施数/套	废水治理设施治理能力/（万吨/日）	废水治理设施运行费用/万元
总　计		329 865	246 532	83 332	713 976	12 846	3 467	1 088 112.6
滇池	云　南	845	658	188	1 943	182	18	7 581.1
巢　湖	安　徽	6 158	5 527	631	25 914	290	156	18 706.2
洞庭湖	江　西	2 125	2 125	0	29 733	128	91	9 489.5
	湖　北	673	673	0	347	12	6	762.8
	湖　南	93 480	91 041	2 439	267 481	3 054	1 112	175 081.6
	广　西	1 481	1 481	0	1 977	112	9	1 767.6
	合　计	97 759	95 320	2 439	299 538	3 306	1 218	187 101.5
鄱阳湖	安　徽	34	34	0	34	19	1	149.1
	江　西	64 378	62 939	1 439	144 708	2 330	715	147 063.1
	合　计	64 412	62 973	1 439	144 743	2 349	716	147 212.2
太　湖	上　海	1 811	152	1 659	1 042	172	6	8 193.4
	江　苏	108 037	61 105	46 931	130 431	3 374	746	558 739.9
	浙　江	40 093	10 172	29 921	81 382	1 988	436	123 954.7
	合　计	149 941	71 429	78 512	212 854	5 534	1 188	690 888.0
丹江口	河　南	3 893	3 893	0	11 260	603	50	16 839.5
	湖　北	2 087	2 087	0	2 185	144	16	4 948.4
	陕　西	4 770	4 646	124	15 540	438	104	14 835.7
	合　计	10 750	10 626	124	28 985	1 185	171	36 623.6

湖泊水库接纳工业废水及处理情况（二）

（2012）

单位：吨

| 流域 | 地 区 名 称 | 工业废水中污染物产生量 | | | | |
		化学需氧量	氨氮	石油类	挥发酚	氰化物
总 计		2 416 975.0	147 262.1	16 827.6	4 071.6	561.5
滇 池	云 南	9 598.3	569.8	63.8	0.8	0.8
巢 湖	安 徽	82 955.8	5 751.4	1 046.1	0.1	3.3
洞庭湖	江 西	23 699.6	302.3	1 067.2	389.8	2.3
	湖 北	16 499.5	72.9	3.2	0.0	0.0
	湖 南	569 029.2	65 901.2	4 388.6	1 173.6	358.5
	广 西	8 758.3	403.2	30.7	0.0	0.0
	合 计	617 986.7	66 679.5	5 489.7	1 563.4	360.8
鄱阳湖	安 徽	1 119.2	5.0	132.7	0.0	0.0
	江 西	444 764.6	32 495.6	4 826.9	2 124.5	64.2
	合 计	445 883.8	32 500.6	4 959.5	2 124.5	64.2
太 湖	上 海	7 405.4	394.5	10.9	0.0	12.8
	江 苏	714 413.3	21 569.6	3 567.7	180.6	113.2
	浙 江	441 349.7	10 488.8	1 002.8	195.3	2.9
	合 计	1 163 168.5	32 452.8	4 581.4	375.9	128.9
丹江口	河 南	29 729.4	981.5	67.6	0.3	1.1
	湖 北	19 005.4	2 385.6	389.9	5.0	1.6
	陕 西	48 647.2	5 940.9	229.6	1.6	0.8
	合 计	97 381.9	9 308.0	687.1	6.8	3.5

湖泊水库接纳工业废水及处理情况（三）

（2012）

单位：吨

流域	地区名称	工业废水中污染物产生量					
		汞	镉	六价铬	总铬	铅	砷
总　计		5.127	802.690	262.109	597.433	822.645	6 792.251
滇　池	云　南	0.000	125.622	0.014	0.014	118.986	451.443
巢　湖	安　徽	0.000	0.058	1.412	1.414	0.589	0.265
洞庭湖	江　西	0.000	0.000	0.000	0.000	0.033	0.036
	湖　北	0.000	0.000	0.000	0.000	0.000	0.098
	湖　南	4.476	203.121	13.996	38.327	494.240	472.109
	广　西	0.000	0.017	0.092	0.092	0.025	0.023
	合　计	4.476	203.139	14.088	38.419	494.299	472.265
鄱阳湖	安　徽	0.000	0.000	0.000	0.000	0.209	0.000
	江　西	0.308	398.938	47.191	51.216	95.181	5 826.440
	合　计	0.308	398.938	47.191	51.216	95.390	5 826.440
太　湖	上　海	0.000	0.000	17.209	17.529	0.104	0.001
	江　苏	0.000	2.263	125.596	246.713	3.444	3.874
	浙　江	0.000	1.307	44.122	228.821	4.010	0.536
	合　计	0.000	3.570	186.927	493.063	7.558	4.410
丹江口	河　南	0.000	0.001	0.769	0.908	0.018	0.006
	湖　北	0.001	0.010	8.067	8.450	0.030	0.630
	陕　西	0.340	71.353	3.642	3.949	105.774	36.791
	合　计	0.342	71.364	12.478	13.307	105.822	37.427

湖泊水库接纳工业废水及处理情况（四）

（2012）

单位：吨

流域	地区名称	工业废水中污染物排放量				
		化学需氧量	氨氮	石油类	挥发酚	氰化物
总　计		388 601.3	46 769.7	1 821.0	35.7	26.4
滇　池	云　南	1 261.5	64.4	46.8	0.1	0.1
巢　湖	安　徽	11 453.6	569.3	44.9	0.0	0.0
洞庭湖	江　西	4 919.9	126.1	32.3	1.2	0.1
	湖　北	1 913.7	55.0	2.3	0.0	0.0
	湖　南	142 748.8	25 533.1	760.2	15.6	11.2
	广　西	3 203.6	244.4	15.7	0.0	0.0
	合　计	152 786.0	25 958.6	810.4	16.8	11.3
鄱阳湖	安　徽	89.9	4.2	5.5	0.0	0.0
	江　西	94 012.7	9 002.3	531.0	7.7	7.7
	合　计	94 102.6	9 006.5	536.5	7.7	7.7
太　湖	上　海	807.0	116.4	9.7	0.0	0.2
	江　苏	65 831.5	4 286.0	178.4	8.9	6.2
	浙　江	34 443.0	3 353.4	41.5	0.5	0.3
	合　计	101 081.4	7 755.8	229.6	9.4	6.7
丹江口	河　南	4 247.3	320.9	2.5	0.0	0.0
	湖　北	9 751.3	768.6	133.8	1.7	0.5
	陕　西	13 917.6	2 325.7	16.4	0.0	0.1
	合　计	27 916.2	3 415.1	152.7	1.7	0.7

湖泊水库接纳工业废水及处理情况（五）

（2012）

单位：吨

流域	地区名称	工业废水中污染物排放量					
		汞	镉	六价铬	总铬	铅	砷
总　计		0.342	16.513	28.439	50.960	47.355	63.974
滇　池	云　南	0.000	0.327	0.000	0.003	0.877	1.412
巢　湖	安　徽	0.000	0.002	0.022	0.024	0.030	0.012
洞庭湖	江　西	0.000	0.000	0.000	0.000	0.012	0.000
	湖　北	0.000	0.000	0.000	0.000	0.000	0.098
	湖　南	0.225	13.454	2.076	18.050	38.475	53.378
	广　西	0.000	0.003	0.002	0.002	0.002	0.002
	合　计	0.225	13.457	2.078	18.052	38.489	53.477
鄱阳湖	安　徽	0.000	0.000	0.000	0.000	0.023	0.000
	江　西	0.088	2.162	17.136	17.301	6.430	8.623
	合　计	0.088	2.162	17.136	17.301	6.453	8.623
太　湖	上　海	0.000	0.000	0.153	0.169	0.020	0.000
	江　苏	0.000	0.005	1.244	3.674	0.123	0.007
	浙　江	0.000	0.013	0.463	4.317	0.189	0.007
	合　计	0.000	0.019	1.860	8.160	0.332	0.014
丹江口	河　南	0.000	0.000	0.045	0.061	0.008	0.003
	湖　北	0.001	0.010	7.237	7.253	0.030	0.014
	陕　西	0.027	0.536	0.061	0.107	1.135	0.419
	合　计	0.029	0.547	7.343	7.420	1.173	0.436

湖泊水库工业污染防治投资情况

（2012）

流域	地区名称	工业废水治理施工项目数/个	工业废水治理竣工项目数/个	工业废水治理项目完成投资/万元	废水治理竣工项目新增处理能力/（万吨/日）
总　计		209	234	122 276.9	33.2
滇　池	云　南	2	3	211.6	0.0
巢　湖	安　徽	20	17	6 315.6	0.6
洞庭湖	江　西	0	0	0.0	0.0
	湖　北	0	0	0.0	0.0
	湖　南	46	55	46 442.1	16.8
	广　西	0	0	0.0	0.0
	合　计	46	55	46 442.1	16.8
鄱阳湖	安　徽	0	0	0.0	0.0
	江　西	39	45	16 574.6	4.1
	合　计	39	45	16 574.6	4.1
太　湖	上　海	0	0	0.0	0.0
	江　苏	40	50	23 397.1	7.4
	浙　江	39	42	16 906.7	2.7
	合　计	79	92	40 303.9	10.1
丹江口	河　南	6	4	1 353.0	0.8
	湖　北	6	7	1 710.5	0.4
	陕　西	11	11	9 365.7	0.3
	合　计	23	22	12 429.2	1.5

湖泊水库农业污染排放情况

（2012）　　　　　　　　　　　　　　　　　　　　　　　单位：万吨

流域	地 区名 称	农业污染物排放/流失总量			
		化学需氧量	氨氮	总氮	总磷
总 计		105.3	12.0	45.2	5.1
滇 池	云 南	0.4	0.1	0.2	0.0
巢 湖	安 徽	5.0	0.4	2.0	0.2
洞庭湖	江 西	0.6	0.1	0.2	0.0
	湖 北	0.4	0.1	0.3	0.0
	湖 南	55.6	6.1	21.6	2.3
	广 西	0.5	0.1	0.4	0.0
	合 计	57.1	6.4	22.5	2.4
鄱阳湖	安 徽	0.0	0.0	0.0	0.0
	江 西	22.6	2.8	10.8	1.3
	合 计	22.6	2.8	10.8	1.3
太 湖	上 海	0.2	0.0	0.1	0.0
	江 苏	5.0	0.5	2.3	0.3
	浙 江	6.1	0.8	2.5	0.3
	合 计	11.3	1.3	4.9	0.6
丹江口	河 南	2.9	0.3	1.2	0.1
	湖 北	1.1	0.1	0.6	0.1
	陕 西	4.8	0.7	2.9	0.3
	合 计	8.8	1.0	4.7	0.5

湖泊水库城镇生活污染排放及处理情况（一）

（2012）

流域	地 区 名 称	城镇人口/ 万人	生活用水总量/ 万吨	污水处理厂数/ 座	污水处理厂设计处理能力/ （万吨/日）	生活污水实际处理量/ 万吨
总 计		9 118	767 643	652	2 153	498 614
滇 池	云 南	329	44 950	10	84	28 069
巢 湖	安 徽	498	43 114	19	107	35 842
洞庭湖	江 西	93	6 075	3	11	2 673
	湖 北	32	2 181	1	2	526
	湖 南	3 030	238 307	131	506	144 809
	广 西	34	2 677	5	7	1 093
	合 计	3 189	249 241	140	525	149 101
鄱阳湖	安 徽	7	338	1	1	256
	江 西	2 008	148 242	114	288	86 776
	合 计	2 015	148 579	115	289	87 032
太 湖	上 海	78	7 038	12	25	6 088
	江 苏	1 533	154 300	250	659	115 278
	浙 江	851	81 227	70	373	52 356
	合 计	2 463	242 565	332	1 056	173 722
丹江口	河 南	135	11 160	6	15	4 290
	湖 北	164	11 311	13	41	9 256
	陕 西	325	16 723	17	34	6 544
	合 计	624	39 194	36	90	20 090

湖泊水库城镇生活污染排放及处理情况（二）

（2012）

流域	地区名称		城镇生活污水排放量/万吨	城镇生活化学需氧量产生量/吨	城镇生活氨氮产生量/吨	城镇生活化学需氧量排放量/吨	城镇生活氨氮排放量/吨
总　计			642 890	2 238 686	277 898	1 215 769	180 474
滇　池	云　南		38 207	89 835	10 038	1 521	2 491
巢　湖	安　徽		36 928	119 040	13 464	60 217	8 816
洞庭湖	江　西		5 771	18 272	2 699	18 796	2 329
	湖　北		1 854	7 328	949	6 278	864
	湖　南		202 930	721 289	92 350	517 876	69 863
	广　西		2 287	7 791	1 028	6 470	839
	合　计		212 843	754 680	97 025	549 421	73 895
鄱阳湖	安　徽		320	1 775	271	1 198	153
	江　西		124 986	502 708	57 119	371 208	47 017
	合　计		125 306	504 483	57 390	372 406	47 170
太　湖	上　海		6 334	21 641	2 762	6 517	1 519
	江　苏		125 807	387 677	50 328	68 553	17 263
	浙　江		63 085	222 044	28 804	58 865	14 493
	合　计		195 226	631 362	81 894	133 935	33 275
丹江口	河　南		8 851	29 237	3 730	18 554	2 805
	湖　北		9 799	37 710	4 713	22 804	3 987
	陕　西		15 731	72 339	9 643	56 912	8 036
	合　计		34 381	139 286	18 086	98 270	14 828

三峡地区接纳工业废水及处理情况（一）

（2012）

流域	地区名称	工业废水排放量/万吨	直接排入环境的	排入污水处理厂的	工业废水处理量/万吨	废水治理设施数/套	废水治理设施治理能力/（万吨/日）	废水治理设施运行费用/万元
	总　计	130 461	121 607	8 854	332 537	7 701	1 756	438 876.3
库　区	湖　北	3 065	2 705	360	2 716	62	14	2 261.9
	重　庆	14 210	13 149	1 061	27 995	856	296	57 355.6
	合　计	17 275	15 854	1 421	30 711	918	310	59 617.5
影响区	湖　北	6 869	6 103	766	6 877	88	27	5 576.7
	重　庆	15 637	15 553	84	15 910	591	93	18 430.0
	四　川	12 448	12 247	201	12 527	654	95	29 160.3
	贵　州	1 523	1 523	0	949	90	5	2 545.2
	合　计	36 478	35 427	1 051	36 263	1 423	220	55 712.2
上游区	重　庆	439	439	0	543	81	4	1 761.8
	四　川	57 536	51 422	6 114	194 928	3 601	905	255 216.5
	贵　州	11 671	11 671	0	31 496	833	159	26 650.3
	云　南	7 063	6 794	268	38 596	845	160	39 918.0
	合　计	76 708	70 326	6 382	265 563	5 360	1 227	323 546.6

三峡地区接纳工业废水及处理情况（二）

（2012）

单位：吨

流域	地区名称	工业废水中污染物产生量				
		化学需氧量	氨氮	石油类	挥发酚	氰化物
	总　计	1 174 192.4	122 377.2	8 966.7	2 700.3	222.6
库　区	湖　北	13 795.3	288.8	3.1	0.0	0.0
	重　庆	162 126.7	23 104.3	1 841.2	675.7	63.5
	合　计	175 922.0	23 393.1	1 844.2	675.7	63.5
影响区	湖　北	21 086.0	2 688.1	375.0	15.9	1.0
	重　庆	121 536.8	4 120.9	866.1	173.9	20.2
	四　川	209 212.8	20 003.5	498.6	53.0	1.9
	贵　州	31 159.5	694.6	0.0	0.0	0.0
	合　计	382 995.1	27 507.0	1 739.6	242.8	23.1
上游区	重　庆	1 696.1	927.0	4.6	0.0	0.8
	四　川	460 330.9	18 434.9	3 482.8	909.2	75.7
	贵　州	70 803.0	27 210.7	881.3	269.7	33.2
	云　南	82 445.4	24 904.5	1 014.2	602.9	26.2
	合　计	615 275.3	71 477.0	5 382.8	1 781.8	136.0

三峡地区接纳工业废水及处理情况（三）

（2012）

单位：吨

流域	地区名称	工业废水中污染物产生量					
		汞	镉	六价铬	总铬	铅	砷
总　计		1.872	231.010	138.008	303.155	400.427	1 125.094
库　区	湖　北	0.000	0.000	0.001	0.051	0.000	0.000
	重　庆	0.000	0.001	65.570	105.988	6.591	1.432
	合　计	0.000	0.001	65.571	106.040	6.591	1.432
影响区	湖　北	0.000	0.000	0.000	0.000	0.000	0.000
	重　庆	0.000	0.000	19.754	57.548	0.399	1.566
	四　川	0.211	0.075	12.999	14.625	0.090	52.381
	贵　州	0.000	0.000	0.000	0.000	0.000	0.000
	合　计	0.211	0.075	32.753	72.173	0.489	53.947
上游区	重　庆	0.000	0.000	0.407	0.407	0.000	0.000
	四　川	0.757	45.294	39.046	120.901	71.495	436.296
	贵　州	0.020	0.471	0.155	2.825	0.685	19.621
	云　南	0.884	185.168	0.075	0.810	321.167	613.798
	合　计	1.661	230.933	39.683	124.943	393.347	1 069.715

三峡地区接纳工业废水及处理情况（四）

（2012）

单位：吨

流域	地区名称	工业废水中污染物排放量				
		化学需氧量	氨氮	石油类	挥发酚	氰化物
总　计		229 263.4	12 289.3	1 270.9	11.6	3.4
库　区	湖　北	5 641.4	237.4	2.7	0.0	0.0
	重　庆	27 411.4	1 750.6	146.7	0.6	0.0
	合　计	33 052.7	1 988.0	149.5	0.6	0.0
影响区	湖　北	6 318.5	1 152.4	2.3	0.0	0.0
	重　庆	20 746.6	803.6	202.6	8.9	0.9
	四　川	36 592.5	1 870.0	109.3	0.1	0.7
	贵　州	10 152.4	306.0	0.0	0.0	0.0
	合　计	73 809.9	4 132.0	314.2	9.1	1.6
上游区	重　庆	787.3	503.5	3.1	0.0	0.6
	四　川	81 785.4	3 721.2	309.5	1.2	0.8
	贵　州	25 589.4	1 402.0	233.9	0.4	0.1
	云　南	14 238.7	542.7	260.7	0.3	0.3
	合　计	122 400.8	6 169.3	807.2	1.9	1.8

三峡地区接纳工业废水及处理情况（五）

（2012）

单位：吨

流域	地区名称	工业废水中污染物排放量					
		汞	镉	六价铬	总铬	铅	砷
总　计		0.087	1.368	1.064	4.077	7.558	8.856
库　区	湖　北	0.000	0.000	0.001	0.051	0.000	0.000
	重　庆	0.000	0.000	0.093	0.374	0.070	1.323
	合　计	0.000	0.000	0.095	0.426	0.070	1.323
影响区	湖　北	0.000	0.000	0.000	0.000	0.000	0.000
	重　庆	0.000	0.000	0.045	0.050	0.003	0.035
	四　川	0.007	0.000	0.224	0.258	0.001	0.134
	贵　州	0.000	0.000	0.000	0.000	0.000	0.000
	合　计	0.007	0.000	0.269	0.308	0.004	0.168
上游区	重　庆	0.000	0.000	0.066	0.082	0.000	0.000
	四　川	0.061	0.127	0.616	3.220	1.482	2.408
	贵　州	0.019	0.103	0.018	0.038	0.112	0.109
	云　南	0.001	1.137	0.001	0.003	5.891	4.847
	合　计	0.081	1.367	0.700	3.343	7.485	7.364

三峡地区工业污染防治投资情况

（2012）

流域	地区名称	工业废水治理施工项目数/个	工业废水治理竣工项目数/个	工业废水治理项目完成投资/万元	废水治理竣工项目新增处理能力/（万吨/日）
总　计		192	306	142 371.6	22.5
库区	湖　北	2	3	5 183.0	0.8
	重　庆	42	51	9 812.8	1.6
	合　计	44	54	14 995.8	2.3
影响区	湖　北	1	1	30.0	0.0
	重　庆	3	19	6 597.0	2.0
	四　川	41	45	24 836.3	2.0
	贵　州	3	58	6 921.1	0.5
	合　计	48	123	38 384.4	4.5
上游区	重　庆	1	3	530.8	0.2
	四　川	53	70	28 810.1	8.5
	贵　州	20	37	5 955.4	5.4
	云　南	26	19	53 695.2	1.7
	合　计	100	129	88 991.4	15.8

三峡地区农业污染排放情况

（2012）

单位：万吨

流域	地区名称	农业污染物排放/流失总量			
		化学需氧量	氨氮	总氮	总磷
总　计		73.9	8.1	34.0	3.8
库区	湖　北	1.0	0.1	0.5	0.1
	重　庆	6.0	0.6	2.7	0.3
	合　计	7.0	0.7	3.2	0.4
影响区	湖　北	0.6	0.1	0.4	0.0
	重　庆	5.6	0.6	2.4	0.3
	四　川	10.0	1.1	4.1	0.4
	贵　州	0.2	0.0	0.1	0.0
	合　计	16.5	1.9	7.0	0.8
上游区	重　庆	0.7	0.1	0.3	0.0
	四　川	44.0	4.6	18.1	2.1
	贵　州	3.3	0.4	2.6	0.2
	云　南	2.4	0.4	2.7	0.2
	合　计	50.4	5.5	23.7	2.6

三峡地区城镇生活污染排放及处理情况（一）

（2012）

流域	地区名称	城镇人口/万人	生活用水总量/万吨	污水处理厂数/座	污水处理厂设计处理能力/（万吨/日）	生活污水实际处理量/万吨
总　计		6 958	502 019	533	1 058	312 674
库区	湖　北	62	3 574	19	14	3 221
	重　庆	1 145	82 548	84	225	64 144
	合　计	1 207	86 122	103	239	67 365
影响区	湖　北	102	6 012	8	29	5 153
	重　庆	459	32 457	45	47	14 888
	四　川	666	39 886	25	53	17 276
	贵　州	43	2 204	4	6	2 141
	合　计	1 271	80 559	82	135	39 458
上游区	重　庆	766	56 801	52	102	32 020
	四　川	699	65 993	34	125	44 681
	贵　州	37	2 490	2	2	679
	云　南	2 977	210 054	260	454	128 472
	合　计	4 480	335 338	348	683	205 851

三峡地区城镇生活污染排放及处理情况（二）

（2012）

流域	地 区 名 称	城镇生活污水排放量/万吨	城镇生活化学需氧量产生量/吨	城镇生活氨氮产生量/吨	城镇生活化学需氧量排放量/吨	城镇生活氨氮排放量/吨
	总 计	424 381	1 713 592	211 324	1 043 361	145 670
库区	湖 北	3 698	14 096	1 762	6 879	1 214
	重 庆	69 388	284 241	36 366	135 527	23 631
	合 计	73 087	298 337	38 128	142 406	24 844
影响区	湖 北	6 187	23 175	2 896	11 473	2 236
	重 庆	27 838	114 035	14 590	83 576	11 932
	四 川	33 513	163 998	20 031	124 760	15 870
	贵 州	2 025	8 943	1 156	6 268	757
	合 计	69 563	310 152	38 673	226 077	30 795
上游区	重 庆	44 034	172 531	21 996	118 995	16 070
	四 川	55 497	177 846	20 621	67 292	11 763
	贵 州	2 271	9 303	1 190	8 559	1 081
	云 南	179 929	745 425	90 717	480 033	61 116
	合 计	281 732	1 105 103	134 523	674 878	90 031

入海陆源接纳工业废水及处理情况（一）

（2012）

流域	地 区 名 称	工业废水排放量/万吨	直接排入环境的	排入污水处理厂的	工业废水处理量/万吨	废水治理设施数/套	废水治理设施治理能力/（万吨/日）	废水治理设施运行费用/万元
	总 计	385 682	264 111	121 571	631 437	14 702	3 187	1 103 079.9
渤 海	天 津	9 532	2 258	7 274	10 204	292	38	88 573.2
	河 北	7 059	4 108	2 951	93 394	238	410	30 963.9
	辽 宁	32 499	30 997	1 502	33 804	410	131	35 082.9
	山 东	29 316	18 303	11 013	57 478	596	463	101 950.2
	合 计	78 407	55 666	22 740	194 880	1 536	1 041	256 570.2
黄 海	辽 宁	4 006	3 503	503	3 149	231	33	5 046.6
	江 苏	24 807	19 719	5 088	34 300	993	163	82 441.4
	山 东	15 672	5 378	10 293	57 818	740	243	40 454.0
	合 计	44 485	28 601	15 884	95 267	1 964	438	127 942.0
东 海	上 海	31 784	18 539	13 245	60 136	928	263	114 682.3
	浙 江	73 053	28 399	44 654	74 246	3 637	404	249 882.6
	福 建	70 517	59 186	11 331	77 081	1 524	279	97 649.5
	广 东	1 736	1 704	31	1 186	112	18	4 500.9
	合 计	177 089	107 828	69 261	212 648	6 201	964	466 715.3
南 海	广 东	75 935	63 561	12 374	95 684	4 548	621	228 187.8
	广 西	5 470	5 068	402	27 061	155	98	16 662.0
	海 南	4 296	3 386	910	5 896	298	26	7 002.7
	合 计	85 701	72 015	13 686	128 642	5 001	744	251 852.5

入海陆源接纳工业废水及处理情况（二）

（2012）

单位：吨

流域	地区名称	工业废水中污染物产生量				
		化学需氧量	氨氮	石油类	挥发酚	氰化物
总 计		3 728 838.2	161 969.3	70 373.3	3 296.8	1 271.5
渤	天 津	54 471.2	3 441.7	652.3	0.2	0.0
	河 北	148 007.8	2 375.8	3 179.4	1 217.9	217.8
	辽 宁	118 651.3	26 129.9	3 631.6	258.6	0.2
海	山 东	362 993.1	10 696.3	21 781.3	104.9	4.5
	合 计	684 123.4	42 643.7	29 244.7	1 581.5	222.5
黄	辽 宁	51 914.9	2 336.3	69.2	0.0	0.5
	江 苏	448 560.4	8 543.1	1 651.9	165.8	43.3
	山 东	134 924.7	5 772.3	1 558.3	8.2	27.0
海	合 计	635 400.0	16 651.7	3 279.3	174.0	70.8
东	上 海	204 890.2	8 014.1	7 344.9	857.0	195.2
	浙 江	826 369.2	55 720.8	19 355.3	80.5	342.8
	福 建	540 773.7	20 535.9	8 750.7	203.8	103.4
海	广 东	8 699.5	164.2	0.0	0.0	0.0
	合 计	1 580 732.5	84 435.0	35 450.9	1 141.3	641.5
南	广 东	668 788.0	15 041.2	1 437.3	8.3	334.2
	广 西	131 816.9	2 027.3	951.2	391.7	2.6
	海 南	27 977.5	1 170.4	9.9	0.0	0.0
海	合 计	828 582.4	18 238.9	2 398.4	400.0	336.7

入海陆源接纳工业废水及处理情况（三）

（2012）

单位：吨

流域	地区名称	工业废水中污染物产生量					
		汞	镉	六价铬	总铬	铅	砷
总 计		0.993	59.638	1 372.781	3 581.136	81.041	252.389
渤	天 津	0.039	0.015	6.786	10.930	15.108	0.948
	河 北	0.000	0.000	1.594	8.720	0.000	0.000
	辽 宁	0.080	34.964	1.900	2.424	32.798	22.634
海	山 东	0.003	0.019	6.012	29.899	1.336	104.653
	合 计	0.121	34.998	16.292	51.972	49.242	128.236
黄	辽 宁	0.000	0.009	2.251	2.253	0.081	0.036
	江 苏	0.020	0.013	59.664	60.064	6.006	0.352
	山 东	0.727	23.749	20.079	31.021	17.819	104.421
海	合 计	0.747	23.771	81.994	93.338	23.906	104.809
东	上 海	0.003	0.026	38.895	50.636	0.694	0.671
	浙 江	0.000	0.192	778.471	1 712.910	0.501	14.903
	福 建	0.004	0.155	396.047	746.231	3.790	0.598
海	广 东	0.000	0.000	0.000	0.000	0.307	0.000
	合 计	0.007	0.372	1 213.413	2 509.778	5.292	16.173
南	广 东	0.114	0.482	61.035	896.989	2.557	3.149
	广 西	0.003	0.001	0.047	12.163	0.041	0.008
	海 南	0.000	0.014		16.896	0.003	0.014
海	合 计	0.117	0.497	61.083	926.048	2.601	3.171

入海陆源接纳工业废水及处理情况（四）

（2012）

单位：吨

流域	地 区 名 称	工业废水中污染物排放量				
		化学需氧量	氨氮	石油类	挥发酚	氰化物
	总　计	424 878.2	30 703.7	2 492.2	62.8	13.0
渤 海	天　津	15 759.8	2 038.4	32.6	0.1	0.0
	河　北	14 056.8	923.8	65.4	0.3	0.1
	辽　宁	27 052.7	2 873.5	216.2	12.6	0.0
	山　东	24 023.1	2 188.4	541.2	7.1	0.6
	合　计	80 892.4	8 024.1	855.3	20.1	0.8
黄 海	辽　宁	14 995.2	1 671.3	30.8	0.0	0.1
	江　苏	39 897.6	3 315.3	222.4	17.9	1.1
	山　东	11 736.9	936.1	57.7	0.3	0.2
	合　计	66 629.7	5 922.7	310.8	18.2	1.5
东 海	上　海	20 116.2	1 322.4	612.4	1.7	1.8
	浙　江	87 010.7	5 941.8	273.6	21.2	3.2
	福　建	44 204.9	2 876.1	235.9	0.1	1.9
	广　东	1 833.8	86.9	0.0	0.0	0.0
	合　计	153 165.5	10 227.1	1 122.0	23.0	6.8
南 海	广　东	101 282.7	5 715.7	140.1	1.0	3.7
	广　西	16 748.9	458.0	60.6	0.5	0.3
	海　南	6 159.0	356.2	3.3	0.0	0.0
	合　计	124 190.6	6 529.8	204.1	1.5	4.0

入海陆源接纳工业废水及处理情况（五）

（2012）

单位：吨

流域	地 区 名 称	工业废水中污染物排放量					
		汞	镉	六价铬	总铬	铅	砷
	总　计	0.016	0.279	11.373	33.121	2.781	0.727
渤 海	天　津	0.004	0.002	0.097	0.326	0.997	0.019
	河　北	0.000	0.000	0.004	0.071	0.000	0.000
	辽　宁	0.000	0.000	0.051	0.091	0.169	0.041
	山　东	0.000	0.001	0.040	3.318	0.004	0.001
	合　计	0.004	0.003	0.192	3.806	1.171	0.060
黄 海	辽　宁	0.000	0.000	0.067	0.069	0.000	0.000
	江　苏	0.001	0.005	1.936	2.211	0.509	0.349
	山　东	0.000	0.000	0.189	0.454	0.025	0.036
	合　计	0.001	0.005	2.191	2.733	0.534	0.386
东 海	上　海	0.000	0.000	0.569	1.291	0.164	0.064
	浙　江	0.000	0.149	4.398	10.387	0.055	0.011
	福　建	0.001	0.018	2.485	11.395	0.314	0.175
	广　东	0.000	0.000	0.000	0.000	0.022	0.000
	合　计	0.001	0.168	7.452	23.073	0.555	0.250
南 海	广　东	0.008	0.103	1.532	3.236	0.518	0.017
	广　西	0.001	0.000	0.006	0.142	0.000	0.001
	海　南	0.000	0.001	0.000	0.131	0.002	0.013
	合　计	0.009	0.103	1.538	3.509	0.520	0.032

入海陆源工业废水污染防治投资情况

（2012）

流域	地区名称	工业废水治理施工项目数/个	工业废水治理竣工项目数/个	工业废水治理项目完成投资/万元	废水治理竣工项目新增处理能力/（万吨/日）
	总 计	310	307	278 772.2	43.9
渤	天 津	41	9	8 319.9	0.2
	河 北	1	2	3 489.2	0.4
	辽 宁	4	5	6 640.0	0.1
海	山 东	31	35	71 751.7	5.5
	合 计	77	51	90 200.8	6.1
黄	辽 宁	0	0	1 740.0	0.4
	江 苏	22	33	13 313.5	4.3
	山 东	7	9	6 775.0	1.4
海	合 计	29	42	21 828.5	6.2
东	上 海	22	21	2 314.3	0.2
	浙 江	55	65	47 988.2	4.4
	福 建	49	39	78 478.2	8.1
海	广 东	0	3	295.0	0.4
	合 计	126	128	129 075.6	13.1
南	广 东	59	60	28 549.2	11.0
	广 西	6	9	4 967.1	0.4
	海 南	13	17	4 150.9	7.1
海	合 计	78	86	37 667.2	18.5

入海陆源农业污染排放情况

（2012）

单位：万吨

流域	地区名称	农业污染物排放/流失总量			
		化学需氧量	氨氮	总氮	总磷
	总 计	108.1	9.5	38.9	5.0
渤	天 津	0.6	0.0	0.2	0.0
	河 北	7.7	0.4	3.4	0.5
	辽 宁	10.8	0.6	2.6	0.4
海	山 东	13.6	0.7	4.7	0.6
	合 计	32.7	1.8	10.9	1.5
黄	辽 宁	7.0	0.3	1.4	0.2
	江 苏	11.5	1.2	5.0	0.5
	山 东	13.6	0.7	3.3	0.5
海	合 计	32.1	2.2	9.7	1.2
东	上 海	2.8	0.3	1.2	0.2
	浙 江	6.7	1.0	3.1	0.3
	福 建	9.4	1.8	4.3	0.6
海	广 东	1.7	0.2	0.6	0.1
	合 计	20.6	3.3	9.2	1.1
南	广 东	13.8	1.5	5.3	0.6
	广 西	1.8	0.1	0.7	0.1
	海 南	7.0	0.6	3.0	0.4
海	合 计	22.6	2.2	9.0	1.1

入海陆源城镇生活污染排放及处理情况（一）

（2012）

流域	地区名称		城镇人口/万人	生活用水总量/万吨	污水处理厂数/座	污水处理厂设计处理能力/（万吨/日）	生活污水实际处理量/万吨
	总	计	9 081	725 205	666	3 011	504 785
渤	天	津	260	12 250	21	58	7 357
	河	北	221	16 665	15	67	13 118
	辽	宁	430	29 205	17	77	16 810
海	山	东	378	25 090	42	106	19 226
	合	计	1 288	83 211	95	308	56 512
黄	辽	宁	280	18 502	15	64	20 262
	江	苏	564	41 489	97	89	18 080
	山	东	915	66 243	54	206	52 389
海	合	计	1 760	126 233	166	359	90 731
东	上	海	846	76 288	24	541	58 547
	浙	江	1 366	112 055	90	489	74 688
	福	建	1 216	93 870	76	265	58 603
海	广	东	106	6 149	6	9	2 777
	合	计	3 533	288 361	196	1 304	194 615
南	广	东	2 142	201 537	158	874	149 936
	广	西	158	12 347	7	40	7 664
	海	南	200	13 516	44	127	5 327
海	合	计	2 500	227 400	209	1 040	162 927

入海陆源城镇生活污染排放及处理情况（二）

（2012）

流域	地区名称		城镇生活污水排放量/万吨	城镇生活化学需氧量产生量/吨	城镇生活氨氮产生量/吨	城镇生活化学需氧量排放量/吨	城镇生活氨氮排放量/吨
	总	计	638 681	2 289 725	297 178	1 060 840	181 152
渤	天	津	10 413	63 935	8 718	36 373	6 601
	河	北	15 356	42 786	5 670	8 280	3 100
	辽	宁	24 948	99 932	14 034	51 567	8 277
海	山	东	23 037	89 145	11 932	27 086	5 688
	合	计	73 754	295 799	40 354	123 306	23 666
黄	辽	宁	15 772	65 360	9 407	31 536	4 587
	江	苏	36 661	141 874	18 481	86 501	14 208
	山	东	56 882	216 771	29 112	48 205	11 796
海	合	计	109 314	424 006	56 999	166 241	30 591
东	上	海	68 660	234 586	29 941	70 638	16 463
	浙	江	95 366	366 891	47 372	167 241	30 396
	福	建	80 631	297 473	38 621	177 542	28 949
海	广	东	5 036	23 885	3 084	20 279	2 801
	合	计	249 693	922 835	119 018	435 699	78 610
南	广	东	184 663	549 896	69 373	253 109	38 938
	广	西	10 223	38 058	4 622	27 432	3 067
	海	南	11 034	59 132	6 812	55 053	6 279
海	合	计	205 920	647 085	80 807	335 594	48 284

12

大气污染防治重点区域废气排放统计

ANNUAL STATISTIC REPORT ON ENVIRONMENT IN CHINA
2012

大气防治重点区域工业废气排放及处理情况（一）

（2012）

区域	地区名称	工业废气排放量/亿米³	废气污染物产生量/吨			废气污染物排放量/吨		
			二氧化硫	氮氧化物	烟（粉）尘	二氧化硫	氮氧化物	烟（粉）尘
总计		341 532	27 829 664	9 119 438	372 929 202	8 991 606	8 276 196	4 169 023
京津冀	北京	3 264	159 493	97 827	4 219 833	59 330	85 331	30 844
	天津	9 032	623 909	297 400	7 758 866	215 481	275 553	59 036
	河北	67 647	3 232 634	1 276 454	62 118 989	1 238 737	1 194 755	1 055 732
	合计	79 943	4 016 036	1 671 681	74 097 688	1 513 548	1 555 639	1 145 613
长三角	上海	13 361	535 339	325 286	7 254 212	193 405	284 764	63 732
	江苏	48 623	3 720 115	1 338 194	39 707 107	959 210	1 133 633	395 978
	浙江	23 967	2 167 547	721 919	26 481 356	610 882	634 950	233 228
	合计	85 952	6 423 001	2 385 399	73 442 675	1 763 497	2 053 347	692 938
珠三角	广东	16 138	1 337 304	572 252	12 487 392	456 170	481 951	147 933
	合计	16 138	1 337 304	572 252	12 487 392	456 170	481 951	147 933
辽宁中部城市群	辽宁	20 771	921 463	430 357	17 082 158	488 220	417 827	396 580
	合计	20 771	921 463	430 357	17 082 158	488 220	417 827	396 580
山东城市群	山东	45 420	5 004 973	1 297 358	62 888 056	1 543 766	1 238 173	525 896
	合计	45 420	5 004 973	1 297 358	62 888 056	1 543 766	1 238 173	525 896
武汉及其周边城市群	湖北	13 619	1 265 081	295 434	12 367 946	321 176	285 534	132 559
	合计	13 619	1 265 081	295 434	12 367 946	321 176	285 534	132 559
长株潭城市群	湖南	4 213	458 473	121 091	9 819 875	104 162	92 750	44 189
	合计	4 213	458 473	121 091	9 819 875	104 162	92 750	44 189
成渝城市群	重庆	8 360	1 285 476	272 968	17 468 228	509 788	272 094	166 142
	四川	16 894	1 565 961	401 435	23 377 219	625 951	374 176	182 872
	合计	25 254	2 851 437	674 404	40 845 447	1 135 740	646 270	349 014
海峡西岸城市群	福建	14 739	976 728	410 160	16 997 420	352 389	360 879	233 989
	合计	14 739	976 728	410 160	16 997 420	352 389	360 879	233 989
山西中北部城市群	山西	14 143	1 348 937	375 217	26 471 771	396 307	357 322	232 497
	合计	14 143	1 348 937	375 217	26 471 771	396 307	357 322	232 497
陕西关中城市群	陕西	8 854	1 618 321	472 475	11 173 341	462 655	399 434	125 002
	合计	8 854	1 618 321	472 475	11 173 341	462 655	399 434	125 002
甘宁城市群	甘肃	5 016	900 480	162 938	6 159 829	175 596	150 606	45 753
	宁夏	2 063	362 115	89 294	4 286 138	105 743	75 090	26 348
	合计	7 080	1 262 594	252 232	10 445 967	281 339	225 697	72 101
新疆乌鲁木齐城市群	新疆	5 405	345 316	161 378	4 809 467	172 637	161 373	70 712
	合计	5 405	345 316	161 378	4 809 467	172 637	161 373	70 712

大气防治重点区域工业废气排放及处理情况（二）

（2012）

区域	地区名称	废气中重金属产生量/千克					
		汞	镉	六价铬	总铬	铅	砷
总计		283	23 283	7 372	9 041	506 515	101 284
京津冀	北京	0	0	0	0	0	1
	天津	156	168	120	156	19 141	10
	河北	1	21 506	3 053	3 053	94 406	45 454
	合计	157	21 673	3 172	3 209	113 547	45 464
长三角	上海	41	0	0	269	261	0
	江苏	10	9	267	315	237 061	103
	浙江	69	55	342	1 253	32 653	48
	合计	121	65	610	1 837	269 974	151
珠三角	广东	0	1	0	1	2	0
	合计	0	1	0	1	2	0
辽宁中部城市群	辽宁	0	1	626	626	236	1
	合计	0	1	626	626	236	1
山东城市群	山东	1	0	2	4	48 549	2
	合计	1	0	2	4	48 549	2
武汉及其周边城市群	湖北	0	0	65	65	49 825	0
	合计	0	0	65	65	49 825	0
长株潭城市群	湖南	0	0	0	0	7 021	0
	合计	0	0	0	0	7 021	0
成渝城市群	重庆	0	0	2 301	2 481	3 763	2
	四川	3	41	95	246	2 636	53 970
	合计	3	41	2 396	2 727	6 399	53 972
海峡西岸城市群	福建	1	227	1	72	8 398	1 344
	合计	1	227	1	72	8 398	1 344
山西中北部城市群	山西	0	0	0	0	0	0
	合计	0	0	0	0	0	0
陕西关中城市群	陕西	1	1 276	0	0	2 565	348
	合计	1	1 276	0	0	2 565	348
甘宁城市群	甘肃	0	0	500	500	0	0
	宁夏	0	0	0	0	0	0
	合计	0	0	500	500	0	0
新疆乌鲁木齐城市群	新疆	0	0	0	0	0	0
	合计	0	0	0	0	0	0

大气防治重点区域工业废气排放及处理情况（三）

（2012）

区域	地区名称	废气中重金属排放量/千克					
		汞	镉	六价铬	总铬	铅	砷
总计		64	349	247	803	20 050	1 426
京津冀	北京	0	0	0	0	0	1
	天津	2	10	3	39	1 428	1
	河北	0	0	63	64	4 988	0
	合计	2	10	66	102	6 416	2
长三角	上海	1	0	0	207	84	0
	江苏	1	9	10	38	3 034	0
	浙江	57	21	43	96	3 390	5
	合计	59	30	52	340	6 507	5
珠三角	广东	0	0	0	0	2	0
	合计	0	0	0	0	2	0
辽宁中部城市群	辽宁	0	0	16	16	56	0
	合计	0	0	16	16	56	0
山东城市群	山东	1	0	1	2	666	2
	合计	1	0	1	2	666	2
武汉及其周边城市群	湖北	0	0	65	65	1 375	0
	合计	0	0	65	65	1 375	0
长株潭城市群	湖南	0	0	0	0	1 534	0
	合计	0	0	0	0	1 534	0
成渝城市群	重庆	0	0	41	49	107	2
	四川	0	4	0	151	134	57
	合计	0	4	41	200	241	59
海峡西岸城市群	福建	0	227	1	72	2 918	1 344
	合计	0	227	1	72	2 918	1 344
山西中北部城市群	山西	0	0	0	0	0	0
	合计	0	0	0	0	0	0
陕西关中城市群	陕西	1	78	0	0	336	14
	合计	1	78	0	0	336	14
甘宁城市群	甘肃	0	0	6	6	0	0
	宁夏	0	0	0	0	0	0
	合计	0	0	6	6	0	0
新疆乌鲁木齐城市群	新疆	0	0	0	0	0	0
	合计	0	0	0	0	0	0

大气防治重点区域工业废气排放及处理情况（四）

（2012）

区域	地区名称	废气治理设施数/套	废气治理设施处理能力/（万米³/时）	废气治理设施运行费用/万元	脱硫治理设施数/套	脱硫治理设施处理能力/（千克/时）	脱硫治理设施运行费用/万元
总计		133 768	973 339	9 291 862	13 880	16 994 610	3 289 878
京津冀	北京	3 004	10 901	87 848	0	382 897	28 095
	天津	3 911	22 382	257 709	1	156 828	90 104
	河北	16 924	182 217	1 304 986	1	2 843 358	369 399
	合计	23 839	215 500	1 650 543	3	3 383 083	487 598
长三角	上海	4 795	32 703	405 020	0	137 711	144 314
	江苏	17 641	99 232	2 243 764	2	2 949 787	473 733
	浙江	17 771	159 773	990 296	1	1 473 955	421 215
	合计	40 207	291 708	4 596 895	4	4 561 452	1 039 262
珠三角	广东	13 335	50 394	605 104	1	245 477	297 929
	合计	13 335	50 394	605 104	1	245 477	297 929
辽宁中部城市群	辽宁	7 048	59 909	417 296	0	510 668	98 204
	合计	7 048	59 909	417 296	0	510 668	98 204
山东城市群	山东	15 915	116 317	1 142 869	3	1 991 431	561 381
	合计	15 915	116 317	1 142 869	3	1 991 431	561 381
武汉及其周边城市群	湖北	2 788	21 962	247 000	0	347 229	90 870
	合计	2 788	21 962	247 000	0	347 229	90 870
长株潭城市群	湖南	1 377	5 175	74 762	0	91 305	31 575
	合计	1 377	5 175	74 762	0	91 305	31 575
成渝城市群	重庆	4 319	19 990	202 828	0	165 532	75 735
	四川	6 355	40 151	315 762	0	4 811 181	113 522
	合计	10 674	60 141	518 590	1	4 976 713	189 256
海峡西岸城市群	福建	7 780	85 065	305 612	0	182 166	128 013
	合计	7 780	85 065	305 612	0	182 166	128 013
山西中北部城市群	山西	5 087	19 713	273 673	2	228 166	158 905
	合计	5 087	19 713	273 673	2	228 166	158 905
陕西关中城市群	陕西	3 171	21 890	191 733	0	344 277	110 708
	合计	3 171	21 890	191 733	0	344 277	110 708
甘宁城市群	甘肃	1 368	10 588	90 716	0	54 241	26 750
	宁夏	419	3 968	66 061	0	51 057	43 258
	合计	1 787	14 556	156 777	0	105 299	70 007
新疆乌鲁木齐城市群	新疆	760	11 010	68 823	0	27 346	26 170
	合计	760	11 010	68 823	0	27 346	26 170

大气防治重点区域工业废气排放及处理情况（五）

（2012）

区域	地区名称	脱硝治理设施数/套	脱硝治理设施处理能力/（千克/时）	脱硝治理设施运行费用/万元	除尘治理设施数/套	除尘治理设施处理能力/（千克/时）	除尘治理设施运行费用/万元
总计		595	383 044	481 289	98 027	289 355 240	3 837 146
京津冀	北京	51	5 686	19 977	2 176	13 183 458	29 421
	天津	9	3 407	10 023	2 378	1 885 382	149 593
	河北	45	35 062	27 707	14 995	48 645 351	874 506
	合计	105	44 155	57 707	19 549	63 714 191	1 053 520
长三角	上海	37	7 678	28 390	2 698	1 778 717	149 908
	江苏	99	116 814	111 155	11 153	43 536 301	414 173
	浙江	61	16 387	83 561	12 036	46 251 165	416 739
	合计	197	140 880	223 106	25 887	91 566 183	980 821
珠三角	广东	101	22 787	79 456	6 422	2 354 048	142 032
	合计	101	22 787	79 456	6 422	2 354 048	142 032
辽宁中部城市群	辽宁	5	3 520	2 104	6 424	17 251 443	291 475
	合计	5	3 520	2 104	6 424	17 251 443	291 475
山东城市群	山东	39	24 006	26 911	12 316	34 045 115	510 105
	合计	39	24 006	26 911	12 316	34 045 115	510 105
武汉及其周边城市群	湖北	19	13 722	5 654	2 496	6 001 229	150 210
	合计	19	13 722	5 654	2 496	6 001 229	150 210
长株潭城市群	湖南	5	2 522	4 125	1 206	4 732 874	31 523
	合计	5	2 522	4 125	1 206	4 732 874	31 523
成渝城市群	重庆	0	0	0	3 338	17 820 269	117 884
	四川	32	81 208	5 297	5 438	25 202 361	176 591
	合计	32	81 208	5 297	8 776	43 022 630	294 474
海峡西岸城市群	福建	45	25 648	27 502	6 812	6 510 672	134 352
	合计	45	25 648	27 502	6 812	6 510 672	134 352
山西中北部城市群	山西	21	5 995	14 881	3 005	7 036 586	96 567
	合计	21	5 995	14 881	3 005	7 036 586	96 567
陕西关中城市群	陕西	16	12 792	14 777	2 704	9 572 552	58 845
	合计	16	12 792	14 777	2 704	9 572 552	58 845
甘宁城市群	甘肃	6	2 927	6 226	1 322	1 966 841	41 379
	宁夏	4	2 883	13 543	384	851 694	9 261
	合计	10	5 810	19 769	1 706	2 818 535	50 640
新疆乌鲁木齐城市群	新疆	0	0	0	724	729 184	42 653
	合计	0	0	0	724	729 184	42 653

大气防治重点区域工业废气排放及处理情况（六）

（2012）

区域	地区名称	工业锅炉数/台				工业锅炉蒸吨数/蒸吨				工业窑炉数/座
		35蒸吨及以上	20(含)~35蒸吨之间	10(含)~20蒸吨之间	10蒸吨以下	35蒸吨及以上	20(含)~35蒸吨之间	10(含)~20蒸吨之间	10蒸吨以下	
总计		7 546	3 492	7 595	39 974	2 754 958	71 517	82 345	127 044	46 831
京津冀	北京	308	268	342	842	30 716	5 575	4 065	2 922	317
	天津	419	307	345	1 131	55 821	6 208	3 683	3 610	702
	河北	753	336	858	4 446	1 382 688	6 985	9 097	15 196	6 377
	合计	1 480	911	1 545	6 419	1 469 225	18 768	16 844	21 728	7 396
长三角	上海	160	70	160	1 732	61 017	1 546	1 721	5 631	1 285
	江苏	943	282	668	5 047	213 860	4 873	6 319	13 929	4 508
	浙江	658	243	891	6 205	143 622	4 691	9 029	19 853	4 789
	合计	1 761	595	1 719	12 984	418 499	11 109	17 069	39 412	10 582
珠三角	广东	364	182	779	4 031	107 434	4 033	8 931	13 823	2 357
	合计	364	182	779	4 031	107 434	4 033	8 931	13 823	2 357
辽宁中部城市群	辽宁	678	380	779	2 055	87 361	8 111	8 787	7 240	4 778
	合计	678	380	779	2 055	87 361	8 111	8 787	7 240	4 778
山东城市群	山东	1 792	482	837	3 416	288 593	10 083	9 282	11 883	4 916
	合计	1 792	482	837	3 416	288 593	10 083	9 282	11 883	4 916
武汉及其周边城市群	湖北	114	47	162	858	39 808	1 014	1 759	2 669	1 144
	合计	114	47	162	858	39 808	1 014	1 759	2 669	1 144
长株潭城市群	湖南	37	14	50	512	13 976	304	600	1 507	887
	合计	37	14	50	512	13 976	304	600	1 507	887
成渝城市群	重庆	112	53	125	1 166	27 637	1 078	1 367	2 708	1 947
	四川	187	160	303	2 702	40 892	3 178	3 354	6 486	5 686
	合计	299	213	428	3 868	68 529	4 256	4 720	9 194	7 633
海峡西岸城市群	福建	153	122	417	2 747	73 999	2 685	5 004	8 983	2 687
	合计	153	122	417	2 747	73 999	2 685	5 004	8 983	2 687
山西中北部城市群	山西	266	167	286	1 638	77 694	3 368	3 060	5 560	1 601
	合计	266	167	286	1 638	77 694	3 368	3 060	5 560	1 601
陕西关中城市群	陕西	221	165	306	898	38 075	3 127	3 050	2 829	1 534
	合计	221	165	306	898	38 075	3 127	3 050	2 829	1 534
甘宁城市群	甘肃	76	84	174	237	17 528	1 860	1 975	1 039	974
	宁夏	84	66	69	119	19 460	1 400	764	598	106
	合计	160	150	243	356	36 988	3 260	2 739	1 637	1 080
新疆乌鲁木齐城市群	新疆	221	64	44	192	34 778	1 400	500	580	236
	合计	221	64	44	192	34 778	1 400	500	580	236

大气防治重点区域工业废气污染防治投资情况（一）

（2012）

区域	地区名称	工业废气治理施工项目数/个			工业废气治理竣工项目数/个		
		脱硫治理项目	脱硝治理项目	其他废气治理项目	脱硫治理项目	脱硝治理项目	其他废气治理项目
总计		338	183	823	334	135	183
京津冀	北京	11	1	15	8	1	1
	天津	9	3	38	5	3	3
	河北	46	16	40	39	12	16
	合计	66	20	93	52	16	20
长三角	上海	10	3	34	10	2	3
	江苏	27	31	108	24	24	31
	浙江	22	26	157	22	17	26
	合计	59	60	299	56	43	60
珠三角	广东	37	12	130	54	11	12
	合计	37	12	130	54	11	12
辽宁中部城市群	辽宁	4	4	14	3	1	4
	合计	4	4	14	3	1	4
山东城市群	山东	87	24	124	79	15	24
	合计	87	24	124	79	15	24
武汉及其周边城市群	湖北	1	4	8	2	3	4
	合计	1	4	8	2	3	4
长株潭城市群	湖南	4	5	11	6	3	5
	合计	4	5	11	6	3	5
成渝城市群	重庆	5	8	23	8	6	8
	四川	6	5	34	7	5	5
	合计	11	13	57	15	11	13
海峡西岸城市群	福建	18	19	49	11	15	19
	合计	18	19	49	11	15	19
山西中北部城市群	山西	22	5	21	26	4	5
	合计	22	5	21	26	4	5
陕西关中城市群	陕西	19	11	10	20	9	11
	合计	19	11	10	20	9	11
甘宁城市群	甘肃	2	2	0	1	1	2
	宁夏	3	0	7	4	2	0
	合计	5	2	7	5	3	2
新疆乌鲁木齐城市群	新疆	5	4	0	5	1	4
	合计	5	4	0	5	1	4

大气防治重点区域工业废气污染防治投资情况（二）

（2012）

区域	地区名称	工业废气治理项目完成投资/万元			废气治理竣工项目新增处理能力/（万米³/时）
		脱硫治理项目	脱硝治理项目	其他废气治理项目	
总计		470 347	593 261	617 128	31 467
京津冀	北京	3 026	210	21 416	266
	天津	21 441	11 965	9 053	721
	河北	105 102	38 907	37 159	3 499
	合计	129 569	51 082	67 628	4 486
长三角	上海	1 091	9 450	44 914	655
	江苏	52 219	122 086	91 433	4 278
	浙江	22 630	82 169	33 623	3 604
	合计	75 941	213 705	169 970	8 537
珠三角	广东	39 641	48 598	71 073	2 776
	合计	39 641	48 598	71 073	2 776
辽宁中部城市群	辽宁	6 340	8 052	10 171	178
	合计	6 340	8 052	10 171	178
山东城市群	山东	136 325	41 345	126 195	8 675
	合计	136 325	41 345	126 195	8 675
武汉及其周边城市群	湖北	7 579	47 689	9 164	1 188
	合计	7 579	47 689	9 164	1 188
长株潭城市群	湖南	3 644	12 840	6 671	797
	合计	3 644	12 840	6 671	797
成渝城市群	重庆	2 942	11 714	1 876	282
	四川	10 894	15 468	11 694	784
	合计	13 835	27 182	13 570	1 066
海峡西岸城市群	福建	11 385	40 853	55 487	2 417
	合计	11 385	40 853	55 487	2 417
山西中北部城市群	山西	14 313	34 028	10 062	436
	合计	14 313	34 028	10 062	436
陕西关中城市群	陕西	6 105	26 257	70 330	512
	合计	6 105	26 257	70 330	512
甘宁城市群	甘肃	3 071	33 290	0	104
	宁夏	5 948	4 680	6 809	294
	合计	9 019	37 970	6 809	398
新疆乌鲁木齐城市群	新疆	16 651	3 660	0	1
	合计	16 651	3 660	0	1

大气防治重点区域城镇生活废气排放情况

（2012）

区域	地区名称	生活煤炭消费量/万吨	生活天然气消费量/万米³	二氧化硫排放量/吨	氮氧化物排放量/吨	烟尘排放量/吨
总计		6 472	2 152 709	802 927	170 422	471 024
京津冀	北京	461	480 152	34 485	11 991	31 868
	天津	231	28 109	8 959	4 447	18 400
	河北	903	75 526	102 366	18 352	65 849
	合计	1 595	583 786	145 810	34 791	116 116
长三角	上海	172	239 963	34 754	19 138	15 454
	江苏	216	128 586	32 468	6 541	18 530
	浙江	123	46 692	14 654	3 256	4 280
	合计	511	415 241	81 877	28 935	38 264
珠三角	广东	57	57 756	8 851	3 945	3 581
	合计	57	57 756	8 851	3 945	3 581
辽宁中部城市群	辽宁	509	18 090	42 428	9 817	32 000
	合计	509	18 090	42 428	9 817	32 000
山东城市群	山东	1 452	105 610	204 847	31 954	119 726
	合计	1 452	105 610	204 847	31 954	119 726
武汉及其周边城市群	湖北	449	14 778	33 594	7 163	26 925
	合计	449	14 778	33 594	7 163	26 925
长株潭城市群	湖南	57	46 178	5 229	1 003	4 945
	合计	57	46 178	5 229	1 003	4 945
成渝城市群	重庆	147	194 906	54 984	4 487	9 139
	四川	377	385 693	55 930	8 465	8 358
	合计	524	580 599	110 914	12 953	17 497
海峡西岸城市群	福建	115	9 441	18 851	2 382	4 941
	合计	115	9 441	18 851	2 382	4 941
山西中北部城市群	山西	586	53 490	57 954	12 936	55 293
	合计	586	53 490	57 954	12 936	55 293
陕西关中城市群	陕西	376	98 369	59 189	17 553	34 863
	合计	376	98 369	59 189	17 553	34 863
甘宁城市群	甘肃	102	32 525	13 895	3 092	3 608
	宁夏	33	69 758	5 613	1 218	2 972
	合计	135	102 283	19 508	4 310	6 580
新疆乌鲁木齐城市群	新疆	105	67 090	13 876	2 682	10 294
	合计	105	67 090	13 876	2 682	10 294

大气防治重点区域机动车废气排放情况

（2012）

区域	地区名称	机动车污染物排放总量/吨			
		总颗粒物	氮氧化物	一氧化碳	碳氢化合物
总计		278 119	3 038 665	16 768 023	2 045 230
京津冀	北京	4 074	79 852	781 137	86 414
	天津	6 587	54 052	447 149	50 922
	河北	53 618	547 971	2 563 286	329 277
	合计	64 278	681 874	3 791 572	466 613
长三角	上海	7 906	97 478	435 342	63 319
	江苏	28 496	338 960	1 813 894	219 724
	浙江	16 315	170 400	1 211 245	140 721
	合计	52 716	606 838	3 460 482	423 764
珠三角	广东	33 848	328 442	1 765 044	203 161
	合计	33 848	328 442	1 765 044	203 161
辽宁中部城市群	辽宁	13 658	138 045	663 495	84 964
	合计	13 658	138 045	663 495	84 964
山东城市群	山东	49 603	468 652	2 357 515	300 915
	合计	49 603	468 652	2 357 515	300 915
武汉及其周边城市群	湖北	8 759	95 360	533 494	62 499
	合计	8 759	95 360	533 494	62 499
长株潭城市群	湖南	3 210	42 888	249 778	27 879
	合计	3 210	42 888	249 778	27 879
成渝城市群	重庆	6 984	106 070	616 026	78 345
	四川	11 858	162 169	979 381	115 980
	合计	18 842	268 239	1 595 407	194 325
海峡西岸城市群	福建	9 440	103 909	651 276	76 989
	合计	9 440	103 909	651 276	76 989
山西中北部城市群	山西	7 698	80 832	416 947	49 758
	合计	7 698	80 832	416 947	49 758
陕西关中城市群	陕西	5 489	98 117	619 469	69 670
	合计	5 489	98 117	619 469	69 670
甘宁城市群	甘肃	2 243	34 607	311 468	37 613
	宁夏	3 700	33 717	141 040	18 546
	合计	5 943	68 324	452 508	56 158
新疆乌鲁木齐城市群	新疆	4 632	57 144	211 036	28 535
	合计	4 632	57 144	211 036	28 535

大气防治重点区域生活垃圾处理厂废气排放情况

（2012）

区域	地区名称	焚烧废气中污染物排放量					
		二氧化硫/吨	氮氧化物/吨	烟尘/吨	汞/千克	镉/千克	铅/千克
总计		674.0	1 248.6	356.6	0.0	0.0	0.0
京津冀	北京	34.0	300.7	41.0	0.0	0.0	0.0
	天津	18.5	150.0	8.9	0.0	0.0	0.0
	河北	78.6	0.0	23.4	0.0	0.0	0.0
	合计	131.1	450.7	73.3	0.0	0.0	0.0
长三角	上海	0.0	0.0	0.0	0.0	0.0	0.0
	江苏	162.2	193.3	35.3	0.0	0.0	0.0
	浙江	85.5	132.6	142.2	0.0	0.0	0.0
	合计	247.7	325.9	177.5	0.0	0.0	0.0
珠三角	广东	0.7	6.1	0.0	0.0	0.0	0.0
	合计	0.7	6.1	0.0	0.0	0.0	0.0
辽宁中部城市群	辽宁	0.0	0.0	0.0	0.0	0.0	0.0
	合计	0.0	0.0	0.0	0.0	0.0	0.0
山东城市群	山东	61.8	128.0	24.8	0.0	0.0	0.0
	合计	61.8	128.0	24.8	0.0	0.0	0.0
武汉及其周边城市群	湖北	0.0	0.0	0.0	0.0	0.0	0.0
	合计	0.0	0.0	0.0	0.0	0.0	0.0
长株潭城市群	湖南	0.0	0.0	0.0	0.0	0.0	0.0
	合计	0.0	0.0	0.0	0.0	0.0	0.0
成渝城市群	重庆	0.0	0.0	0.0	0.0	0.0	0.0
	四川	228.7	330.1	79.0	0.0	0.0	0.0
	合计	228.7	330.1	79.0	0.0	0.0	0.0
海峡西岸城市群	福建	4.1	7.7	2.0	0.0	0.0	0.0
	合计	4.1	7.7	2.0	0.0	0.0	0.0
山西中北部城市群	山西	0.0	0.0	0.0	0.0	0.0	0.0
	合计	0.0	0.0	0.0	0.0	0.0	0.0
陕西关中城市群	陕西	0.0	0.0	0.0	0.0	0.0	0.0
	合计	0.0	0.0	0.0	0.0	0.0	0.0
甘宁城市群	甘肃	0.0	0.0	0.0	0.0	0.0	0.0
	宁夏	0.0	0.0	0.0	0.0	0.0	0.0
	合计	0.0	0.0	0.0	0.0	0.0	0.0
新疆乌鲁木齐城市群	新疆	0.0	0.0	0.0	0.0	0.0	0.0
	合计	0.0	0.0	0.0	0.0	0.0	0.0

大气防治重点区域危险（医疗）废物处置厂废气排放情况

（2012）

区域	地区名称	焚烧废气中污染物排放量					
		二氧化硫/吨	氮氧化物/吨	烟尘/吨	汞/千克	镉/千克	铅/千克
总计		700.3	963.7	494.0	7.1	2.9	60.6
京津冀	北京	0.6	20.5	2.3	2.8	0.0	0.1
	天津	62.7	22.6	29.1	0.0	0.0	9.8
	河北	19.6	32.2	58.8	2.5	1.6	5.6
	合计	83.0	75.3	90.2	5.2	1.6	15.5
长三角	上海	59.2	235.5	52.5	0.0	0.4	3.3
	江苏	126.7	281.4	164.4	0.4	0.6	16.3
	浙江	144.5	110.5	59.5	0.1	0.3	2.1
	合计	330.4	627.4	276.5	0.5	1.2	21.7
珠三角	广东	105.2	71.8	21.9	0.0	0.0	0.0
	合计	105.2	71.8	21.9	0.0	0.0	0.0
辽宁中部城市群	辽宁	2.1	5.7	3.8	0.0	0.0	0.1
	合计	2.1	5.7	3.8	0.0	0.0	0.1
山东城市群	山东	131.9	68.7	20.2	0.0	0.0	0.0
	合计	131.9	68.7	20.2	0.0	0.0	0.0
武汉及其周边城市群	湖北	19.8	16.0	12.3	0.0	0.0	0.0
	合计	19.8	16.0	12.3	0.0	0.0	0.0
长株潭城市群	湖南	0.0	1.2	0.1	0.0	0.0	0.0
	合计	0.0	1.2	0.1	0.0	0.0	0.0
成渝城市群	重庆	4.4	19.2	4.7	0.2	0.0	0.2
	四川	6.7	16.6	27.9	0.0	0.0	0.0
	合计	11.1	35.9	32.5	0.2	0.0	0.2
海峡西岸城市群	福建	7.8	32.0	6.3	1.1	0.0	23.0
	合计	7.8	32.0	6.3	1.1	0.0	23.0
山西中北部城市群	山西	1.0	11.8	1.1	0.0	0.0	0.0
	合计	1.0	11.8	1.1	0.0	0.0	0.0
陕西关中城市群	陕西	4.3	12.4	1.5	0.0	0.0	0.0
	合计	4.3	12.4	1.5	0.0	0.0	0.0
甘宁城市群	甘肃	0.3	1.0	0.1	0.0	0.0	0.0
	宁夏	2.7	1.9	0.6	0.0	0.0	0.0
	合计	3.0	2.9	0.7	0.0	0.0	0.0
新疆乌鲁木齐城市群	新疆	0.7	2.6	26.9	0.0	0.0	0.0
	合计	0.7	2.6	26.9	0.0	0.0	0.0

13

环境管理统计

ANNUAL STATISTIC REPORT ON ENVIRONMENT IN CHINA
2012

各地区环保机构情况（一）

（2012）

単位：个

地　区 名　称	环保系统 机构总数	国家级、 省级机 构数	环保行政主管 部门机构数	环境监察 机构数	环境监测 站机构数	辐射监测 机构数	科研机 构数	宣教机 构数	信息机 构数	应急机 构数	其他机 构数
总　计	13 225	439	32	51	47	28	34	31	24	10	182
国家级	45	45	1	12	1	0	3	1	1	1	25
北　京	103	13	1	1	1	0	1	1	1	0	7
天　津	83	15	1	1	1	1	1	1	1	1	7
河　北	779	13	1	2	1	1	1	1	1	0	5
山　西	609	19	1	1	2	1	1	1	1	0	11
内蒙古	405	12	1	3	1	0	1	1	0	0	5
辽　宁	488	20	1	1	1	1	1	1	1	0	13
吉　林	255	11	1	1	1	1	1	1	1	1	3
黑龙江	441	16	1	1	7	1	1	1	1	0	3
上　海	69	8	1	1	1	1	1	1	1	0	1
江　苏	583	20	1	4	2	1	1	1	0	1	9
浙　江	579	12	1	1	2	1	1	1	1	0	4
安　徽	393	11	1	1	1	1	1	1	1	0	4
福　建	848	13	1	1	3	1	1	1	1	1	3
江　西	427	11	1	1	1	1	1	1	1	0	4
山　东	724	12	1	1	1	1	2	1	0	0	5
河　南	786	8	1	1	1	1	1	1	0	0	2
湖　北	493	10	1	1	1	1	1	1	1	1	2
湖　南	495	15	1	1	3	1	1	1	1	1	5
广　东	740	11	1	0	1	1	1	1	1	0	5
广　西	374	13	1	1	2	1	1	1	0	1	5
海　南	183	8	1	1	0	1	1	1	1	0	2
重　庆	207	11	1	1	1	1	0	1	1	0	5
四　川	720	13	1	1	1	1	1	1	1	0	6
贵　州	398	13	1	1	1	0	1	1	0	1	7
云　南	502	10	1	1	1	1	1	1	0	0	4
西　藏	136	8	1	1	1	1	0	0	1	0	3
陕　西	378	15	1	2	1	1	1	1	1	0	7
甘　肃	396	14	1	1	1	1	1	1	1	1	6
青　海	126	10	1	1	2	1	1	1	1	0	2
宁　夏	69	13	1	2	1	1	1	1	0	0	6
新　疆	391	16	1	2	2	1	2	1	1	0	6

各地区环保机构情况（二）

（2012）

单位：个

地 区 名 称	地市级环 保机构数	环保行政主管 部门机构数	环境监察 机构数	环境监测 站机构数	辐射监测 机构数	科研机 构数	宣教机 构数	信息机 构数	应急机 构数	其他机 构数
总　计	2 102	333	344	351	73	190	115	126	43	527
国家级	—	—	—	—	—	—	—	—	—	—
北　京	—	—	—	—	—	—	—	—	—	—
天　津	—	—	—	—	—	—	—	—	—	—
河　北	84	11	17	11	2	11	7	7	0	18
山　西	103	11	12	15	7	11	8	10	4	25
内蒙古	76	12	11	13	2	7	4	1	1	25
辽　宁	133	14	14	14	0	13	12	8	4	54
吉　林	64	9	9	9	3	2	7	3	3	19
黑龙江	89	13	15	16	5	11	6	4	0	19
上　海	—	—	—	—	—	—	—	—	—	—
江　苏	97	13	15	14	4	9	6	6	5	25
浙　江	75	11	12	11	0	7	4	2	1	27
安　徽	88	16	16	15	2	9	3	13	0	14
福　建	63	9	9	10	3	9	5	9	2	7
江　西	73	11	12	11	3	10	6	2	0	18
山　东	128	17	19	18	2	11	7	5	4	45
河　南	112	17	18	18	10	11	5	3	4	26
湖　北	77	13	14	13	3	9	3	3	0	19
湖　南	94	14	16	16	2	11	2	5	0	28
广　东	140	21	11	22	0	16	10	12	2	46
广　西	61	14	14	14	2	8	1	3	2	3
海　南	12	3	2	2	0	2	0	2	0	1
重　庆	—	—	—	—	—	—	—	—	—	—
四　川	91	21	21	25	4	1	2	10	0	7
贵　州	62	9	11	9	3	2	3	3	5	17
云　南	76	16	14	17	2	6	4	1	0	16
西　藏	27	7	7	7	0	0	0	0	0	6
陕　西	64	10	13	10	1	3	6	1	0	20
甘　肃	99	14	14	15	11	6	1	7	5	26
青　海	29	8	8	6	0	1	0	1	1	4
宁　夏	20	5	5	6	0	1	0	1	0	2
新　疆	65	14	15	14	2	3	3	4	0	10

各地区环保机构情况（三）

（2012）

单位：个

地 区 名 称	县级环保 机构数	环保行政主管 部门机构数	环境监察 机构数	环境监测 站机构数	辐射监测 机构数	科研机 构数	宣教机 构数	信息机 构数	应急机 构数	其他机 构数	乡镇环保 机构数
总 计	8 807	2 812	2 497	2 327	42	96	94	85	42	812	1 877
国家级	—	—	—	—	—	—	—	—	—	—	—
北 京	85	16	16	18	4	3	1	0	0	27	5
天 津	66	16	16	16	0	0	0	0	0	18	2
河 北	527	172	148	140	0	8	8	6	1	44	155
山 西	429	119	130	116	7	3	5	10	0	39	58
内蒙古	315	102	95	92	0	0	1	0	1	24	2
辽 宁	334	100	92	80	0	2	2	0	0	58	1
吉 林	173	57	52	51	0	0	5	1	0	7	7
黑龙江	335	114	103	94	0	0	0	0	0	24	1
上 海	61	17	17	17	0	2	2	0	0	6	0
江 苏	364	101	102	92	4	9	3	2	9	42	102
浙 江	324	90	81	72	0	4	3	3	4	67	168
安 徽	282	104	85	65	0	1	0	7	0	20	12
福 建	280	84	82	81	2	2	1	5	4	19	492
江 西	336	100	98	101	0	2	0	1	0	34	7
山 东	508	137	127	120	0	5	9	6	6	98	76
河 南	493	159	130	121	12	14	2	2	7	46	173
湖 北	329	99	93	83	2	12	6	0	0	34	77
湖 南	363	121	105	102	3	4	0	0	1	27	23
广 东	400	117	46	105	0	22	20	19	2	69	189
广 西	291	105	91	78	0	0	0	1	3	13	9
海 南	52	20	18	13	0	0	0	0	0	1	111
重 庆	140	38	38	40	2	1	10	3	3	5	56
四 川	567	181	180	171	0	1	3	10	0	21	49
贵 州	316	88	87	86	1	0	8	7	1	38	7
云 南	381	129	114	120	0	1	3	1	0	13	35
西 藏	101	74	11	15	0	0	0	0	0	1	0
陕 西	299	107	106	85	0	0	0	0	0	1	0
甘 肃	238	86	87	58	5	0	1	0	0	1	45
青 海	87	43	32	9	0	0	0	0	0	3	0
宁 夏	36	17	9	9	0	0	0	0	0	1	0
新 疆	295	99	106	77	0	0	1	1	0	11	15

各地区环保机构情况（四）

（2012）

单位：人

地 区 名 称	环保系统 人员总数	国家级、省 级人员数	环保行政 主管部门 人员数	环境监 察机构 人员数	环境监 测站人 员数	辐射监 测机构 人员数	科研机 构人员 数	宣教机 构人员 数	信息机 构人员 数	应急机 构人员 数	其他机 构人员 数
总 计	205 334	16 993	3 102	1 742	3 347	857	2 988	557	300	132	3 968
国家级	2 839	2 839	357	481	184	0	514	35	34	35	1 199
北 京	2 684	873	132	65	163	0	237	48	15	0	213
天 津	1 888	629	101	56	176	28	145	22	20	6	75
河 北	18 146	401	78	60	81	27	47	16	12	0	80
山 西	11 595	474	93	47	105	21	45	9	10	0	144
内蒙古	5 742	306	66	44	60	0	22	15	0	0	99
辽 宁	9 141	515	73	29	89	22	102	12	19	0	169
吉 林	5 376	365	84	16	94	24	69	18	13	9	38
黑龙江	5 134	418	82	31	167	17	74	11	8	0	28
上 海	2 199	643	103	60	169	31	206	27	15	0	32
江 苏	10 508	725	131	52	167	39	191	14	0	24	107
浙 江	6 845	435	78	38	135	67	74	7	6	0	30
安 徽	5 538	373	86	41	88	20	75	11	12	0	40
福 建	4 402	407	88	40	89	47	43	9	7	17	67
江 西	5 308	340	62	43	62	23	92	10	13	0	35
山 东	14 168	491	93	34	89	40	73	68	0	0	94
河 南	21 951	364	109	46	66	30	40	22	0	0	51
湖 北	7 476	327	84	37	63	15	75	17	7	8	21
湖 南	9 887	1 011	69	23	174	10	130	23	5	10	567
广 东	11 009	615	146	0	136	47	43	17	9	0	217
广 西	3 808	306	54	23	96	38	37	16	0	5	37
海 南	1 555	227	96	22	0	14	77	1	12	0	5
重 庆	3 085	566	136	69	193	29	0	24	27	0	88
四 川	9 227	508	100	33	103	69	140	11	9	0	43
贵 州	3 702	375	73	51	73	0	80	9	0	9	80
云 南	5 425	490	98	55	89	30	150	15	0	0	53
西 藏	833	156	55	14	39	14	0	0	7	0	27
陕 西	5 884	436	99	78	81	44	43	14	16	0	61
甘 肃	4 340	404	102	25	81	18	79	13	9	9	68
青 海	901	200	49	23	73	23	0	7	7	0	18
宁 夏	924	305	56	53	62	32	0	23	0	0	79
新 疆	3 814	469	69	53	100	38	85	13	8	0	103

各地区环保机构情况（五）

（2012）

单位：人

地 区 名 称	地市级环保 机构人员数	环保行政 主管部门 人员数	环境监察机 构人员数	环境监测 站人员数	辐射监测机 构人员数	科研机构 人员数	宣教机构 人员数	信息机构 人员数	应急机构 人员数	其他机构 人员数
总　计	45 203	9 920	9 390	16 015	414	3 328	823	761	270	4 282
国家级	—	—	—	—	—	—	—	—	—	—
北　京	—	—	—	—	—	—	—	—	—	—
天　津	—	—	—	—	—	—	—	—	—	—
河　北	2 711	529	601	829	27	271	85	55	0	314
山　西	1 899	377	412	562	19	168	39	82	61	179
内蒙古	1 537	294	255	542	63	129	28	4	13	209
辽　宁	3 413	508	596	990	0	552	102	69	32	564
吉　林	1 073	287	270	277	18	35	42	11	17	116
黑龙江	1 512	392	316	475	12	119	44	15	0	139
上　海	—	—	—	—	—	—	—	—	—	—
江　苏	2 693	525	665	1 038	30	193	44	26	32	140
浙　江	1 585	299	357	623	0	128	28	18	3	129
安　徽	1 546	340	411	595	9	66	14	59	0	52
福　建	1 175	191	292	415	19	144	33	43	10	28
江　西	1 255	260	308	396	17	111	39	14	0	110
山　东	3 070	729	554	1 010	6	193	70	32	26	450
河　南	3 369	684	823	1 162	71	201	46	28	40	314
湖　北	1 700	326	370	635	13	157	24	23	0	152
湖　南	2 175	437	448	856	7	222	18	31	0	156
广　东	3 724	1 134	416	1 112	0	308	60	57	8	629
广　西	1 375	311	302	629	13	72	11	23	0	14
海　南	273	54	58	91	0	15	0	31	0	24
重　庆	—	—	—	—	—	—	—	—	—	—
四　川	2 259	545	501	1 030	23	66	22	46	0	26
贵　州	908	226	200	340	10	13	21	20	11	67
云　南	1 297	360	227	504	12	70	17	3	0	104
西　藏	287	178	41	53	0	0	0	0	0	15
陕　西	1 448	280	301	618	5	28	29	11	0	176
甘　肃	1 167	272	189	495	37	28	0	35	17	94
青　海	200	69	53	67	0	4	0	7	0	0
宁　夏	355	77	105	158	0	0	0	8	0	7
新　疆	1 197	236	319	513	3	35	7	10	0	74

各地区环保机构情况（六）

（2012）

单位：人

地 区 名 称	县级环 保机构 人员数	环保行 政主管 部门人 员数	环境监 察机构 人员数	环境监 测站人 员数	辐射监 测机构 人员数	科研机 构人员 数	宣教机 构人员 数	信息机 构人员 数	应急机 构人员 数	其他机构 人员数	乡镇环 保机构 人员数
总 计	135 628	40 264	49 949	37 192	169	882	424	368	142	6 238	7 510
国家级	—	—	—	—	—	—	—	—	—	—	—
北 京	1 787	413	273	361	22	18	16	0	0	684	24
天 津	1 236	336	237	484	0	0	0	0	0	179	23
河 北	13 673	5 075	4 616	3 111	0	141	57	65	10	598	1 361
山 西	8 601	1 471	3 615	2 930	46	43	24	83	0	389	621
内蒙古	3 872	1 273	1 464	980	0	0	2	0	1	152	27
辽 宁	5 202	1 223	2 146	1 258	0	15	5	0	0	555	11
吉 林	3 913	744	2 014	1 016	0	0	24	4	0	111	25
黑龙江	3 204	846	1 409	828	0	0	0	0	0	121	0
上 海	1 556	265	467	741	0	8	13	0	0	62	0
江 苏	6 920	2 008	2 337	2 311	13	62	7	1	19	162	170
浙 江	4 345	1 141	1 501	1 395	0	20	2	6	12	268	480
安 徽	3 581	990	1 577	842	0	0	0	20	0	152	38
福 建	2 719	649	1 079	901	5	0	2	35	5	43	101
江 西	3 683	1 328	1 235	981	0	20	0	3	0	116	30
山 东	9 786	3 131	2 904	2 732	0	34	99	33	21	832	821
河 南	16 810	4 045	7 885	4 215	53	152	16	9	39	396	1 408
湖 北	5 053	1 635	1 852	1 283	2	89	33	0	0	159	396
湖 南	6 548	2 069	2 498	1 792	8	21	0	0	6	154	153
广 东	5 674	2 466	601	1 704	0	165	54	62	10	612	996
广 西	2 112	798	656	614	0	0	0	2	9	33	15
海 南	689	303	250	135	0	0	0	0	0	1	366
重 庆	2 370	617	823	840	3	0	34	14	10	29	149
四 川	6 403	1 900	2 262	1 941	0	91	5	22	0	182	57
贵 州	2 385	842	845	571	3	0	10	6	0	108	34
云 南	3 540	1 569	884	1 015	0	3	12	0	0	57	98
西 藏	390	339	25	11	0	0	0	0	0	15	0
陕 西	4 000	866	2 116	1 014	0	0	0	0	0	4	0
甘 肃	2 665	926	1 257	461	14	0	6	0	0	1	104
青 海	501	189	193	113	0	0	0	0	0	6	0
宁 夏	264	67	82	107	0	0	0	0	0	8	0
新 疆	2 146	740	846	505	0	0	3	3	0	49	2

各地区环境信访与环境法制情况（一）

（2012）

单位：件

地 区名 称	当年颁布地方性法规	当年颁布地方政府规章	现行有效的地方性法规总数	现行有效的地方政府规章总数	当年受理行政复议案件数	本级行政处罚案件数	电话/网络投诉数	电话/网络投诉办结数
总 计	23	27	348	347	427	117 308	892 348	888 836
国家级	—	—	—	—	79	11	1 554	1 554
北 京	2	0	6	3	3	7 166	17 985	17 985
天 津	1	1	4	10	0	194	19 762	19 711
河 北	0	1	10	13	1	4 742	26 015	26 039
山 西	0	0	11	6	0	1 920	12 757	12 742
内蒙古	2	1	17	9	1	1 293	9 039	8 881
辽 宁	0	0	15	18	1	13 293	32 277	34 258
吉 林	0	0	10	8	2	1 227	15 698	15 691
黑龙江	2	5	18	26	4	37 942	15 337	14 556
上 海	0	1	3	10	11	963	34 960	34 792
江 苏	2	2	23	22	12	5 002	122 825	122 224
浙 江	0	0	19	21	28	9 743	42 563	41 713
安 徽	2	1	12	6	28	665	25 548	25 406
福 建	0	0	19	7	42	2 413	32 071	29 909
江 西	3	1	3	3	11	1 717	7 295	7 297
山 东	1	0	20	10	7	4 858	46 707	46 699
河 南	0	2	7	11	15	2 048	37 309	37 231
湖 北	0	1	8	8	2	1 209	30 706	30 681
湖 南	1	0	6	6	10	2 141	21 468	21 131
广 东	4	1	36	17	95	10 028	134 214	132 606
广 西	0	1	6	29	6	753	23 596	23 124
海 南	0	0	5	4	6	183	8 593	8 482
重 庆	0	0	15	11	32	804	72 927	72 927
四 川	0	1	12	11	10	1 596	29 832	31 932
贵 州	0	1	8	4	0	994	5 222	5 248
云 南	2	6	28	54	2	865	18 705	18 657
西 藏	0	0	2	6	0	13	343	343
陕 西	0	1	5	2	5	1 277	28 368	28 391
甘 肃	0	0	5	1	1	331	4 802	4 563
青 海	0	0	3	0	0	84	1 444	1 438
宁 夏	1	0	4	7	0	290	3 491	3 489
新 疆	0	0	8	4	13	1 543	8 935	9 136

各地区环境信访与环境法制情况（二）

（2012）

地 区 名 称	当年备案的 地方环境标 准数/件	累积备案的 地方环境标 准总数/件	来信总数/ 件	来访总数/ 批次	来访总数 （人数）/ 人	来信、来访 已办结数量/ 件	承办的人大 建议数/件	承办的政协 提案数/件
总 计	19	95	107 120	43 260	96 145	159 283	6 093	13 374
国家级	—	—	2 470	490	1 210	2 688	293	233
北 京	1	28	877	129	256	1 006	88	81
天 津	1	3	217	259	463	476	15	26
河 北	4	4	2 297	2 089	4 058	4 983	177	291
山 西	0	0	1 850	672	1 601	2 654	134	172
内蒙古	0	0	8 400	2 123	4 167	9 012	98	165
辽 宁	0	1	2 959	2 310	5 098	6 847	192	263
吉 林	0	0	1 593	1 659	3 289	3 840	47	55
黑龙江	0	2	2 246	2 241	5 061	4 340	152	120
上 海	1	9	1 992	731	4 120	2 708	66	88
江 苏	0	0	8 165	2 101	5 270	14 972	460	621
浙 江	3	4	4 852	1 817	3 896	6 447	655	716
安 徽	0	0	5 125	1 373	2 728	6 677	190	377
福 建	2	4	3 758	1 170	2 720	4 797	320	390
江 西	0	0	3 419	1 872	4 050	5 912	182	217
山 东	4	14	8 383	1 076	2 915	11 025	268	476
河 南	2	4	3 261	1 987	4 394	4 997	266	423
湖 北	0	0	6 010	2 212	4 855	10 058	315	452
湖 南	0	0	4 178	3 238	8 174	6 772	300	412
广 东	0	13	13 279	2 180	4 770	16 018	416	519
广 西	0	0	2 076	1 457	2 738	3 538	146	183
海 南	0	0	145	78	239	201	21	11
重 庆	0	6	3 073	868	1 689	4 379	214	232
四 川	0	0	7 601	3 139	7 128	10 584	278	5 567
贵 州	0	0	1 968	1 121	1 959	2 763	119	149
云 南	0	0	2 118	1 761	4 016	3 862	253	327
西 藏	0	0	35	75	179	117	21	17
陕 西	1	2	1 368	1 124	1 849	2 258	97	158
甘 肃	0	1	1 528	581	1 149	2 252	100	104
青 海	0	0	281	109	152	263	49	19
宁 夏	0	0	110	446	579	228	12	356
新 疆	0	0	1 486	772	1 373	2 609	149	154

各地区环境保护能力建设投资情况（一）

（2012）

单位：万元

地区名称	本级环保能力建设资金使用总额	监测能力建设	监察能力建设	核与辐射安全监管能力建设	固体废物管理能力建设	环境应急能力建设	环境信息能力建设	环境宣教能力建设	环境监管运行保障
总　计	1 148 349.1	551 243.1	139 308.8	33 427.7	15 621.6	30 548.0	47 198.0	24 731.4	248 969.0
北　京	30 383.5	14 736.8	4 520.8	2 983.0	2.0	143.3	1 417.2	561.0	5 583.6
天　津	14 149.1	7 380.7	1 675.8	857.0	600.0	141.0	342.0	561.3	2 384.1
河　北	83 304.9	48 567.8	9 510.8	1 734.7	330.6	1 840.2	2 590.8	1 123.3	4 609.8
山　西	36 364.0	22 751.6	4 558.9	360.1	287.7	341.2	932.8	314.9	1 827.6
内蒙古	57 445.8	28 051.0	6 653.8	920.9	352.2	781.1	2 561.0	1 253.5	7 780.6
辽　宁	50 810.0	16 447.5	3 607.5	767.5	81.3	2 971.3	794.2	525.8	22 973.5
吉　林	15 284.3	10 429.3	2 750.2	192.4	27.0	106.8	119.2	91.7	1 102.6
黑龙江	19 445.3	10 040.3	4 108.7	694.3	71.8	284.1	188.1	644.9	3 036.1
上　海	23 968.6	11 588.4	297.0	1 395.9	106.6	2 429.5	1 421.5	465.5	5 649.9
江　苏	105 168.3	49 091.4	11 891.8	3 222.1	1 417.5	1 313.8	5 139.8	2 794.2	27 805.2
浙　江	83 588.2	41 721.5	7 196.1	3 453.6	1 597.2	2 367.3	3 302.8	1 813.6	21 878.8
安　徽	26 755.5	10 117.7	4 921.8	578.0	1 764.5	1 994.8	1 169.7	671.0	5 526.9
福　建	37 801.3	12 774.2	5 899.8	1 280.3	86.8	552.8	512.5	800.4	15 894.5
江　西	36 291.4	14 601.4	5 715.6	218.0	162.2	726.5	2 917.3	926.8	6 243.9
山　东	35 871.9	15 221.6	3 425.6	793.5	133.8	1 769.9	2 358.1	677.1	10 879.5
河　南	42 188.4	23 407.7	7 955.6	2 338.6	222.6	639.0	1 028.0	847.7	3 977.6
湖　北	22 748.8	9 064.6	4 528.3	554.9	366.0	1 597.1	1 164.2	330.0	4 651.6
湖　南	36 615.5	15 715.5	7 553.4	606.9	787.0	1 016.3	2 016.9	1 345.9	6 841.6
广　东	114 294.5	51 125.1	6 142.9	3 681.3	4 830.4	2 803.6	4 418.4	2 908.0	33 210.3
广　西	29 168.5	16 942.5	2 208.6	146.0	124.0	2 234.8	629.6	230.7	5 521.6
海　南	2 731.6	2 404.2	85.7	11.0	0.0	48.0	7.7	21.7	153.2
重　庆	27 983.0	10 654.0	5 777.5	664.0	162.0	1 075.8	2 074.9	992.7	4 664.4
四　川	51 304.3	18 027.4	6 909.9	1 026.0	452.7	510.5	2 938.8	734.5	19 831.2
贵　州	18 329.0	10 819.9	2 799.0	142.0	51.0	356.7	553.1	345.6	2 377.2
云　南	31 119.5	14 829.8	5 089.0	1 399.0	772.5	1 076.7	504.2	1 830.2	3 448.1
西　藏	7 976.6	2 456.1	543.0	510.0	0.0	11.0	751.0	21.9	3 540.0
陕　西	37 414.6	19 060.8	4 128.3	1 194.3	374.0	818.0	4 404.6	921.0	5 883.7
甘　肃	14 128.8	8 561.6	2 582.1	428.3	189.1	210.9	197.3	275.8	1 283.3
青　海	12 750.8	8 370.7	1 088.7	729.0	66.0	132.0	61.5	243.6	1 940.0
宁　夏	2 299.3	1 108.8	199.9	0.0	0.0	0.0	152.0	66.0	479.0
新　疆	40 663.8	25 173.2	4 982.7	544.8	202.7	254.0	528.8	391.1	7 989.6

各地区环境保护能力建设投资情况（二）

（2012）

单位：万元

地 区 名 称	资金来源			
	国家拨付	省级拨付	地市拨付	县（区）拨付
总 计	199 318.4	313 643.2	314 704.5	320 949.4
北 京	1 852.0	19 980.3	0.0	8 551.2
天 津	1 374.8	10 837.3	0.0	1 937.0
河 北	33 294.0	12 353.7	21 482.8	16 201.4
山 西	6 403.8	9 777.2	11 949.3	8 233.7
内蒙古	11 090.6	24 437.6	14 323.1	7 594.5
辽 宁	10 930.7	15 573.9	20 466.8	3 913.1
吉 林	4 713.0	4 478.9	3 889.0	2 203.4
黑龙江	4 923.5	6 089.7	5 507.1	2 925.0
上 海	140.0	7 643.1	2 668.0	13 517.5
江 苏	3 583.9	16 424.5	28 507.5	56 652.4
浙 江	3 405.7	21 061.6	17 852.1	41 268.8
安 徽	6 421.8	5 550.2	8 834.3	6 052.2
福 建	3 565.7	6 473.7	15 629.7	12 132.2
江 西	4 452.0	18 016.4	6 437.6	7 385.4
山 东	2 360.1	8 648.5	12 952.5	11 910.8
河 南	6 650.4	11 531.8	10 198.7	13 807.5
湖 北	4 474.9	4 885.7	8 906.8	4 481.4
湖 南	7 566.5	9 063.8	11 067.6	8 917.6
广 东	10 191.5	27 271.8	45 404.9	31 488.2
广 西	10 141.7	10 326.1	5 503.8	3 196.9
海 南	1 017.0	63.3	630.5	1 020.8
重 庆	5 459.3	6 932.8	4 479.6	11 111.3
四 川	0.0	15 661.0	14 309.9	21 333.4
贵 州	3 315.5	3 702.1	5 729.7	5 581.7
云 南	8 675.8	8 037.6	4 493.9	9 912.2
西 藏	4 706.8	2 540.4	545.5	183.9
陕 西	10 864.2	13 862.0	11 264.2	1 424.2
甘 肃	7 956.8	1 646.0	1 954.1	2 571.9
青 海	6 453.0	4 778.9	709.0	809.9
宁 夏	923.3	592.2	424.3	359.5
新 疆	12 410.1	5 401.1	18 582.2	4 270.4

各地区环境污染源控制与管理情况（一）

（2012）

地 区 名 称	清洁生产审核 当年完成企业 数/家	强制性审核 当年完成数	应开展监测的重 金属污染防控重 点企业数/家	重金属排放达标 的重点企业数	已发放危险废物 经营许可证数/ 个	具有医疗废物经营 范围的许可证数
总　计	11 971	8 490	4 525	3 786	1 528	245
国家级	—	—	—	—	16	2
北　京	25	8	4	4	10	2
天　津	136	70	37	37	14	1
河　北	632	632	127	127	65	13
山　西	120	97	46	39	13	6
内蒙古	51	36	51	49	35	9
辽　宁	448	384	89	84	27	5
吉　林	121	99	37	32	65	4
黑龙江	60	53	11	11	22	5
上　海	322	153	21	19	33	2
江　苏	1 910	1 412	54	54	320	27
浙　江	1 675	522	269	161	76	14
安　徽	274	175	64	48	45	13
福　建	496	394	185	182	29	8
江　西	86	64	211	191	52	10
山　东	504	409	41	41	94	20
河　南	293	268	592	541	73	17
湖　北	94	72	100	89	27	13
湖　南	229	172	912	826	118	12
广　东	2 783	2 020	507	470	133	19
广　西	217	179	129	117	18	4
海　南	33	24	9	7	8	2
重　庆	653	634	57	47	17	4
四　川	322	262	218	93	47	17
贵　州	55	50	31	20	19	3
云　南	157	103	358	278	42	6
西　藏	2	0	37	0	0	0
陕　西	100	65	84	44	43	7
甘　肃	99	93	196	139	23	13
青　海	17	14	17	11	7	1
宁　夏	25	14	21	17	14	3
新　疆	32	12	10	8	23	9

各地区环境污染源控制与管理情况（二）

（2012）

地 区 名 称	地表水集中式饮 用水水源取水量/ 万米³	地表水集中式饮 用水水源地达标供 水量/万米³	地下水集中式饮 用水水源取水量/ 万米³	地下水集中式饮 用水水源地达标供 水量/万米³	受省级环保部门 委托的机动车环 保检验机构数/ 个	受省级环保部门 委托的机动车环 保检验机构人数/ 人
总 计	53 666 001	53 525 884	805 090	790 437	1 622	23 385
北 京	31 900	31 900	23 456	23 388	41	2 674
天 津	57 919	57 919	2 227	2 227	30	180
河 北	26 668	26 668	25 134	25 134	38	559
山 西	13 620	12 821	36 831	33 331	77	720
内蒙古	30 182	30 182	50 119	46 128	54	609
辽 宁	98 296	98 296	31 479	29 016	34	582
吉 林	74 392	73 942	13 720	13 587	72	487
黑龙江	70 245	70 004	29 167	29 167	87	555
上 海	322 751	296 409	12 807	12 778	96	2 612
江 苏	490 346	487 708	8 296	8 296	134	1 635
浙 江	2 201 451	2 133 020	443	443	76	1 066
安 徽	105 005	104 568	11 846	11 846	46	504
福 建	149 988	149 978	1 282	1 282	126	1 700
江 西	76 604	76 604	1 744	1 714	86	940
山 东	137 298	136 496	42 817	42 816	151	2 037
河 南	71 758	71 758	350 903	350 903	42	609
湖 北	213 759	213 757	1 430	1 414	12	83
湖 南	176 756	154 775	3 067	2 237	8	192
广 东	738 789	729 587	4 598	4 598	183	2 306
广 西	108 408	107 132	7 009	6 959	7	174
海 南	45 699	43 719	1 881	1 881	24	331
重 庆	48 005 034	48 002 727	5 443	5 336	62	807
四 川	172 763	172 723	18 551	18 551	1	
贵 州	39 687	39 581	4 508	4 508	38	639
云 南	78 254	75 876	2 246	2 204	6	83
西 藏	1 527	1 116	3 428	3 413	2	30
陕 西	42 654	42 654	13 817	13 817	26	332
甘 肃	57 101	56 926	17 863	17 782	38	372
青 海	2 922	2 922	14 472	14 472	5	118
宁 夏	5 997	5 997	11 183	11 183	16	229
新 疆	18 228	18 118	53 325	50 027	4	220

各地区环境监测情况（一）

（2012）

地 区 名 称	监测用房面积/米²	监测业务经费/万元	环境监测—监测仪器设备台套数/台套	环境监测—监测仪器设备原值总值/万元	环境空气监测点位数/个	国控监测点位数	酸雨监测点位/个	沙尘天气影响环境质量监测点位数/个
总　计	2 634 680	458 556	233 932	1 247 624.0	3 189	1 436	1 672	220
国家级	10 998	7 719	706	30 636.7	—	—	440	82
北　京	29 511	6 211	6 786	38 314.4	35	12	5	1
天　津	24 165	3 862	3 056	15 377.6	27	15	9	1
河　北	140 301	30 774	10 536	42 311.7	51	53	21	3
山　西	86 224	8 938	9 165	35 045.6	113	58	21	2
内蒙古	117 614	8 004	9 300	34 616.8	81	44	32	20
辽　宁	103 340	10 143	10 271	53 753.5	103	77	59	17
吉　林	60 162	9 344	3 473	15 297.0	163	33	22	5
黑龙江	61 694	4 306	4 513	17 439.6	186	57	30	9
上　海	54 232	6 190	4 818	37 195.2	74	10	18	1
江　苏	174 153	69 057	16 400	178 883.3	140	72	111	22
浙　江	129 645	44 864	13 763	60 128.1	159	47	85	0
安　徽	73 897	41 715	4 860	17 177.3	67	68	42	3
福　建	64 832	5 735	8 882	27 836.9	80	37	36	0
江　西	101 607	8 051	5 219	10 723.9	60	60	23	0
山　东	126 087	16 746	14 435	247 088.8	167	74	67	7
河　南	94 410	12 939	10 031	27 911.6	102	75	50	1
湖　北	85 179	10 063	5 932	18 789.1	91	51	53	0
湖　南	106 226	11 601	9 030	24 351.1	156	78	61	0
广　东	163 596	28 534	17 311	90 959.7	477	102	71	3
广　西	59 646	7 387	5 172	19 118.6	67	50	37	0
海　南	24 645	1 260	1 797	8 468.7	39	7	26	3
重　庆	69 993	14 883	9 798	29 604.7	93	17	51	0
四　川	164 740	18 314	16 091	60 891.9	187	94	123	4
贵　州	53 022	4 212	3 188	13 183.6	91	33	21	0
云　南	188 081	5 236	12 484	24 688.6	109	40	49	0
西　藏	18 243	871	666	2 912.5	16	18	8	0
陕　西	82 283	20 526	6 241	16 997.8	55	50	29	4
甘　肃	61 207	32 582	2 695	13 469.2	62	33	26	9
青　海	13 777	1 501	1 623	6 990.5	22	11	9	3
宁　夏	14 619	684	1 264	3 215.0	20	19	7	4
新　疆	76 551	6 304	4 426	54 881.7	96	41	30	16

各地区环境监测情况（二）

（2012）

单位：个

地区名称	地表水水质监测断面（点位）数	国控断面（点位）数	饮用水水源地监测点位数	地表水监测点位数	地下水监测点位数	近岸海域监测点位数	近岸海域环境功能区点位数	近岸海域环境质量点位数
总　计	8 173	972	2 995	2 125	870	645	418	312
北　京	199	8	6	2	4	0	0	0
天　津	256	11	1	1	0	22	12	10
河　北	259	32	51	10	41	20	10	10
山　西	111	15	48	11	37	0	0	0
内蒙古	141	34	67	7	60	0	0	0
辽　宁	250	37	94	24	70	74	51	28
吉　林	206	41	49	27	22	0	0	0
黑龙江	163	68	61	29	32	0	0	0
上　海	226	3	43	35	8	6	6	10
江　苏	963	83	110	83	27	24	13	16
浙　江	717	40	183	164	19	93	63	50
安　徽	303	73	122	83	39	0	0	0
福　建	272	17	87	81	6	92	79	35
江　西	215	38	38	37	1	0	0	0
山　东	457	50	120	76	44	114	66	48
河　南	208	47	62	17	45	0	0	0
湖　北	488	51	121	116	5	0	0	0
湖　南	137	39	91	81	10	0	0	0
广　东	389	20	159	142	17	119	68	52
广　西	142	17	54	33	21	8	8	22
海　南	115	12	70	49	21	73	42	31
重　庆	383	11	577	512	65	0	0	0
四　川	251	21	160	123	37	0	0	0
贵　州	148	8	131	102	29	0	0	0
云　南	620	69	180	173	7	0	0	0
西　藏	28	12	23	17	6	0	0	0
陕　西	134	23	35	10	25	0	0	0
甘　肃	61	12	76	39	37	0	0	0
青　海	31	8	23	6	17	0	0	0
宁　夏	47	5	21	6	15	0	0	0
新　疆	253	67	132	29	103	0	0	0

各地区环境监测情况（三）

（2012）

单位：个

地 区 名 称	开展环境噪声 监测的监测点 位数	区域环境噪声 监测点位数	道路交通噪声 监测点位数	功能区环境噪 声检测点位数	开展生态监测的 监测点位数	开展污染源监督性 监测的重点企业数
总 计	246 349	180 554	49 813	17 384	87	57 136
北 京	776	191	581	4	0	364
天 津	1 338	802	463	73	0	609
河 北	9 282	6 497	2 572	693	34	3 779
山 西	9 590	7 367	1 514	709	0	1 083
内蒙古	8 174	5 377	2 568	229	8	1 065
辽 宁	10 270	7 495	2 133	766	0	2 328
吉 林	6 652	4 932	1 416	304	1	609
黑龙江	10 639	7 167	2 232	1 240	0	1 494
上 海	635	380	199	56	1	1 243
江 苏	15 442	11 206	3 301	935	0	4 906
浙 江	14 618	10 577	2 971	1 070	0	4 275
安 徽	5 128	3 633	1 311	611	1	1 538
福 建	9 454	7 654	1 739	188	0	2 815
江 西	4 741	3 315	1 093	333	0	1 358
山 东	20 935	16 203	3 586	1 146	22	5 240
河 南	11 124	8 149	1 944	1 031	0	2 078
湖 北	10 896	7 620	2 142	1 134	0	2 189
湖 南	12 962	10 042	2 348	572	1	2 259
广 东	17 252	12 101	4 294	851	0	7 573
广 西	6 889	4 667	952	1 270	0	1 825
海 南	1 744	1 380	360	4	0	297
重 庆	5 218	4 077	875	266	0	944
四 川	16 157	13 169	2 178	810	2	2 286
贵 州	6 676	5 501	855	320	0	846
云 南	6 963	4 960	1 426	827	0	1 158
西 藏	290	228	53	9	0	14
陕 西	7 666	5 236	1 560	870	0	1 034
甘 肃	5 843	3 811	1 362	670	0	479
青 海	587	438	137	12	16	306
宁 夏	1 180	817	272	91	0	220
新 疆	7 228	5 562	1 376	290	1	922

各地区污染源自动监控情况

（2012）　　　　　　　　　　　　　　　　　　　　　　　　　　　　　　　　　单位：个

地区名称	已实施自动监控国家重点监控企业数	已实施自动监控国家重点监控企业中排放口数		已实施自动监控国家重点监控企业中监控设备与环保部门稳定联网数			
		水排放口数	气排放口数	COD监控设备与环保部门稳定联网数	NH3-N监控设备与环保部门稳定联网数	SO2监控设备与环保部门稳定联网数	NOx监控设备与环保部门稳定联网数
总　计	9 215	7 293	6 765	4 503	3 194	4 314	4 106
北　京	33	24	11	18	18	10	0
天　津	61	43	94	43	43	94	92
河　北	555	522	686	301	171	424	402
山　西	762	533	771	112	100	148	139
内蒙古	244	136	345	136	112	345	345
辽　宁	279	191	217	149	138	209	192
吉　林	172	115	167	65	46	32	31
黑龙江	197	135	165	122	33	128	125
上　海	76	63	43	54	48	43	43
江　苏	773	635	357	383	246	167	165
浙　江	349	249	159	249	154	159	159
安　徽	247	207	211	166	102	125	127
福　建	283	249	121	130	94	85	80
江　西	203	157	141	126	71	100	73
山　东	727	502	450	105	60	105	108
河　南	597	610	459	354	254	459	459
湖　北	422	356	217	356	179	178	168
湖　南	428	367	207	330	195	27	25
广　东	476	410	243	0	0	174	160
广　西	515	427	242	200	204	137	137
海　南	37	30	25	30	25	25	25
重　庆	132	98	77	87	67	44	34
四　川	442	337	204	337	292	203	204
贵　州	144	161	168	135	133	168	160
云　南	158	159	136	96	71	97	90
陕　西	413	251	295	192	155	230	191
甘　肃	118	109	176	89	62	129	132
青　海	55	29	64	29	29	48	34
宁　夏	96	55	88	55	55	81	73
新　疆	221	133	226	54	37	140	133

各地区排污费征收情况

（2012）

地 区 名 称	排污费解缴入库户数/户	排污费解缴入库户金额/万元
总 计	351 326	1 889 204
北 京	1 718	3 103
天 津	3 673	19 214
河 北	21 735	166 539
山 西	10 261	126 524
内 蒙 古	4 664	98 363
辽 宁	22 645	128 649
吉 林	12 985	33 578
黑 龙 江	9 553	43 321
上 海	4 710	18 961
江 苏	30 684	190 582
浙 江	24 584	86 262
安 徽	11 574	56 879
福 建	13 873	34 095
江 西	8 094	84 307
山 东	14 591	147 638
河 南	17 812	104 390
湖 北	7 577	40 146
湖 南	12 872	60 083
广 东	57 653	87 351
广 西	7 195	26 996
海 南	969	3 530
重 庆	6 798	36 342
四 川	9 448	55 326
贵 州	7 725	51 209
云 南	7 416	36 399
陕 西	4 934	54 654
甘 肃	5 253	23 296
青 海	878	6 760
宁 夏	1 845	17 796
新 疆	7 607	46 912

各地区自然生态保护与建设情况（一）

（2012）

地 区 名 称	自然保护区个数/个	国家级	省级	自然保护区面积/公顷	国家级	省级
总 计	2 669	363	876	149 787 347	94 145 640	40 908 533
北 京	20	2	12	133 966	26 403	71 413
天 津	8	3	5	91 115	37 862	53 253
河 北	43	13	25	692 709	253 653	414 274
山 西	46	6	40	1 161 022	107 351	1 053 671
内蒙古	184	25	62	13 688 964	4 048 938	6 974 624
辽 宁	105	14	30	2 673 537	1 006 522	883 969
吉 林	39	16	15	2 329 251	1 002 724	1 289 846
黑龙江	224	28	85	6 751 783	2 715 749	2 699 805
上 海	4	2	2	93 821	66 175	27 646
江 苏	30	3	10	567 126	336 211	85 401
浙 江	32	10	9	196 904	146 542	15 455
安 徽	104	7	29	524 494	139 221	287 728
福 建	93	13	26	463 611	218 034	136 140
江 西	200	11	35	1 259 848	184 078	386 992
山 东	86	7	37	1 081 950	219 828	538 345
河 南	34	11	21	734 658	426 316	306 942
湖 北	65	13	22	954 997	282 983	416 256
湖 南	129	18	33	1 285 259	517 856	467 562
广 东	368	13	64	3 552 658	294 768	528 955
广 西	78	17	49	1 452 941	333 833	878 216
海 南	50	9	24	2 735 320	106 526	2 614 554
重 庆	57	5	17	850 236	241 387	253 065
四 川	167	27	65	8 974 274	2 832 955	3 117 044
贵 州	129	8	4	951 766	243 539	56 965
云 南	159	19	38	2 854 278	1 480 385	712 721
西 藏	47	9	14	41 368 882	37 153 065	4 209 486
陕 西	57	17	33	1 163 080	527 121	539 823
甘 肃	59	17	38	7 346 761	4 910 058	2 321 803
青 海	11	5	6	21 822 201	20 252 490	1 569 711
宁 夏	14	6	8	535 570	426 916	108 654
新 疆	27	9	18	21 494 365	13 606 151	7 888 214

各地区自然生态保护与建设情况（二）

（2012）

单位：个

地 区 名 称	生态市、县建设 个数	国家级生态市 个数	国家级生态县（区） 个数	省级生态市 个数	省级生态县（区） 个数
总　计	287	4	51	45	187
北　京	13	0	2	9	2
天　津	1	0	1	0	0
河　北	0	0	0	0	0
山　西	2	0	0	0	2
内蒙古	6	0	0	0	6
辽　宁	11	0	5	1	5
吉　林	0	0	0	0	0
黑龙江	22	0	0	7	15
上　海	1	0	1	0	0
江　苏	27	3	19	1	4
浙　江	55	0	6	9	40
安　徽	14	0	3	1	10
福　建	27	0	0	0	27
江　西	12	0	0	1	11
山　东	12	0	3	3	6
河　南	2	0	0	0	2
湖　北	0	0	0	0	0
湖　南	8	0	0	0	8
广　东	12	1	4	1	6
广　西	2	0	0	1	1
海　南	5	0	0	3	2
重　庆	4	0	0	0	4
四　川	24	0	4	4	16
贵　州	3	0	0	0	3
云　南	0	0	0	0	0
西　藏	0	0	0	0	0
陕　西	6	0	2	0	4
甘　肃	8	0	0	4	4
青　海	0	0	0	0	0
宁　夏	0	0	0	0	0
新　疆	10	0	1	0	9

各地区自然生态保护与建设情况（三）

（2012）

地 区 名 称	农村生态示范建设 个数	国家级生态乡镇 个数	国家级生态村 个数	国家有机食品生产 基地数量
总　计	1 797	1 559	238	138
北　京	72	70	2	1
天　津	15	15	0	0
河　北	37	26	11	0
山　西	7	4	3	0
内蒙古	36	32	4	4
辽　宁	85	77	8	15
吉　林	26	23	3	6
黑龙江	30	27	3	1
上　海	53	51	2	2
江　苏	282	238	44	11
浙　江	283	274	9	30
安　徽	91	70	21	11
福　建	21	18	3	0
江　西	50	41	9	3
山　东	223	217	6	16
河　南	35	28	7	1
湖　北	50	28	22	0
湖　南	92	65	27	0
广　东	50	44	6	1
广　西	3	2	1	2
海　南	3	2	1	0
重　庆	5	5	0	3
四　川	105	98	7	3
贵　州	34	26	8	1
云　南	19	16	3	5
西　藏	0	0	0	0
陕　西	36	25	11	2
甘　肃	11	5	6	0
青　海	0	0	0	2
宁　夏	12	8	4	0
新　疆	31	24	7	18

各地区环境影响评价（一）

（2012）

地区名称	当年开工建设的建设项目数/个	执行环境影响评价制度的建设项目数量	当年审批的建设项目投资总额/万元	当年审批的建设项目环保投资总额/万元
总　计			2 968 125 433	84 389 902
国家级		256	139 260 486	5 592 617
北　京	9 480	9 480	52 114 580	967 329
天　津	3 264	3 264	67 108 516	1 093 310
河　北	23 864	23 864	326 357 853	3 362 904
山　西			66 501 448	1 683 978
内蒙古	114	114	78 248 379	4 609 339
辽　宁	15 333	15 333	60 049 267	1 469 135
吉　林	4 515	4 515	59 661 111	2 063 693
黑龙江	6 914	6 914	37 888 850	981 514
上　海	12 830	12 830	58 804 420	5 121
江　苏	26 389	26 389	380 162 150	13 183 515
浙　江	28 321	28 321	170 116 000	3 140 300
安　徽	13 392	13 284	99 469 397	3 465 832
福　建	13 921	13 921	82 273 134	2 853 031
江　西	7 504	7 499	49 989 223	1 633 727
山　东	28 612	18 612	213 152 670	8 438 135
河　南	8 506	8 506	110 829 191	2 405 520
湖　北	8 469	8 459	95 978 454	3 130 969
湖　南	9 773	9 773	40 163 287	1 150 859
广　东			231 113 697	4 997 048
广　西	6 163	6 163	60 317 378	1 943 879
海　南	998	978	27 966 746	825 258
重　庆	5 299	5 299	53 054 551	1 323 050
四　川	14 564	14 564	85 863 100	1 964 829
贵　州	10 890	10 873	55 521 278	3 052 104
云　南	16 457	16 457	70 085 166	2 177 006
西　藏			4 417 907	212 048
陕　西	4 291	4 291	54 203 418	1 837 667
甘　肃	4 870	4 870	22 487 819	454 228
青　海	1 567	1 567	25 353 994	986 987
宁　夏	2 120	2 120	28 931 805	1 045 099
新　疆	7 993	7 862	60 680 158	2 339 870

各地区环境影响评价（二）

（2012）

单位：个

地 区 名 称	当年审批的建设项目环境 影响评价文件数量	编制报告书的项目数量	填报报告表的项目数量	编制登记表的项目数量
总 计	427 573	29 171	179 202	219 233
国家级	246	228	10	8
北 京	9 480	433	3 152	5 895
天 津	3 264	439	2 301	524
河 北	23 453	924	10 587	11 942
山 西	7 935	979	3 984	2 975
内蒙古	8 756	818	3 657	4 311
辽 宁	15 976	984	6 836	8 156
吉 林	2 923	440	1 808	675
黑龙江	6 914	660	3 236	3 018
上 海	12 830	534	6 117	6 179
江 苏	37 455	2 443	16 949	18 063
浙 江	31 152	2 052	15 162	13 938
安 徽	13 971	1 275	6 344	6 352
福 建	13 921	1 072	6 756	6 093
江 西	7 016	734	2 949	3 333
山 东	35 702	1 827	15 416	18 459
河 南	17 061	965	7 452	8 644
湖 北	9 086	1 158	4 170	3 758
湖 南	9 827	1 024	3 087	5 716
广 东	56 040	1 723	31 093	23 224
广 西	15 013	659	4 211	10 143
海 南	3 006	260	1 247	1 499
重 庆	5 617	773	2 400	2 444
四 川	14 564	2 746	4 547	7 271
贵 州	12 666	832	3 488	8 346
云 南	16 571	985	3 253	12 333
西 藏	4 919	68	471	4 380
陕 西	5 191	510	1 876	2 805
甘 肃	4 967	375	1 288	3 304
青 海	1 998	141	454	1 403
宁 夏	2 484	258	928	1 298
新 疆	17 569	852	3 973	12 744

各地区建设项目竣工环境保护验收情况（一）

（2012）

地 区 名 称	当年完成环保验收 项目数/个	环保验收一次合格 项目数	经限期改正验收合格 项目数	当年完成环保验收 项目总投资/万元
总　计	132 310	128 758	3 618	987 20 9 186
国家级	221	215	6	195 572 892
北　京	3 578	3 573	5	9 130 299
天　津	1 319	1 319	0	11 525 082
河　北	7 334	7 211	123	17 635 014
山　西	1 525	1 521	4	8 602 574
内蒙古	2 040	2 030	10	12 756 397
辽　宁	1 619	1 565	27	17 651 929
吉　林	3 474	3 451	23	10 574 790
黑龙江	2 356	2 327	29	7 325 953
上　海	4 505	4 499	6	24 654 540
江　苏	9 548	9 240	308	382 340 275
浙　江	12 245	11 691	554	40 205 389
安　徽	3 757	3 659	98	20 526 073
福　建	5 532	4 952	610	20 747 337
江　西	1 921	1 851	70	8 628 138
山　东	13 792	13 539	253	37 497 177
河　南	4 254	4 208	46	12 082 425
湖　北	3 057	2 986	71	14 380 681
湖　南	3 328	3 168	160	5 964 156
广　东	15 980	15 275	705	40 365 850
广　西	5 948	5 805	149	13 149 165
海　南	718	713	5	3 575 633
重　庆	2 290	2 290	0	15 457 009
四　川	4 172	4 129	43	14 714 895
贵　州	3 089	3 083	6	3 333 929
云　南	4 882	4 786	153	9 400 779
西　藏	125	107	18	1 372 312
陕　西	1 392	1 376	16	5 648 983
甘　肃	1 456	1 449	7	6 080 839
青　海	516	494	22	2 877 701
宁　夏	567	556	11	5 393 800
新　疆	5 770	5 690	80	8 037 171

各地区建设项目竣工环境保护验收情况（二）

（2012）

单位：万元

地 区 名 称	当年完成环保验收项目环保投资	废水治理环保投资	废气治理环保投资	噪声治理环保投资	固体废物治理环保投资	绿化及生态环保投资
总　计	26 903 521					
国家级	3 266 550					
北　京	263 791	52 229	71 298	39 516	31 257	69 493
天　津	398 772	108 975	111 763	36 308	60 570	81 156
河　北	929 586	238 147	224 697	63 492	54 526	78 967
山　西	557 956	125 960	272 161	20 256	51 475	72 785
内蒙古	821 512	233 265	272 871	63 598	52 227	181 464
辽　宁	3 318 714	111 438	134 181	23 825	22 604	28 436
吉　林	370 627	50 505	49 332	15 780	9 869	85 340
黑龙江	509 931	304 917	89 562	18 620	20 659	31 510
上　海	572 522					
江　苏	2 378 588	914 445	498 898	182 206	148 592	293 056
浙　江	1 834 198	828 866	547 493	125 776	114 141	217 922
安　徽	664 671	271 318	192 419	50 565	34 651	115 718
福　建	764 943	237 139	210 230	39 170	46 292	135 091
江　西	386 426	108 162	60 914	16 963	22 964	51 902
山　东	2 255 833	904 700	499 411	142 168	219 400	350 018
河　南	548 345	145 826	142 152	35 052	25 815	29 248
湖　北	807 807	353 922	192 577	44 225	35 485	117 240
湖　南	421 302	152 981	113 722	30 506	29 256	94 838
广　东	1 291 177	508 942	355 957	58 001	39 569	219 961
广　西	723 320	216 198	223 489	18 743	29 429	90 539
海　南	164 150	31 194	12 884	7 059	6 694	80 653
重　庆	565 781	184 238	103 943	27 986	39 690	119 607
四　川	669 332	254 497	48 183	7 663	93 780	233 269
贵　州	207 020	26 703	41 999	10 558	8 228	11 979
云　南	639 832	96 200	110 291	17 478	21 995	75 951
西　藏	30 366					
陕　西	301 552	119 069	59 878	26 554	32 994	44 040
甘　肃	302 921	63 228	81 950	5 697	18 724	18 198
青　海	96 371	9 680	34 970	1 827	5 075	44 819
宁　夏	319 323	158 792	92 234	19 045	15 270	17 601
新　疆	520 304	136 707	71 284	13 006	24 316	274 991

各地区突发环境事件

（2012）

单位：次

地 区 名 称	突发环境事件 次数	特别重大环境 事件次数	重大环境事件次数	较大环境事件次数	一般环境事件次数
总 计	542	0	5	5	532
北 京	21	0	0	0	21
天 津	5	0	0	0	5
河 北	10	0	0	0	10
山 西	0	0	0	0	0
内蒙古	10	0	0	0	10
辽 宁	15	0	1	1	13
吉 林	1	0	0	0	1
黑龙江	0	0	0	0	0
上 海	192	0	0	0	192
江 苏	77	0	0	0	77
浙 江	23	0	0	0	23
安 徽	20	0	0	0	20
福 建	4	0	0	0	4
江 西	1	0	0	0	1
山 东	3	0	0	0	3
河 南	14	0	0	0	14
湖 北	4	0	0	0	4
湖 南	3	0	1	1	1
广 东	23	0	1	0	22
广 西	20	0	1	3	16
海 南	2	0	0	0	2
重 庆	25	0	0	0	25
四 川	16	0	0	0	16
贵 州	4	0	1	0	3
云 南	1	0	0	0	1
西 藏	0	0	0	0	0
陕 西	23	0	0	0	23
甘 肃	8	0	0	0	8
青 海	4	0	0	0	4
宁 夏	0	0	0	0	0
新 疆	13	0	0	0	13

各地区环境宣教情况

（2012）

地 区 名 称	当年开展的社会环境宣传教育活动数/次	当年开展的社会环境宣传教育活动人数/人	环境教育基地数/个
总 计	10 209	46 643 567	1 845
国家级	70	2 028 208	0
北 京	790	358 065	28
天 津	18	206 000	3
河 北	769	10 158 600	50
山 西	99	295 798	24
内蒙古	234	178 325	138
辽 宁	367	405 300	68
吉 林	92	304 410	12
黑龙江	135	389 905	61
上 海	394	494 450	69
江 苏	491	1 187 957	316
浙 江	518	491 650	181
安 徽	205	173 695	22
福 建	84	448 840	47
江 西	65	40 930	12
山 东	722	763 557	212
河 南	273	780 666	73
湖 北	379	539 956	142
湖 南	204	1 580 667	19
广 东	506	6 595 091	119
广 西	141	123 028	16
海 南	64	19 603	4
重 庆	874	15 122 488	39
四 川	862	1 175 321	40
贵 州	79	10 366	20
云 南	803	981 795	78
西 藏	82	3 097	6
陕 西	108	845 948	13
甘 肃	128	527 224	7
青 海	385	300 328	16
宁 夏	54	1 782	7
新 疆	214	110 517	3

14

附　表

ANNUAL STATISTIC REPORT ON ENVIRONMENT IN CHINA
2012

全国行政区划

（2012 年底）

单位：个

省级区划名称	地级区划数	地级市	县级区划数	县级市	市辖区	县	自治县
全　国	333	285	2 852	368	860	1 453	117
北京市			16		14	2	
天津市			16		13	3	
河北省	11	11	172	22	37	107	6
山西省	11	11	119	11	23	85	
内蒙古自治区	12	9	101	11	21	17	
辽宁省	14	14	100	17	56	19	8
吉林省	9	8	60	20	20	17	3
黑龙江省	13	12	128	18	64	45	1
上海市			17		16	1	
江苏省	13	13	102	23	55	24	
浙江省	11	11	90	22	32	35	1
安徽省	16	16	105	6	43	56	
福建省	9	9	85	14	26	45	
江西省	11	11	100	11	19	70	
山东省	17	17	138	30	48	60	
河南省	17	17	159	21	50	88	
湖北省	13	12	103	24	38	38	2
湖南省	14	13	122	16	35	64	7
广东省	21	21	121	23	56	39	3
广西壮族自治区	14	14	109	7	34	56	12
海南省	3	3	20	6	4	4	6
重庆市			38		19	15	4
四川省	21	18	181	14	45	118	4
贵州省	9	6	88	7	13	56	11
云南省	16	8	129	11	13	76	29
西藏自治区	7	1	74	1	1	72	
陕西省	10	10	107	3	24	80	
甘肃省	14	12	86	4	17	58	7
青海省	8	1	43	2	4	30	7
宁夏回族自治区	5	5	22	2	9	11	
新疆维吾尔自治区	14	2	101	22	11	62	6
香港特别行政区							
澳门特别行政区							
台湾省							

注：数据摘自《中国统计摘要》（中国统计出版社），以下同。

国民经济与社会发展总量指标摘要

指标	单位	2005 年	2006 年	2007 年	2008 年	2009 年	2010 年	2011 年	2012 年
人口									
年底总人口	万人	130 756	131 448	132 129	132 802	133 450	134 091	134 735	135 404
其中：城镇人口	万人	56 212	58 288	60 633	62 403	64 512	66 978	69 079	71 182
乡村人口	万人	74 544	73 160	71 496	70 399	68 938	67 113	65 656	64 222
国民经济核算									
国内生产总值	亿元	184 937.4	216 314.4	265 810.3	314 045.4	340 902.8	401 512.8	471 563.7	518 942.1
其中：第一产业	亿元	22 420.0	24 040.0	28 627.0	33 702.0	35 226.0	40 533.6	47 712.0	52 373.6
第二产业	亿元	87 598.1	103 719.5	125 831.4	149 003.4	157 638.8	187 383.2	220 591.6	235 162.0
第三产业	亿元	77 230.8	88 554.9	111 351.9	131 340.0	148 038.0	173 596.0	203 260.1	231 406.5
人均国内生产总值	元/人	14 185	16 500	20 169	23 708	25 608	30 015	35 083	
支出法国内生产总值	亿元	187 437.4	222 711.9	266 556.7	315 977.2	348 771.3	402 818.7	465 998.7	529 238.4
其中：最终消费	亿元	99 357.5	113 103.8	132 232.9	153 422.5	169 274.8	194 115.0	224 740.8	261 832.8
资本形成总额	亿元	77 856.8	92 954.1	110 943.2	138 325.3	164 463.2	193 603.9	229 102.4	252 773.2
固定资产投资									
全社会固定资产投资总额	亿元	88 773.6	109 998.2	137 323.9	172 828.4	224 599	278 121.9	311 021.9	374 694.7
城镇	亿元	75 095.1	93 368.7	117 464.5	148 738.3	193 920	241 430.9	301 932.8	364 854.1
其中：房地产开发	亿元	15 909.2	19 422.9	25 288.8	31 203.2	36 242	48 259.4	61 739.8	71 803.8
农村	亿元	13 678.5	16 629.5	19 859.5	24 123.0	30 678	36 724.9		
主要农业、工业产品产量									
粮食	万吨	48 402.2	49 804	50 160.3	52 870.9	53 082	54 647.7	57 121	58 958.0
棉花	万吨	571.4	753.3	762.4	749.2	638	596.1	658.9	683.6
油料	万吨	3 077.1	2 640.3	2 568.7	2 952.8	3 154	3 230.1	3 306.8	3 436.8
肉类	万吨	6 938.9	7 089.0	6 865.7	7 278.3	7 650	7 925.8	7 957.8	8 387.2
原煤	亿吨	23.50	25.29	26.92	28.02	29.73	32.35	35.20	36.50
原油	万吨	18 135.29	18 476.57	18 632	19 043.1	18 949	20 241.4	20 287.6	20 748
发电量	亿千瓦·时	25 002.60	28 657.26	32 816	34 957.6	37 146.5	42 071.6	47 000.7	49 876
粗钢	万吨	35 324.0	41 914.9	48 929	50 305.8	57 218	63 723.0	68 388.3	72 388
水泥	万吨	106 884.8	123 676.5	136 117	142 355.7	164 398	188 191.2	208 500.0	220 984

国民经济与社会发展总量指标摘要（续表）

指标	单位	2005 年	2006 年	2007 年	2008 年	2009 年	2010 年	2011 年	2012 年
国内商业和对外贸易									
社会消费品零售总额	亿元	68 352.6	79 145.2	93 571.6	114 830.1	132 678.4	156 998	183 918.6	210 307.0
进出口总额	亿美元	14 219	17 604	21 765.7	25 632.6	22 075.4	29 740.0	36 420.6	38 671.2
其中：出口额	亿美元	7 620	9 689	12 204.6	14 306.9	12 016.1	15 777.5	18 986.0	20 487.1
进口额	亿美元	6 600	7 915	9 561.1	11 325.6	10 059.2	13 962.4	17 434.6	18 184.1
利用外资									
实际利（使用）用外资额	亿美元	638.1	735.2	783.4	952.6	918.0	1 088.2	1 177.0	1 132.9
其中：外商直接投资	亿美元	603.3	694.7	747.7	924.0	900.3	1 057.3	1 160.1	1 117.2
外商其他投资	亿美元	34.8	40.6	35.7	28.6	17.7	30.9	16.9	15.8
旅游									
入境过夜旅游者人数	万人次	4 681	4 991	5 472.0	5 304.9	5 087.5	5 566.5	5 758.1	5 772.5
国内旅游人数	亿人次	12.1	13.9	16.1	17.1	19.0	21.0	26.41	
国际旅游外汇收入	亿美元	293.0	339.5	419.2	408.4	396.8	458.1	484.6	500.3
国内旅游总收入（国内旅游总花费	亿元	5 285.9	6 229.7	7 770.6	8 749.3	10 183.7	12 579.8	19 305.4	
教育、科技、文化、卫生									
普通高等学校学生数	万人	1 561.8	1 738.8	1 884.9	2 021.0	2 144.7	2 231.8	2 308.5	2 391.3
普通中等学校学生数	万人	8 581	8 451.9	8 243.3	8 050.4	7 876.9	7 703.2	7 519.0	7 228.4
小学学生数	万人	10 864	10 711.5	10 564.0	10 331.5	10 071.5	9 940.7	9 926.4	9 695.9
研究与试验发展经费支出	亿元	2 450	3 003.1	3 710.2	4 570.0	5 791.9	7 063	8 610	10 298.4
技术市场成交额	亿元	1 551	1 818.2	2 227	2 665	3 039	3 907	4 764	6 437.1
图书总印数	亿册（张）	64.7	64.0	62.9	70.6	70.4	71.7	76.8	79.2
杂志总印数（期刊）	亿册	27.6	28.5	30.4	31.0	31.5	32.2	32.7	
报纸总印数	亿份	412.6	424.5	438.0	442.9	439.1	452.1	466.8	
医院、卫生院数（医疗卫生机构数）	万个	88.2	91.8	91.2	89.1	91.7	93.7	95.4	95.0
医生数	万人	204.2	209.9	212.3	220.2	232.9	241.3	246.6	261.6
医院、卫生院床位数（医疗卫生机构床位数	万张	336.8	351.2	370.1	403.9	441.7	478.7	516.0	416.1
城市公用事业									
城市房屋集中供热面积	亿米²	25.2	26.6				43.6	47.4	51.8
年末供水综合生产能力	万米³/日	24 720	26 962						
年末污水处理能力	万米³/日	7 990	9 734						
年末城市道路长度	千米	247 015	241 351						

注：① 由于计算误差的影响，支出法国内生产总值不等于按生产法计算的国内生产总值。

② 本表价值量指标均按当年价格计算。

国民经济与社会发展结构指标摘要

单位：%

指标	2005 年	2006 年	2007 年	2008 年	2009 年	2010 年	2011 年	2012 年
总人口	100	100	100	100	100	100	100	100
其中：城镇	43.0	44.3	45.9	47.0	48.3	50.0	51.3	51.3
乡村	57.0	5 537	54.1	53.0	51.7	50.0	48.7	48.7
国内生产总值	100	100	100	100	100	100	100	100
其中：第一产业	12.1	11.1	10.8	10.7	10.3	10.1	10.1	33.6
第二产业	47.4	47.9	47.3	47.4	46.2	46.7	46.8	30.3
第三产业	40.5	40.9	41.9	41.8	43.4	43.2	43.1	36.1
固定资产投资总额	100	100	100	100	100	100	100	100
城镇	84.6	84.9	85.6	86.0	86.3	86.8		
农村	15.4	15.1	14.4	14.0	13.7	13.2		
财政收入	100	100	100	100	100	100	100	100
其中：中央	52.3	52.8	54.1	53.3	52.4	51.1	49.5	47.9
地方	47.7	47.2	45.9	46.7	47.6	48.9	50.5	52.1
财政支出	100	100	100	100	100	100	100	100
其中：中央	25.9	24.7	23.1	21.4	20.0	17.8	15.2	14.9
地方	74.1	75.3	76.9	78.6	80.0	82.2	84.8	85.1
农、林、牧、渔业产值	100	100	100	100	100	100	100	100
其中：农业	49.7	51.1	50.4	55.7	50.7	53.3	51.6	52.5
林业	3.6	3.9	3.8	3.8	3.9	3.7	3.8	3.9
牧业	33.7	29.4	33.0	29.7	32.3	30.0	31.7	30.4
渔业	10.2	9.6	9.1	10.9	9.3	9.3	9.3	9.7
农、林、牧、渔、服务业	2.8	6.0	3.7		3.8	3.7		
工业总产值	100	100	100	100	100	100	100	100
其中：轻工业	31.4	30.5						
重工业	68.6	69.5						

注：2002 年起轻重工业结构为国有及规模以上非国有工业企业口径。

自然资源状况

项目	单位	2005 年	2006 年	2007 年	2008 年	2009 年	2010 年	2011 年	2012 年
国土面积	万千米²	960	960	960	960	960	960	960	960
海域面积	万千米²	473	473	473	473	473	473	473	473
大陆岸线长度	万千米	1.80	1.80	1.80	1.80	1.80	1.80	1.8	1.8
岛屿面积	万千米²	3.87	3.87	3.87	3.87	3.87	3.87	3.87	3.87
降水量									
台湾中部山区	毫米	≥4 000	≥4 000	≥4 000	≥4 000	≥4 000	≥4 000	≥4 000	≥4 000
华南沿海	毫米	1 600～2 000	1 600～2 000	1 600～2 000	1 600～2 000	1 600～2 000	1 600～2 000	1 600～2 000	1 600～2 000
长江流域	毫米	1 000～1 500	1 000～1 500	1 000～1 500	1 000～1 500	1 000～1 500	1 000～1 500	1 000～1 500	1 000～1 500
华北、东北	毫米	400～800	400～800	400～800	400～800	400～800	400～800	400～800	400～800
西北内陆	毫米	100～200	100～200	100～200	100～200	100～200	100～200	100～200	100～200
塔里木盆地、吐鲁番盆地和柴达木盆地	毫米	≤25	≤25	≤25	≤25	≤25	≤25	≤25	≤25
耕地面积	万公顷	13 004	13 004	13 004	12 178	12 178	12 172	12 172	12 172
荒地面积	万公顷	10 800	10 800	10 800	10 800	10 800			
林业用地面积	万公顷	26 329	28 493	28 493	28 493	30 378	30 590	30 590	30 590
草地面积	万公顷	40 000	40 000	40 000	40 000	40 000			
活立木总蓄积量	亿米³	136.2	136.2	136.2	136.2	136.2	149.1	149.1	149.1
森林面积	亿公顷	1.75	1.75	1.75	1.75	1.95	1.95	1.95	1.95
森林覆盖率	%	18.21	18.21	18.21	18.21	20.36	20.36	20.36	20.36
水资源总量	亿米³	28 124	27 430	25 567	24 696	23 763	29 658	24 022	24 022
水力资源蕴藏量	亿千瓦	6.76	6.76	6.76	6.76	6.76	6.76	6.76	6.76
其中：可开发量	亿千瓦	3.79	3.79	3.79	3.79	3.79	3.79	3.79	3.79
内陆水域总面积	万公顷	1 747	1 747	1 747	1 747	1 747	1 747	1 747	1 747
其中：可养殖面积	万公顷	675	675	675	675	675	675	675	675
海水可养殖面积	万公顷	260	260	260	260	260	260	260	260
其中：已养殖面积	万公顷	109	109	109	109	109	109	109	109

注：① 森林资源为 1994—1998 年调查数；草地资源为 1991 年调查数；水面资源为 1988 年数；水力资源为 1999 年公报数；耕地面积为 1996 年农业普查数。

② 本表除国土面积、海域面积和降水量外，其余指标均未包括香港特别行政区、澳门特别行政区和台湾省数据。

《中国环境统计年报·2012》各章编写作者

第1章	统计调查企业基本情况	董广霞
第2章	废水	周冏、封雪
第3章	废气	吕卓
第4章	工业固体废物	赵银慧、刘通浩
第5章	环境污染治理投资	王鑫
第6章	环境管理	李治国
数据表汇总及整理		周冏、封雪、吕卓、谢光轩、黄英志

15

主要统计指标解释

ANNUAL STATISTIC REPORT ON ENVIRONMENT IN CHINA
2012

1. 工业企业污染排放及处理利用情况

【工业用水量】指报告期内企业厂区内用于工业生产活动的水量，它等于取水量与重复用水量之和。

【取水量】指报告期内企业厂区内用于工业生产活动的水量中从外部取水的量。根据《工业企业产品取水定额编制通则》（GB/T 18220—2002），工业生产的取水量，包括取自地表水（以净水厂供水计量）、地下水、城镇供水工程，以及企业从市场购得的其他水（如其他企业回用水量）或水的产品（如蒸汽、热水、地热水等），不包括企业自取的海水和苦咸水等以及企业为外供给市场的水的产品（如蒸汽、热水、地热水等）而取用的水量。

工业生产活动用水包括主要工业生产用水、辅助生产（包括机修、运输、空压站等）用水和附属生产（包括厂内绿化、职工食堂、非营业的浴室及保健站、厕所等）用水；不包括①非工业生产单位的用水，如厂内居民家庭用水和企业附属幼儿园、学校、对外营业的浴室、游泳池等的用水量；②生活用水单独计量且生活污水不与工业废水混排的水量。

【重复用水量】指报告期内企业生产用水中重复再利用的水量，包括循环使用、一水多用和串级使用的水量（含经处理后回用量）。

指企业内部对工业生产活动排放的废水直接利用或经过处理后回收再利用的水量，不包括从城市污水处理厂回用的水量。每重复利用一次，则计算一次重复用水量。但锅炉、循环冷却系统等封闭式系统内的循环水不能计算重复用水量。重复用水量的计算原则：

（1）开放原则。即水的循环在开放系统进行，循环一次计算一次，但锅炉、循环冷却系统等封闭式系统内的循环水不能计算重复用水量。

（2）"源头"计算原则。对循环水来说，使用后的水，又回流到系统的取水源头，流经源头一次，计算一次。循环系统中的中间环节用水不得计算重复用水量。

（3）异地原则。对于非循环系统，根据不同工艺对不同水质的要求，在一个地方（工艺）使用过的水，在另外一个地方（工艺）中又进行使用，使用一次，计算一次。在同一地方（容器）多次使用的水，不得计算重复用水量。

（4）经过净化处理后的水重复再用，在任何情况下都按照重复用水计算。

【煤炭消耗量】指报告期内企业所用煤炭的总消耗量。

【燃料煤消耗量】指报告期内企业厂区内用作燃料的煤炭消耗量（实物量），包括企业厂区内生产、生活用燃料煤，也包括砖瓦、石灰等产品生产用的内燃煤，不包括在生产工艺中用作原料并能转换成新的产品实体的煤炭消耗量。如转换为水泥、焦炭、煤气、碳素、活性炭、氮肥的煤炭。

【燃料油消耗量（不含车船用）】指报告期内企业用作燃料的原油、汽油、柴油、煤油等各种油料总消耗量，不包括车船交通用油量。

【焦炭消耗量】指报告期内企业消耗的焦炭总量。

【天然气消耗量】指报告期内企业用作燃料的天然气消耗量。

【其他燃料消耗量】指报告期内企业除了煤炭、燃油、天然气等以外，用作燃料的其他燃料消耗量。其他燃料应根据当地的折标系数折算为标准煤后统一填报。

各类能源的参考折标系数表

能源种类		折标系数	能源种类		折标系数
原煤		0.714 3	煤焦油		1.142 9
洗精煤		0.900 0	粗苯		1.428 6
其他洗煤	洗中煤	0.285 7	原油		1.428 6
	煤泥	0.285 7～0.428 6	汽油		1.471 4
型 煤		0.5～0.7	煤油		1.471 4
焦 炭		0.971 4	柴油		1.457 1
焦炉煤气		0.571 4～0.614 3 千克标准煤/米³	燃料油		1.428 6
高炉煤气		0.128 6 千克标准煤/米³	热力		0.034 12 千克标准煤/百万千焦 0.142 86 千克标准煤/1 000 千卡
天 然 气		1.330 0 千克标准煤/米³	电力		0.122 9 千克标准煤/千瓦·时
液化天然气		1.757 2	生物质能	大豆秆、棉花秆	0.543
液化石油气		1.714 3		稻秆	0.429
炼厂干气		1.571 4		麦秆	0.500
其他煤气	发生炉煤气	0.178 6 千克标准煤/米³		玉米秆	0.529
	重油催化裂解煤气	0.657 1 千克标准煤/米³		杂草	0.471
	重油热裂解煤气	1.214 3 千克标准煤/米³		树叶	0.500
	焦炭制气	0.557 1 千克标准煤/米³		薪柴	0.571
	压力气化煤气	0.514 3 千克标准煤/米³		沼气	0.714 千克标准煤/米³
	水煤气	0.357 1 千克标准煤/米³		—	—

注：除表中标注单位的能源外，其余能源折标系数单位均为千克标准煤/千克。

各地的能源折标系数由当地环保部门协调统计部门提供。调查对象也可根据燃料品质分析报告，自行折标填报。

【用电量】指报告期内企业的用电量，包括动力用电和照明用电。

【工业锅炉数】指报告期内企业厂区内用于生产和生活的大于 1 蒸吨（含 1 蒸吨）的蒸汽锅炉、热水锅炉总台数和总蒸吨数，包括燃煤、燃油、燃气和燃电的锅炉，不包括茶炉。

【其中 35 蒸吨及以上的】指报告期内企业厂区内用于生产和生活的大于 35 蒸吨（含 35 蒸吨）的蒸汽锅炉、热水锅炉总台数和总蒸吨数。

【其中 20（含）～35 蒸吨的】指报告期内企业厂区内用于生产和生活的大于 20 蒸吨（含 20 蒸吨）小于 35 蒸吨的蒸汽锅炉、热水锅炉总台数和总蒸吨数。

【其中 10（含）～20 蒸吨的】指报告期内企业厂区内用于生产和生活的大于 10 蒸吨（含 10 蒸吨）小于 20 蒸吨的蒸汽锅炉、热水锅炉总台数和总蒸吨数。

【其中 10 蒸吨以下的】指报告期内企业厂区内用于生产和生活的小于 10 蒸吨的蒸汽锅炉、热水锅炉总台数和总蒸吨数。

【工业炉窑数】指报告期内企业生产用的炉窑总数，如炼铁高炉、炼钢炉、冲天炉、烘干炉窑、锻造加热炉、水泥窑、石灰窑等。

【废水治理设施数】指报告期内企业用于防治水污染和经处理后综合利用水资源的实有设施（包括构筑物）数，以一个废水治理系统为单位统计。附属于设施内的水治理设备和配套设备不单独计算。备用的、报告期内未运行的、已经报废的设施不统计在内。

只填报企业内部的废水治理设施,工业废水排入的城镇污水处理厂、集中工业废水处理厂不能算作企业的废水治理设施。企业内的废水治理设施包括一级、二级和三级处理的设施,如企业有 2 个排污口,1 个排污口为一级处理(隔油池、化粪池、沉淀池等),另 1 个排污口为二级处理(如生化处理),则该企业有 2 套废水治理设施;若该企业只有 1 个排污口,经由该排污口的废水先经过一级处理,再经二级(甚至三级)处理后外排,则该企业视为 1 套废水治理设施。即针对同一股废水的所有水治理设备均视为 1 套治理设施,针对不同废水的水治理设备可视为多套治理设施。

【废水治理设施处理能力】指报告期内企业内部的所有废水治理设施实际具有的废水处理能力。

【废水治理设施运行费用】指报告期内企业维持废水治理设施运行所发生的费用。包括能源消耗、设备维修、人员工资、管理费、药剂费及与设施运行有关的其他费用等。

【工业废水处理量】指经各种水治理设施(含城镇污水处理厂、工业废水处理厂)实际处理的工业废水量,包括处理后外排的和处理后回用的工业废水量。虽经处理但未达到国家或地方排放标准的废水量也应计算在内。计算时,如遇有车间和厂排放口均有治理设施,并对同一废水分级处理时,不应重复计算工业废水处理量。

【工业废水排放量】指报告期内经过企业厂区所有排放口排到企业外部的工业废水量。包括生产废水、外排的直接冷却水、废气治理设施废水、超标排放的矿井地下水和与工业废水混排的厂区生活污水,不包括独立外排的间接冷却水(清浊不分流的间接冷却水应计算在内)。

直接冷却水:在生产过程中,为满足工艺过程需要,使产品或半成品冷却所用与之直接接触的冷却水(包括调温、调湿使用的直流喷雾水)。

间接冷却水:在工业生产过程中,为保证生产设备能在正常温度下工作,用来吸收或转移生产设备的多余热量,所使用的冷却水(此冷却用水与被冷却介质之间由热交换器壁或设备隔开)。

【直接排入环境的】指废水经过工厂的排污口或经过下水道直接排入环境中,包括排入海、河流、湖泊、水库、蒸发地、渗坑以及农田等。对应的排水去向代码为 A、B、C、D、F、G、K。

【排入污水处理厂的】指企业产生的废水直接或间接经市政管网排入污水处理厂的废水量,包括排入城镇污水处理厂、集中工业废水处理厂以及其他单位的污水处理设施的废水量。对应的排水去向代码为 E、L、H。

【工业废水中污染物产生量】指报告期调查对象生产过程中产生的未经过处理的废水中所含的化学需氧量、氨氮、石油类、挥发酚、氰化物等污染物和砷、铅、汞、镉、六价铬、总铬等重金属本身的纯质量。它可采用产排污系数根据生产的产品产量或原辅料用量计算求得,也可以通过工业废水产生量和其中污染物的浓度相乘求得,计算公式是:

污染物产生量(纯质量)= 工业废水产生量×废水处理设施入口污染物的平均浓度(无处理设施可使用排口浓度)

计算砷、铅、汞、镉、六价铬、总铬等重金属污染物时,上述计算公式中"工业废水产生量"为产生重金属废水的车间年实际产生的废水量,"废水处理设施入口污染物的平均浓度"为该车间废水处理设施入口的年实际加权平均浓度,如没有设施则为车间排口的年实际加权平均浓度。

【工业废水中污染物排放量】指报告期内企业排放的工业废水中所含化学需氧量、氨氮、石油类、挥发酚、氰化物等污染物和砷、铅、汞、镉、六价铬等重金属本身的纯质量。它可采用产排污系数根据生产的产品产量或原辅料用量计算求得,也可以通过工业废水排放量和其中污染物的浓度相乘求得,计算公式是:

污染物排放量(纯质量)= 工业废水排放量×排放口污染物的平均浓度

(1)如企业排出的工业废水经城镇污水处理厂或工业废水处理厂集中处理的,计算化学需氧量、氨氮、石油类、挥发酚、氰化物等污染物时,上述计算公式中"排放口污染物的平均浓度"即为污水处理厂排放口的年实际加权平均浓度。如果厂界排放浓度低于污水处理厂的排放浓度，以污水处理厂的排放浓度为准。

（2）计算砷、铅、汞、镉、六价铬等重金属污染物时，上述计算公式中"工业废水排放量"为车间排放口的年实际废水量，"排放口污染物的平均浓度"为车间排放口的年实际加权平均浓度。

【工业废气排放量】指报告期内企业厂区内燃料燃烧和生产工艺过程中产生的各种排入空气中含有污染物的气体的总量，以标准状态（273 K，101 325 Pa）计。

工业废气排放总量＝燃料燃烧过程中废气排放量＋生产工艺过程中废气排放量

【废气治理设施数】指报告期末企业用于减少在燃料燃烧过程与生产工艺过程中排向大气的污染物或对污染物加以回收利用的废气治理设施总数，以一个废气治理系统为单位统计。包括除尘、脱硫、脱硝及其他的污染物的烟气治理设施。备用的、报告期内未运行的、已报废的设施不统计在内。锅炉中的除尘装置属于"三同时"设备，应统计在内。

【除尘设施数】指专门设计、建设的去除废气烟（粉）尘的设施。

【脱硫设施数】脱硫设施指专门设计、建设的去除废气二氧化硫的设施，具有兼性脱硫效果的设施，如湿法除尘等治理设施等其他可能具有脱硫效果的废气治理设施不计入脱硫设施。具有脱硫效果的生产装置，如制酸、水泥生产等不作为脱硫设施。

【脱硝设施数】指在治理设施中采用选择性催化还原技术（SCR）、选择性非催化还原技术（SNCR）及其联合技术或采用活性炭吸附进行烟气脱销的设施。具有脱硝效果的生产装置，如水泥生产等不作为脱硝设施。

【废气治理设施处理能力】指报告期末企业实有的废气治理设施的实际废气处理能力。

【废气脱硫设施处理能力】指报告期末企业实有的废气脱硫设施的实际废气处理能力。

【废气脱硝设施处理能力】指报告期末企业实有的废气脱硝设施的实际废气处理能力。

【废气除尘设施处理能力】指报告期末企业实有的废气除尘设施的实际废气处理能力。

【废气治理设施运行费用】指报告期内维持废气治理设施运行所发生的费用。包括能源消耗、设备折旧、设备维修、人员工资、管理费、药剂费及与设施运行有关的其他费用等。

【二氧化硫产生量】指当年全年调查对象生产过程中产生的未经过处理的废气中所含的二氧化硫总质量。

【二氧化硫排放量】指报告期内企业在燃料燃烧和生产工艺过程中排入大气的二氧化硫总质量。工业中二氧化硫主要来源于化石燃料（煤、石油等）的燃烧，还包括含硫矿石的冶炼或含硫酸、磷肥等生产的工业废气排放。

【氮氧化物产生量】指当年全年调查对象生产过程中产生的未经过处理的废气中所含的氮氧化物总质量。

【氮氧化物排放量】指报告期内企业在燃料燃烧和生产工艺过程中排入大气的氮氧化物总质量。

【烟（粉）尘产生量】烟尘是指通过燃烧煤、石煤、柴油、木柴、天然气等产生的烟气中的尘粒。通过有组织排放的，俗称烟道尘。工业粉尘指在生产工艺过程中排放的能在空气中悬浮一定时间的固体颗粒。如钢铁企业耐火材料粉尘、焦化企业的筛焦系统粉尘、烧结机的粉尘、石灰窑的粉尘、建材企业的水泥粉尘等。烟（粉）尘产生量指当年全年调查对象生产过程中产生的未经过处理的废气中所含的烟尘及工业粉尘的总质量之和。

【烟（粉）尘排放量】指报告期内企业在燃料燃烧和生产工艺过程中排入大气的烟尘及工业粉尘的总质量之和。烟尘或工业粉尘排放量可以通过除尘系统的排风量和除尘设备出口烟尘浓度相乘求得。

【重金属产生量】指报告期调查对象生产过程中产生的未经过处理的废气中分别所含的砷、铅、汞、镉、铬及其化合物的总质量（以元素计）。

【重金属排放量】指报告期内企业在燃料燃烧和生产工艺过程中分别排入大气的砷、铅、汞、镉、铬及其化合物的总质量（以元素计）。

【一般工业固体废物产生量】是指未被列入《国家危险废物名录》或者根据国家规定的危险废物鉴别标准（GB 5085）、固体废物浸出毒性浸出方法（GB 5086）及固体废物浸出毒性测定方法（GB/T 15555）鉴别方法判定不具有危险特性的工业固体废物。根据其性质分为两种：

（1）第Ⅰ类一般工业固体废物：按照 GB 5086 规定方法进行浸出试验而获得的浸出液中，任何一种污染物的浓度均未超过 GB 8978 最高允放排放浓度，且 pH 值在 6～9 范围之内的一般工业固体废物；

（2）第Ⅱ类一般工业固体废物：按照 GB 5086 规定方法进行浸出试验而获得的浸出液中，有一种或一种以上的污染物浓度超过 GB 8978 最高允许排放浓度，或者是 pH 值在 6～9 范围之外的一般工业固体废物。

主要包括：

代码	名称	代码	名称
SW01	冶炼废渣	SW07	污泥
SW02	粉煤灰	SW08	放射性废物
SW03	炉渣	SW09	赤泥
SW04	煤矸石	SW10	磷石膏
SW05	尾矿	SW99	其他废物
SW06	脱硫石膏		

不包括矿山开采的剥离废石和掘进废石（煤矸石和呈酸性或碱性的废石除外）。酸性或碱性废石是指采掘的废石其流经水、雨淋水的 pH 值小于 4 或 pH 值大于 10.5 者。

冶炼废渣：指在冶炼生产中产生的高炉渣、钢渣、铁合金渣等，不包括列入《国家危险废物名录》中的金属冶炼废物。

粉煤灰：指从燃煤过程产生烟气中收捕下来的细微固体颗粒物，不包括从燃煤设施炉膛排出的灰渣。主要来自电力、热力的生产和供应行业和其他使用燃煤设施的行业，又称飞灰或烟道灰。主要从烟道气体收集而得，应与其烟尘去除量基本相等。

炉渣：指企业燃烧设备从炉膛排出的灰渣，不包括燃料燃烧过程中产生的烟尘。

煤矸石：指与煤层伴生的一种含碳量低、比煤坚硬的黑灰色岩石，包括巷道掘进过程中的掘进矸石、采掘过程中从顶板、底板及夹层里采出的矸石以及洗煤过程中挑出的洗矸石。主要来自煤炭开采和洗选行业。

尾矿：指矿山选矿过程中产生的有用成分含量低、在当前的技术经济条件下不宜进一步分选的固体废物，包括各种金属和非金属矿石的选矿。主要来自采矿业。

脱硫石膏：指废气脱硫的湿式石灰石/石膏法工艺中，吸收剂与烟气中 SO_2 等反应后生成的副产物。

污泥：指污水处理厂污水处理中排出的、以干泥量计的固体沉淀物。

放射性废物：指含有天然放射性核素，并其比活度大于 $2×10^4$ Bq/kg 的尾矿砂、废矿石及其他放射性固体废物（指放射性浓度或活度或污染水平超过规定下限的固体废物）。

赤泥：指从铝土矿中提炼氧化铝后排出的污染性废渣，一般还氧化铁量大，外观与赤色泥土相似。

磷石膏：指在磷酸生产中用硫酸分解磷矿时产生的二水硫酸钙、酸不溶物，未分解磷矿及其他杂质的混合物。主要来自磷肥制造业。

其他废物：指除上述 10 类一般工业固体废物以外的未列入《国家危险废物名录》中的固体废物，如机械工业切削碎屑、研磨碎屑、废砂型等；食品工业的活性炭渣；硅酸盐工业和建材工业的砖、瓦、碎砾、混凝土碎块等。

一般工业固体废物产生量＝（一般工业固体废物综合利用量－其中：综合利用往年贮存量）＋
一般工业固体废物贮存量＋（一般工业固体废物处置量－其中：
处置往年贮存量）＋一般工业固体废物倾倒丢弃量

【一般工业固体废物综合利用量】 指报告期内企业通过回收、加工、循环、交换等方式，从固体废物中提取或者使其转化为可以利用的资源、能源和其他原材料的固体废物量（包括当年利用的往年工业固体废物累计贮存量）。如用作农业肥料、生产建筑材料、筑路等。综合利用量由原产生固体废物的单位统计。

工业固体废物综合利用的主要方式有：

序号	综合利用方式	序号	综合利用方式
1	铺路	9	再循环/再利用不是用作溶剂的有机物
2	建筑材料	10	再循环/再利用金属和金属化合物
3	农肥或土壤改良剂	11	再循环/再利用其他无机物
4	矿渣棉	12	再生酸或碱
5	铸石	13	回收污染减除剂的组分
6	其他	14	回收催化剂组分
7	作为燃料（直接燃烧除外）或以其他方式产生能量	15	废油再提炼或其他废油的再利用
8	溶剂回收/再生（如蒸馏、萃取等）	16	其他有效成分回收

【一般工业固体废物贮存量】 指报告期内企业以综合利用或处置为目的，将固体废物暂时贮存或堆存在专设的贮存设施或专设的集中堆存场所内的量。专设的固体废物贮存场所或贮存设施必须有防扩散、防流失、防渗漏、防止污染大气、水体的措施。

粉煤灰、钢渣、煤矸石、尾矿等的贮存量是指排入灰场、渣场、矸石场、尾矿库等贮存的量。

专设的固体废物贮存场所或贮存设施指符合环保要求的贮存场，即选址、设计、建设符合《一般工业固体废物贮存、填埋场污染控制标准》（GB 18599—2001）等相关环保法律法规要求，具有防扩散、防流失、防渗漏、防止污染大气和水体措施的场所和设施。

工业固体废物的贮存方式主要有：

序号	贮存方式
1	灰场堆放
2	渣场堆放
3	尾矿库堆放
4	其他贮存（不包括永久性贮存）

【一般工业固体废物处置量】 指报告期内企业将工业固体废物焚烧和用其他改变工业固体废物的物理、化学、生物特性的方法，达到减少或者消除其危险成分的活动，或者将工业固体废物最终置于符合环境保护规定要求的填埋场的活动中，所消纳固体废物的量。

处置方式如：填埋、焚烧、专业贮存场（库）封场处理、深层灌注、回填矿井及海洋处置（经海洋管理部门同意投海处置）等。

处置量包括本单位处置或委托给外单位处置的量。还包括当年处置的往年工业固体废物贮存量。

工业固体废物的主要处置方式有：

处置方式
围隔堆存（属永久性处置）
填埋
置放于地下或地上（如填埋、填坑、填浜）
特别设计填埋
海洋处置
经海洋管理部门同意的投海处置
埋入海床
焚化
陆上焚化
海上焚化
水泥窑共处置（指在水泥生产工艺中使用工业固体废物或液态废物作为替代燃料或原料，消纳处理工业固体或液态废物的方式）
固化
其他处置（属于未在上面5种指明的处置作业方式外的处置）
废矿井永久性堆存（包括将容器置于矿井）
土地处理（属于生物降解，适合于液态固废或污泥固废）
地表存放（将液态固废或污泥固废放入坑、氧化塘、池中）
生物处理
物理化学处理
经环保管理部门同意的排入海洋之外的水体（或水域）
其他处理方法

【一般工业固体废物倾倒丢弃量】指报告期内企业将所产生的固体废物倾倒或者丢弃到固体废物污染防治设施、场所以外的量。倾倒丢弃方式如：

（1）向水体排放的固体废物；

（2）在江河、湖泊、运河、渠道、海洋的滩场和岸坡倾倒、堆放和存贮废物；

（3）利用渗井、渗坑、渗裂隙和溶洞倾倒废物；

（4）向路边、荒地、荒滩倾倒废物；

（5）未经环保部门同意作填坑、填河和土地填埋固体废物；

（6）混入生活垃圾进行堆置的废物；

（7）未经海洋管理部门批准同意，向海洋倾倒废物；

（8）其他去向不明的废物；

（9）深层灌注。

一般工业固体废物倾倒丢弃量计算公式是：

一般工业固体废物倾倒丢弃量＝一般工业固体废物产生量－一般工业固体废物贮存量－（一般工

业固体废物综合利用量－其中：综合利用往年贮存量）－（一般工业固体废物处置量－其中：处置往年贮存量）

【危险废物产生量】指当年全年调查对象实际产生的危险废物的量。危险废物指列入国家危险废物名录或者根据国家规定的危险废物鉴别标准和鉴别方法认定的，具有爆炸性、易燃性、易氧化性、毒性、腐蚀性、易传染性疾病等危险特性之一的废物。按《国家危险废物名录》（环境保护部、国家发展和改革委员会 2008 部令第 1 号）填报。

【危险废物综合利用量】指当年全年调查对象从危险废物中提取物质作为原材料或者燃料的活动中消纳危险废物的量。包括本单位利用或委托、提供给外单位利用的量。

【危险废物贮存量】指将危险废物以一定包装方式暂时存放在专设的贮存设施内的量。

专设的贮存设施指对危险废物的包装、选址、设计、安全防护、监测和关闭等符合《危险废物贮存污染控制标准》（GB 18597—2001）等相关环保法律法规要求，具有防扩散、防流失、防渗漏、防止污染大气和水体措施的设施。

【危险废物处置量】指报告期内企业将危险废物焚烧和用其他改变工业固体废物的物理、化学、生物特性的方法，达到减少或者消除其危险成分的活动，或者将危险废物最终置于符合环境保护规定要求的填埋场的活动中，所消纳危险废物的量。处置量包括处置本单位或委托给外单位处置的量。

【危险废物倾倒丢弃量】指报告期内企业将所产生的危险废物未按规定要求处理处置的量。

2. 工业企业污染防治投资情况

【污染治理项目名称】指以治理老污染源的污染、"三废"综合利用为主要目的的工程项目名称，或本年完成建设项目"三同时"环境保护竣工验收的项目名称。

【项目类型】按照不同的项目性质，污染治理项目分为 3 类，并给予不同的代码。

1——老工业污染源治理在建项目；2——老工业污染源治理本年竣工项目；

3——建设项目"三同时"环境保护竣工验收本年完成项目。

【治理类型】按照不同的企业污染治理对象，污染治理项目分为 14 类：

1——工业废水治理；2——工业废气脱硫治理；3——工业废气脱硝治理；4——其他废气治理；5——一般工业固体废物治理；6——危险废物治理（企业自建设施）；7——噪声治理（含振动）；8——电磁辐射治理；9——放射性治理；10——工业企业土壤污染治理；11——矿山土壤污染治理；12——污染物自动在线监测仪器购置；13——污染治理搬迁；14——其他治理（含综合防治）。

【本年完成投资及资金来源】指在报告期内，企业实际用于环境治理工程的投资额。投资额中的资金来源，是指投资单位在本年内收到的用于污染治理项目投资的各种货币资金，包括排污费补助、政府其他补助、企业自筹。各种来源的资金均为报告期投入的资金，不包括以往历年的投资。

本年污染治理资金合计＝排污费补助＋政府其他补助＋企业自筹

【竣工项目设计或新增处理能力】设计能力是指设计中规定的主体工程（或主体设备）及相应的配套的辅助工程（或配套设备）在正常情况下能够达到的处理能力。报告期内竣工的污染治理项目，属新建项目的填写设计文件规定的处理、利用"三废"能力；属改扩建、技术改造项目的填写经改造后新增加的处理利用能力，不包括改扩建之前原有的处理能力；只更新设备或重建构筑物，处理利用"三废"能力没有改变的则不填。

工业废水设计处理能力的计量单位为 t/d；工业废气设计处理能力的计量单位为（标态）m^3/h；工业固体废物设计处理能力的计量单位为 t/d；噪声治理（含振动）设计处理能力以降低分贝数表示；电磁辐射治理设计处理能力以降低电磁辐射强度表示（电磁辐射计量单位有电场强度单位：V/m、磁场强度单位：A/m、功率密度单位：W/m^2）。放射性治理设计处理能力以降低放射性浓度表示，废水计量单位为 Bq/L，固体废物计量单位为 Bq/kg。

3．农业源污染排放及处理利用情况

【治污设施累计完成投资】养殖场/小区当前可用于养殖污染治理设施的累计投资总额。

【新增固定资产】指报告期内交付使用的固定资产价值。对于新建治污设施，本年新增固定资产投资等于总投资；对于改、扩建治污设施，本年新增固定资产投资仅指报告期内交付使用的改、扩建部分的固定资产投资，属于累计完成投资的一部分。

4．城镇生活污染排放及处理情况

调查范围：

城镇生活包括"住宿业与餐饮业、居民服务和其他服务业、医院和独立燃烧设施以及城镇生活污染源"。

根据《关于统计上划分城乡的暂行规定》（国统字[2006]60号），城镇是指在我国市镇建制和行政区划的基础上，经以下规定划定的区域。城镇包括城区和镇区。

（1）城区是指在市辖区和不设区的市中，包括：①街道办事处所辖的居民委员会地域；②城市公共设施、居住设施等连接到的其他居民委员会地域和村民委员会地域。

（2）镇区是指在城区以外的镇和其他区域中，包括：①镇所辖的居民委员会地域；②镇的公共设施、居住设施等连接到的村民委员会地域；③常住人口在 3 000 人以上独立的工矿区、开发区、科研单位、大专院校、农场、林场等特殊区域。

城镇生活源的基本调查单位为地（市、州、盟），其所属的区、县城以及镇区数据包含在所在地（市、州、盟）数据中。

【城镇人口】是指居住在城镇范围内的全部常住人口。相关数据以统计部门公布数据为准。

【生活煤炭消费量】指报告期内调查区域用作城镇生活的煤炭总量，包括批发和零售贸易业、餐饮业、居民生活以及生活供热三个部分，以统计部门的能源平衡表数据为准。其中，批发和零售贸易业、餐饮业和居民生活煤炭消费量从能源平衡表直接获取；生活供热煤炭消费量需由能源平衡表中的供热总煤耗扣减工业供热煤耗得到。生活煤炭消费量计算公式为：

$$生活煤炭消费量 = 批发和零售贸易业、餐饮业煤炭消费量 + 居民生活煤炭消费量 +$$
$$生活供热煤炭消费量$$

$$生活供热煤炭消费量 = 供热总煤耗 - 工业供热煤耗（环境统计工业调查中 4 430 行业$$
$$煤炭消费总量）$$

【生活用水总量】指报告期内调查区域住宿业与餐饮业、居民服务和其他服务业、医院以及城镇生活等的用水总量（含自来水和自备水），包括居民家庭用水和公共服务用水的总量两个部分（包括自来水和自备水），以城市供水管理部门的统计数据为准。

【居民家庭用水总量】指城镇范围内所有居民家庭的日常生活用水。包括城镇居民、农民家庭、公共供水站用水，以城市供水管理部门的统计数据为准。

【公共服务用水总量】指为城镇社会公共服务的用水。包括行政事业单位、部队营区和公共设施服务、社会服务业、批发零售贸易业、旅馆饮食业等单位的用水。

【城镇生活污水污染物产生量】指报告期内各类生活源从贮存场所排入市政管道、排污沟渠和周边环境的量。

【城镇生活污水污染物排放量】指报告期内最终排入外环境生活污水污染物的量，即生活污水污染物产生量扣减经集中污水处理设施去除的生活污水污染物量。

【城镇生活污水排放量】用人均系数法测算。

如果辖区内的城镇污水处理厂未安装再生水回用系统，无再生水利用量，则

城镇生活污水排放量＝城镇生活污水排放系数×城镇人口数×365

反之，辖区内的城镇污水处理厂配备再生水回用系统，其再生水利用量已经污染减排核查确认，则

城镇生活污水排放量＝城镇生活污水排放系数×城镇人口数×365－城镇污水处理厂再生水利用量

5. 机动车

调查范围：

以直辖市、地区（市、州、盟）为单位调查机动车保有量，以及其中分车型、分年度登记数量；机动车调查对象包括汽车和摩托车。

机动车保有量分类登记数量信息由直辖市、地区（市、州、盟）环保部门协调同级公安交通管理部门负责提供，调查表由各级环保部门填写。

大型载客汽车：车长大于等于 6 米或者乘坐人数大于等于 20 人。乘坐人数可变的，以上限确定。乘坐人数包括驾驶员；

中型载客汽车：车长小于 6 米，乘坐人数大于 9 人且小于 20 人；

小型载客汽车：车长小于 6 米，乘坐人数小于等于 9 人；

微型载客汽车：车长小于等于 3.5 米，发动机汽缸总排量小于等于 1 升；

重型载货汽车：车长大于等于 6 米，总质量大于等于 12 000 公斤；

中型载货汽车：车长大于等于 6 米，总质量大于等于 4 500 公斤且小于 12 000 公斤；

轻型载货汽车：车长小于 6 米，总质量小于 4 500 公斤；

微型载货汽车：车长小于等于 3.5 米，载货质量小于等于 1 800 公斤；

三轮汽车（原三轮农用运输车）：以柴油机为动力，最高设计车速小于等于 50 公里/小时，最大设计总质量不大于 2 000 公斤，长小于等于 4.6 米，宽小于等于 1.6 米，高小于等于 2 米，具有三个车轮的货车；采用方向盘转向，由操纵杆传递动力、有驾驶室且驾驶员车厢后有物品存放的长小于等于 5.2 米，宽小于等于 2.8 米，高小于等于 2.2 米；

低速载货汽车（原四轮农用运输车）：以柴油机为动力，最高设计车速小于 70 公里/小时，最大设计总质量小于等于 4 500 公斤，长小于等于 6 米，宽小于等于 2 米，高小于等于 2.5 米，具有四个车轮的货车；

普通摩托车：最大设计时速大于 50 公里/小时或者发动机汽缸总排量大于 50 毫升；

轻便摩托车：最大设计时速小于等于 50 公里/小时，发动机汽缸总排量小于等于 50 毫升。

6. 污水处理厂

填报范围：

城镇污水处理厂及其他社会化运营的集中污水处理单位，包括专业化的工业废水集中处理设施或单位填报本套表。

污水处理厂包括城镇污水处理厂、工业废（污）水集中处理设施和其他污水处理设施。

城镇污水处理厂：指城镇污水（生活污水、工业废水）通过排水管道集于一个或几个处所，并利用由各种处理单元组成的污水处理系统进行净化处理，最终使处理后的污水和污泥达到规定要求后排放或再利用的设施。

工业废（污）水集中处理设施：指提供社会化有偿服务、专门从事为工业园区、联片工业企业或周边企业处理工业废水（包括一并处理周边地区生活污水）的集中设施或独立运营的单位。不包括企

业内部的污水处理设施。

其他污水处理设施：指对不能纳入城市污水收集系统的居民区、风景旅游区、度假村、疗养院、机场、铁路车站以及其他人群聚集地排放的污水进行就地集中处理的设施。

污水处理厂调查中，不包括氧化塘、渗水井、化粪池、改良化粪池、无动力地埋式污水处理装置和土地处理系统处理工艺。

【污水处理厂累计完成投资】指至当年末，调查对象建设实际完成的累计投资额，不包括运行费用。

【新增固定资产】指报告期内交付使用的固定资产价值。对于新建污水处理厂，本年新增固定资产投资等于总投资；对于改、扩建污水处理厂，本年新增固定资产投资仅指报告期内交付使用的改、扩建部分的固定资产投资，属于累计完成投资的一部分。

【运行费用】指报告期内维持污水处理厂（或处理设施）正常运行所发生的费用。包括能源消耗、设备维修、人员工资、管理费、药剂费及与污水处理厂（或处理设施）运行有关的其他费用等，不包括设备折旧费。

【污水设计处理能力】指截至当年末调查对象设计建设的设施正常运行时每天能处理的污水量。

【污水实际处理量】指调查对象报告期内实际处理的污水总量。

【再生水利用量】指调查对象报告期内处理后的污水中再回收利用的水量。该水量仅指总量减排核定 COD 或氨氮削减量的用于工业冷却、洗涤、冲渣和景观用水、生活杂用的水量。

【工业用水量】指调查对象报告期内污水再生水利用量中用于工业冷却用水等工业方面的水量。

【市政用水（杂用水）】指调查对象报告期内污水再生水利用量中用于消防、城市绿化等市政方面的水量。

【景观用水量】指调查对象报告期内污水再生水利用量中用于营造城市景观水体和各种水景构筑物的水量。

【污泥产生量】指调查对象报告期内在整个污水处理过程中最终产生污泥的质量。折合为 80% 含水率的湿泥量填报。污泥指污水处理厂（或处理设施）在进行污水处理过程中分离出来的固体。

【污泥处置量】指报告期内采用土地利用、填埋、建筑材料利用和焚烧等方法对污泥最终消纳处置的质量。

【土地利用量】指报告期内将处理后的污泥作为肥料或土壤改良材料，用于园林、绿化或农业等场合的处置方式处置的污泥质量。

【填埋处置量】指报告期内采取工程措施将处理后的污泥集中堆、填、埋于场地内的安全处置方式处置的污泥质量。

【建筑材料利用量】指报告期内将处理后的污泥作为制作建筑材料的部分原料的处置方式处置的污泥质量。

【焚烧处置量】指报告期内利用焚烧炉使污泥完全矿化为少量灰烬的处置方式处置的污泥质量。

【污泥倾倒丢弃量】指报告期内不作处理、处置而将污泥任意倾倒弃置到划定的污泥堆放场所以外的任何区域的质量。

7. 生活垃圾处理厂（场）

填报范围：

所有垃圾处理厂（场）填报本表。

垃圾处理厂（场）包括垃圾填埋厂（场）、堆肥厂（场）、焚烧厂（场）和其他方式处理垃圾的处理厂（场）。其中，垃圾焚烧厂（场）不包括垃圾焚烧发电厂，垃圾焚烧发电厂纳入工业源调查。

【生活垃圾处理厂（场）累计完成投资】指至当年末调查对象建设实际完成的累计投资额，不包括运行费用。

【本年新增固定资产】指报告期内交付使用的固定资产价值。对于新建垃圾处理厂（场），本年新增固定资产投资等于总投资；对于改、扩建垃圾处理厂（场），本年新增固定资产投资仅指报告期内交付使用的改、扩建部分的固定资产投资，属于累计完成投资的一部分。

【本年运行费用】指报告期内维持垃圾处理厂（场）正常运行所发生的费用。包括能源消耗、设备维修、人员工资、管理费及与垃圾处理厂（场）运行有关的其他费用等，不包括设备折旧费。

【本年实际处理量】指报告期内对垃圾采取焚烧、填埋、堆肥或其他方式处理的垃圾总质量。

【渗滤液污染物排放量】指报告期内排放的渗滤液中所含的化学需氧量、氨氮、石油类、总磷、挥发酚、氰化物、砷和汞、镉、铅、铬等重金属污染物本身的纯质量。按年排放量填报。

表中所列出的污染物，调查对象要根据监测结果或产排污系数如实填报。

【焚烧废气污染物排放量】指报告期内垃圾焚烧过程中排放到大气中的废气（包括处理过的、未经过处理的）中所含的二氧化硫、氮氧化物、烟尘和汞、镉、铅等重金属及其化合物（以重金属元素计）的固态、气态污染物的纯质量。按年排放量填报。

表中所列出的污染物，调查对象要根据监测结果或产排污系数如实填报。

8．危险废物（医疗废物）集中处理（置）厂

填报范围：

所有危险废物（医疗废物）集中处（理）置厂填报本表。集中式危险废物处（理）置厂指统筹规划建设并服务于一定区域专营或兼营危险废物处（理）置且有危险废物经营许可证的危险废物处置厂。

【危险废物（医疗废物）集中处（理）置厂累计完成投资】指至当年末，调查对象建设实际完成的累计投资额，不包括运行费用。

【本年新增固定资产】指报告期内交付使用的固定资产价值。对于新建危险废物（医疗废物）处置厂，本年新增固定资产投资等于总投资；对于改、扩建危险废物（医疗废物）处置厂，本年新增固定资产投资仅指报告期内交付使用的改、扩建部分的固定资产投资，属于累计完成投资的一部分。

【本年运行费用】指报告期内维持危险废物处置厂正常运行所发生的费用。包括能源消耗、设备维修、人员工资、管理费及与危险废物处置厂运行有关的其他费用等，不包括设备折旧费。

【危险废物设计处置能力】指调查对象设计建设的每天可能处置危险废物的量。

【本年实际处置危险废物量】指报告期内调查对象将危险废物焚烧和用其他改变危险废物的物理、化学、生物特性的方法，达到减少已产生的危险废物数量、缩小危险废物体积、减少或者消除其危险成分的活动，或者将危险废物最终置于符合环境保护规定要求的填埋场的活动中，所消纳危险废物的量。

【处置工业危险废物量】指调查对象报告期内采用各种方式处置的工业危险废物的总量。医疗废物集中处置厂不得填写该项指标。

【处置医疗废物量】指调查对象报告期内采用各种方式处置的医疗废物的总量。

【处置其他危险废物量】指调查对象报告期内采用各种方式处置的除工业危险废物和医疗废物以外其他危险废物的总质量，如教学科研单位实验室、机械电器维修、胶卷冲洗、居民生活等产生的危险废物。医疗废物集中处置厂不得填写该项指标。

【本年危险废物综合利用量】指报告期内调查对象从危险废物中提取物质作为原材料或者燃料的活动中消纳危险废物的量。

【渗滤液污染物排放量】指报告期内排放的渗滤液中所含的汞、镉、铅、总铬等重金属和砷、氰化物、挥发酚、石油类、化学需氧量、氨氮、总磷等污染物本身的纯质量。按年排放量填报。

表中所列出的污染物，调查对象要根据监测结果或产排污系数核算结果如实填报。

【焚烧废气污染物排放量】指报告期内危险废物焚烧过程中排放到大气中的废气（包括处理过的、未经过处理的）中所含的二氧化硫、氮氧化物、烟尘和汞、镉、铅等等重金属及其化合物（以重金属元素计）的固态、气态污染物的纯质量。按年排放量填报。

9. 环境管理

【环保系统机构数/人数】指环保系统行政主管部门及其所属事业单位、社会团体设置情况，包括机构总数和在编人员总数。

【环保行政主管部门机构数/人数】指各级人民政府环境保护行政主管部门设置情况，不包括各类开发区等非行政区环保主管部门。

【环境监测站机构/人数】统计独立设置的核与辐射环境监测机构以外的环境监测机构设置情况。在机构数中应注明加挂了核与辐射环境监测站牌子的机构数，以及未加挂牌子但承担了核与辐射环境监测职能的机构数；在总人数中注明承担核与辐射环境监测职能的人数。

【环境监察机构数/人数】指独立设置的环境执法监察机构设置情况。

【核与辐射环境监测站机构数/人数】统计独立设置的核与辐射环境监测站机构设置情况。

【科研机构数/人数】指独立设置的环境科研机构设置情况。

【宣教机构数/人数】指独立设置的环境宣传教育机构设置情况。

【信息机构数/人数】指独立设置的环境信息机构设置情况。

【环境应急（救援）机构数/人数】指环保部门建立或与其他部门共同建立的专兼职环境应急救援机构数、人数。

【当年受理行政复议案件数】指调查年度内环保部门受理的所有行政复议案件数（含非环保案件），包括已受理但未办结的案件；但不包括非本统计年受理而在本统计年内办理或办结的案件。

【本级行政处罚案件数】

【电话/网络投诉数；电话/网络投诉办结数】开通12369环保举报热线电话和网络的地区，按受理情况填报；未开通地区，可根据当地环保部门的公开举报电话和网络信箱的投诉情况填报。只填报本级受理的投诉数，不填报上级转到下级处理的投诉数。

【来访总数】指各级环保部门接待上访人员的数量。同时统计批次、人次。

【来信、来访已办结数量】

【承办的人大建议】指国家、省、市、县环保部门承办的本年度本级人大代表建议数总和。

【承办的政协提案数】指国家、省、市、县环保部门承办的本年度本级政协提案数总和。

【本级环保能力建设资金使用总额】是指本级使用的、当年已完成的用于提高环境保护监督管理能力和环境科技发展能力的基本建设和购置固定资产的投资。具体包括：各级环境保护行政主管部门、各类环境保护事业单位、各类环境监测站点等的基本建设投资、购置固定资产的投资及监测、监察等环境监管运行保障经费（含办公业务用房和科研用实验室的建设费用；不含行政运行费用、生活福利设施的建设费用）。

环境保护事业单位包括：环境监测、环境监察、环境应急、固体废物管理、环境信息、环境宣传、核与辐射、环境科研以及环保部门驻外和派出机构等。

其中：

（1）监测能力建设资金使用总额是指各级环境监测机构、各类环境监测站点等当年已完成的基本建设投资和购置固定资产的投资总额。

（2）监察能力建设资金使用总额是指各级环境监察机构、污染源监控中心、污染源自动监控设施等当年已完成的基本建设投资和购置固定资产的投资总额。

（3）核与辐射安全监管能力建设资金使用总额是指各级核与辐射安全监管机构、核与辐射自动监测站点等当年已完成的基本建设投资和购置固定资产的投资总额。

（4）固体废物管理能力建设资金使用总额是指各级固体废物管理机构当年已完成的基本建设投资和购置固定资产的投资总额。

（5）环境应急能力建设资金使用总额是指各级环境应急管理机构当年已完成的基本建设投资和购置固定资产的投资总额。

（6）环境信息能力建设资金使用总额是指各级环境信息机构、环境信息网络当年已完成的基本建设投资和购置固定资产的投资总额。

（7）环境宣教能力建设资金使用总额是指各级环境宣教机构、培训基地等当年已完成的基本建设投资和购置固定资产的投资总额。

（8）环境监管运行保障资金使用总额是指各级监测、监察、核与辐射安全、宣教等机构开展污染源与总量减排监管、环境监测与评估、环境信息等业务发生的环境监管运行保障经费。

【清洁生产审核当年完成企业数】是指本地区在调查年度完成清洁生产审核并通过评估或验收的企业数量。各地区在清洁生产审核的管理一般有两种情况：评估和验收分开实施，则统计该地区当年完成评估的企业数量；评估和验收同时进行，则统计地区当年完成验收的企业数量。

【强制性审核当年完成数】是指本地区在调查年度完成强制性清洁生产审核并通过评估或验收的企业数量。各地区在强制性清洁生产审核的管理一般有两种情况：评估和验收分开实施，则统计该地区当年完成评估的企业数量；评估和验收同时进行，则统计该地区当年完成验收的企业数量。

【应开展监测的重金属污染防控重点企业数】指按照《重金属污染综合防治"十二五"规划》要求开展监测的重金属污染防控重点企业家数。重金属污染防控重点企业指《重金属污染综合防治"十二五"规划》中的 4 452 家重金属排放企业，主要考核重点企业废气、废水中重点重金属污染物达标排放情况，若企业关闭则不纳入统计基数。《重金属污染综合防治"十二五"规划》要求各地对 4 452 家重点企业每两个月开展一次监督性监测，重点企业应实现稳定达标排放。

【重金属排放达标的重点企业数】指重点重金属污染物达标排放的重点企业数。由各地环保部门按照《重金属污染综合方式"十二五"规划》要求对重点企业实施每两个月一次的监督性监测，涉及企业实际排放的铅、汞、镉、铬、砷等 5 种重点重金属中的一种或多种按排放标准评价其达标排放情况。若一次监测不达标则视为该企业不达标；未开展重金属监督性监测视为不达标。

【已发放危险废物经营许可证数】截至报告期末，由各级环境保护行政主管部门依法审批发放并处于有效期内的危险废物经营许可证数量，包括综合经营许可证和收集经营许可证。

【具有医疗废物经营范围的许可证数】指在"已发放危险废物经营许可证数"指标的统计范围内，核准经营危险废物类别包括 HW01 医疗废物的危险废物经营许可证数。

【监测用房面积】指开展环境监测工作所需的实验室用房、监测业务用房、大气、水质自动监测用房等的面积。

【监测业务经费】指完成常规监测、专项监测、应急监测、质量保证、报告编写、信息统计等工作所需经费，不含自动监测、信息系统运行费和仪器设备购置费等。

【监测仪器设备台套数及原值总值】指基本仪器设备、应急环境监测仪器设备和专项监测仪器设备等的数量。监测仪器设备的原值总值，是指通过政府采购、公开招投标或其他采购方式购买的各类监测仪器设备的购置金额。

【空气监测点位数】指位于本辖区、为监测所代表地区的空气质量而设置的常年运行的例行监测点位的数量。

【其中：国控监测点位数】指位于本辖区、由国家批准纳入国家城市环境空气质量监测网络的空气监测点位数。

【酸雨监测点位数】指位于本辖区、为了掌握当地、区域或全国酸雨状况，了解酸雨变化趋势而设立的例行采集降水样品的点位数。

【沙尘天气影响空气质量监测点位数】指位于本辖区、为反映监测沙尘天气对我国城市环境空气质量的影响，由环保部专门设立的环境空气质量监测点位数。

【地表水水质监测断面数】指位于本辖区、为反映地表水水质状况而设置的监测点位数。一般包括背景断面、对照断面、控制断面、削减断面等。

【其中：国控断面数】指位于本辖区、由国家组织实施监测的、为反映水体水质状况而设置的监测点位数。

【地表水集中式饮用水水源地监测点位数】指位于本辖区、为反映地表水集中式饮用水水源地水质状况而设置的监测点位数。

【近岸海域环境功能区点位数】指位于本辖区、为反映近岸海域功能区环境质量而布设的监测点位数。

【近岸海域环境质量点位数】指位于本辖区、为反映近岸海域环境质量而布设的环境监测点位数。

【开展污染源监督性监测的重点企业数】指按照相关要求开展污染源监督性监测的重点企业家数，以省、市、县本级实施监督性监测的企业数为准，委托监测的企业由受托方统计。

重点企业：由各级政府部门监控的占辖区内主要污染物排放负荷达到一定比例以上的，以及其他根据环境保护规划或者专项环境污染防治需要而列入监控范围的排污单位，包括国控、省控、市控企业和其他企业数。

污染源监督性监测：环境保护行政主管部门所属环境监测机构对辖区内的污染源实施的不定期监测，包括对污染源污染物排放的抽查监测和对自动监测设备的比对监测。

【已实施自动监控国家重点监控企业数】根据污染源自动监控工作进展情况，至本报告期末在环保部门污染源监控中心已经实现自动监控的国家重点监控企业数。

【已实施自动监控国家重点监控企业中水排放口数】已实施自动监控的国家重点监控企业中，环保部门污染源监控中心能够实施自动监控的水排放口数。

【已实施自动监控国家重点监控企业中气排放口数】已实施自动监控的国家重点监控企业中，环保部门污染源监控中心能够实施自动监控的气排放口数。

【COD 监控设备与环保部门稳定联网数】已实施自动监控的国家重点监控企业中，其 COD 自动监控设备正常运行、自动监控数据（浓度和排放量）能通过数据采集与传输设备与环保部门污染源监控中心稳定联网报送的企业数。

【NH_3-N 监控设备与环保部门稳定联网数】已实施自动监控的国家重点监控企业中，其 NH_3-N 自动监控设备正常运行、自动监控数据（浓度和排放量）能通过数据采集与传输设备与环保部门污染源监控中心稳定联网报送的企业数。

【SO_2 监控设备与环保部门稳定联网数】已实施自动监控的国家重点监控企业中，其 SO_2 自动监控设备正常运行、自动监控数据（浓度和排放量）能通过数据采集与传输设备与环保部门污染源监控中心稳定联网报送的企业数。

【NO_x 监控设备与环保部门稳定联网数】已实施自动监控的国家重点监控企业中，其 NO_x 自动监控设备正常运行、自动监控数据（浓度和排放量）能通过数据采集与传输设备与环保部门污染源监控中心稳定联网报送的企业数。

【排污费解缴入库户数】指调查年度经对账、实际解缴国库的排污费所对应的户数，同一排污者分期分批计征或解缴排污费的不重复计算户数。

【排污费解缴入库户金额】指调查年度经对账、实际解缴国库的排污费所累计金额。

【自然保护区个数】指全国（不含香港特别行政区、澳门特别行政区和台湾地区）建立的各种类型、不同级别的自然保护区的总数相加之和，应包括国家级、省级和市（县）级自然保护区。

【自然保护区个数国家级】至报告期末，经国务院批准建立的自然保护区的个数。

【自然保护区个数省级】至报告期末，经省（自治区、直辖市）人民政府批准建立的自然保护区的个数。

【自然保护区面积】指全国（不含香港特别行政区、澳门特别行政区和台湾地区）建立的各种类型、不同级别的自然保护区的总面积相加之和，应包括国家级、省级和市（县）级自然保护区。

【自然保护区面积国家级】至报告期末，国家级自然保护区的面积总和。

【自然保护区面积省级】至报告期末，省（自治区、直辖市）级自然保护区的面积总和。

【生态市、县建设个数】

【国家级生态市个数】至报告期末，环境保护部对外正式公告的国家级生态市个数。

【国家级生态县个数】至报告期末，环境保护部对外正式公告的国家级生态县个数。

【省级生态市个数】至报告期末，获得省政府或省环境保护主管部门命名的省级生态市个数。

【省级生态县个数】至报告期末，获得省政府或省环境保护主管部门命名的省级生态市个数。

【农村生态示范建设个数】

【国家级生态乡镇个数】指环境保护部公告命名的国家级生态乡镇数量。

【国家级生态村个数】指环境保护部公告命名的国家级生态村数量。

【国家有机食品生产基地数量】指符合《国家有机食品生产基地考核管理规定》相关要求的有机食品生产基地的数量。

【当年开工建设的建设项目数量】指一切基本建设项目和技术改造项目，包括饮食服务等三产项目（区域开发已列入规划环评管理，不再计入建设项目）。由省级环保部门根据统计部门数据填报。

【执行环境影响评价制度的建设项目数量】指调查年度开工并依法履行环境影响评价制度的建设项目数。由省级环保部门填报。

【当年审批的建设项目环境影响评价文件数量】指调查年度审批的建设项目数（包括新建项目、改扩建项目）。按审批权限统计，分别由国家、省级、地市级、县级环保部门对审批管理的项目进行统计。委托下级审批的项目，由受委托部门统计。

【当年审批的建设项目投资总额】指调查年度审批的建设项目工程总投资的汇总数额。

【当年审批的建设项目环保投资总额】指调查年度审批的建设项目环保投资的汇总数额。

【当年审查的规划环境影响评价文件数量】指调查年度审查的规划环境影响评价文件数量。按审查权限统计，分别由国家、省级、地市级环保部门对负责审查的规划环境影响评价文件进行统计。

【当年完成环保验收项目数】指调查年度完成环保验收的建设项目数。按审批权限统计，分别由国家、省级、地市级、县级环保部门对负责验收管理的项目进行统计。委托下级验收的项目，由受委托部门统计。

【环保验收一次合格项目数】指执行了环保"三同时"制度，经环保部门验收一次合格的建设项目数。

【经限期改正验收合格项目数】指未执行环保"三同时"制度，经限期改正后通过环保验收的建设项目数。

【工业企业当年完成环保验收项目总投资】指调查年度工业企业完成环保验收项目工程总投资的汇总数额。

【工业企业当年完成环保验收项目环保投资】指调查年度工业企业完成环保验收项目环保投资的汇总数额。

【突发环境事件次数】指调查年度发生突发性环境事件的次数。若突发环境事件涉及两个以上省（自治区、直辖市），由事发地省级环保部门负责上报。特别重大、重大、较大、一般环境事件的级别划分参考《突发环境事件信息报告办法》（环保部 17 号令）。

【当年开展的社会环境宣传教育活动数/人数】指以环保部、省级和地市级环保部门或环境宣教机构本级名义主办的面向社会的环境宣传教育活动，主要包括重要纪念日环保宣传纪念活动、公众参与环保宣传活动、环保主题大型文艺演出、环保影视作品公映等形式。人数为参加该活动的人数。

【环境教育基地数】环保部、省级和地市级环保部门或环境宣教机构命名的环境教育基地。